FOREST INFLUENCES

FOREST INFLUENCES

The Effects of Woody Vegetation on Climate, Water, and Soil, with Applications to the Conservation of Water and the Control of Floods and Erosion

By JOSEPH KITTREDGE
Late Professor of Forestry
University of California

DOVER PUBLICATIONS, INC.
NEW YORK

Published in Canada by General Publishing Company, Ltd., 30 Lesmill Road, Don Mills, Toronto, Ontario.
Published in the United Kingdom by Constable and Company, Ltd., 10 Orange Street, London WC 2.

This Dover edition, first published in 1973, is an unabridged and unaltered republication of the work originally published by McGraw-Hill Book Company, Inc., in 1948.

International Standard Book Number: 0-486-20942-3
Library of Congress Catalog Card Number: 73-75874

Manufactured in the United States of America
Dover Publications, Inc.
180 Varick Street
New York, N. Y. 10014

This new edition is dedicated
to the memory of

JOSEPH KITTREDGE
(November 26, 1890 — August 5, 1971)

as a tribute to his contribution to
education and forest ecology in
the United States

PREFACE

Forests and rainfall, forests and stream flow, and forests and floods have been subjects of intermittent controversy between foresters and others for many years. Yet many foresters today have only a hazy conception of what is meant by forest influences, watershed management, and the protection forest. Parts of the subject have been referred to briefly in courses dealing primarily with other subjects. There has been no textbook in English, and the considerable literature covering specific phases has been widely scattered in many publications, including some rarely consulted by foresters. Moreover, the treatment of pertinent data has often been planned with different objectives and interpretations. The fact that the subject is broad and necessarily overlaps the related fields of meteorology, hydrology, and soil science is additional evidence of the need for a book that might collect and unify these diverse materials and serve as introduction, text, or source of reference for foresters and others. If the presentation is somewhat lacking in popular appeal, perhaps that defect may be partially compensated by the content of substantial information.

As in all the natural sciences, every unit of land with its flora and fauna has peculiarities differentiating it from every other unit. Consequently, for an area of land containing many unit complexes, the possibility of generalizing conclusions is often dubious. The attempts at broad generalizations based on inadequate information have been the commonest sources of the controversies of past and present. In many cases the information needed to resolve these conflicts or to clarify the problems must be quantitative. Consequently every effort has been made to bring together numerical data in support of conclusions and, wherever possible, to express relations in equations as the most useful form of quantitative generalization. Admittedly, many of these equations or the empirical constants require confirmation or further study of the variations associated with other factors.

The subject matter of forest influences proper is summarized in a series of statements at the end of each chapter. The attempt has been made to make them definite and conclusive, perhaps more so than the present state of our knowledge justifies. The summaries do not recapitulate the introductory parts of many of the chapters, which are devoted to the basic physical considerations essential to an understanding of the influences and which have not been sufficiently covered in the earlier preparation of most forestry students.

The knowledge of the subject, besides being widely scattered, is expanding rapidly so that omissions or outdated statements will be noted early and with increasing frequency as time goes on. Moreover, it is too much to hope that among so many numerical details some errors have not been detected. The author will consider it a favor to be notified of any such inaccuracies so that they may be corrected promptly.

The encouragement and suggestions of Walter Mulford and the help of H. W. Anderson in many ways during the preparation of the manuscript is gratefully acknowledged.

JOSEPH KITTREDGE

BERKELEY, CALIF.
January, 1948

CONTENTS

CHAPTER VIII

CHAPTER IX

CHAPTER X

CHAPTER XI

CHAPTER XII

CHAPTER XIII

CHAPTER XIV

CHAPTER XV

CHAPTER XVI

CHAPTER XVII

CHAPTER XVIII

CHAPTER XIX

CHAPTER XX

CHAPTER XXI

CHAPTER XXII

CHAPTER XXIII

CHAPTER XXIV

CHAPTER XXV

CHAPTER I

INTRODUCTION

1. Definitions.—Forest influences may be defined for most purposes as including all effects resulting from the presence of forest or brush upon climate, soil water, runoff, stream flow, floods, erosion, and soil productivity. Health and economic conditions have been included in the older definitions, but their addition expands the scope of the subject beyond its usual limits and beyond the physical and biological aspects that will be covered in this book. More briefly and less specifically, forest influences may be conceived as the effects of natural vegetation on climate, water, and soil. These definitions mention forests, brush, and natural vegetation, terms that are evidently not synonymous. In the present treatment the aim will be to cover the influences of woody vegetation quite thoroughly without attempting to include with equal completeness the many important effects of range vegetation on water and soil. At the same time, reference will be made to herbaceous and grassy vegetation whenever it seems essential to complete the picture, as is often the case where forests and grasses and herbs occur together or alternately with successional or retrogressional changes on the same land.

Another definition might be derived from that for ecological reactions, which comprise all effects of climate or community on the habitat. Thus forest influences would be the ecological reactions of the forest or associated vegetation. However, this would not include the applications in forestry just as ecology does not include silviculture.

Conservation of water and soil is a phrase frequently used to designate some of the applied aspects of forest influences. When conservation is defined as the protection and development of the full usefulness of natural resources, including forests, waters, minerals, and lands, the relation to applied forest influences is evident. It should be noted, however, that conservation denotes not merely preservation but also use. Thus it might also be stated that forest influences are the means through which forests are or can be made effective in the conservation of water and soil.

The important phases of forest influences concerned with water, such as precipitation, soil water, stream flow, and floods, might appropriately be called "forest hydrology," inasmuch as hydrology is defined as

. . . the science that treats of the phenomena of water in all its states; of the distribution and occurrence of water in the atmosphere, on the earth's surface, and in the soil and rock strata and of the relation of these phenomena to the life and activities of man (251).

Finally, some of the applications of forest influences are indicated by the concept of a protection forest as one "whose chief value is to regulate stream flow, prevent erosion, hold shifting sand, or exert any other indirect beneficial effect" (328). Of similar import is the definition of watershed management as the administration and regulation of the aggregate resources of a drainage basin for the production of water and the control of erosion, stream flow, and floods (329).

2. Scope.—In simplest form, forest influences is concerned with sun, wind, water, and soil. The vegetation with its canopy and roots modifies the effects of these elements just above, at the surface, and in the soil. The relations are indicated in Fig. 1. Wind may be modified by the crowns as to velocity, direction, and turbulence. The sun's radiation is partly reflected and partly absorbed by the canopy, and partly penetrates to the forest floor. At the surface of the forest floor there is another division between reflection and absorption. Penetration of the floor and into the mineral soil results in temperature effects.

The canopy will vary in density and thickness, and its effects on the atmospheric elements and on the accumulation of forest floor by the deposition of litter will vary accordingly.

The circulation of water between atmosphere and soil constitutes a cycle of several phases. Precipitation is intercepted by the canopy, and one phase of the cycle is completed by evaporation of intercepted moisture from the crowns. Part of the precipitation passes through the crowns and into the forest floor where again evaporation completes a phase of the cycle. A part passes into the mineral soil and is subject to evaporation. Some of the soil moisture is taken up by the roots and returned to the atmosphere in the process of transpiration. Some of the moisture reaching the floor and soil passes down over the surface as surface runoff to a lake, reservoir, or to the sea. Part of the water passes downward, or laterally above an impermeable layer, to the water table or to a lake or to the sea. There is an interchange of water between bodies of water and the adjacent soil. Any surface moisture in channels or water bodies is also subject to evaporation. The understanding of the role of vegetation in these various transformations and relations and the magnitude of the effects constitutes the subject of forest influences. Obviously the field is broad and involves many interrelations, not to say complications.

3. Relations to Other Subjects.—Forest influences is related to and draws material from several associated fields of knowledge, including the silvics, silviculture, protection, and management phases of forestry;

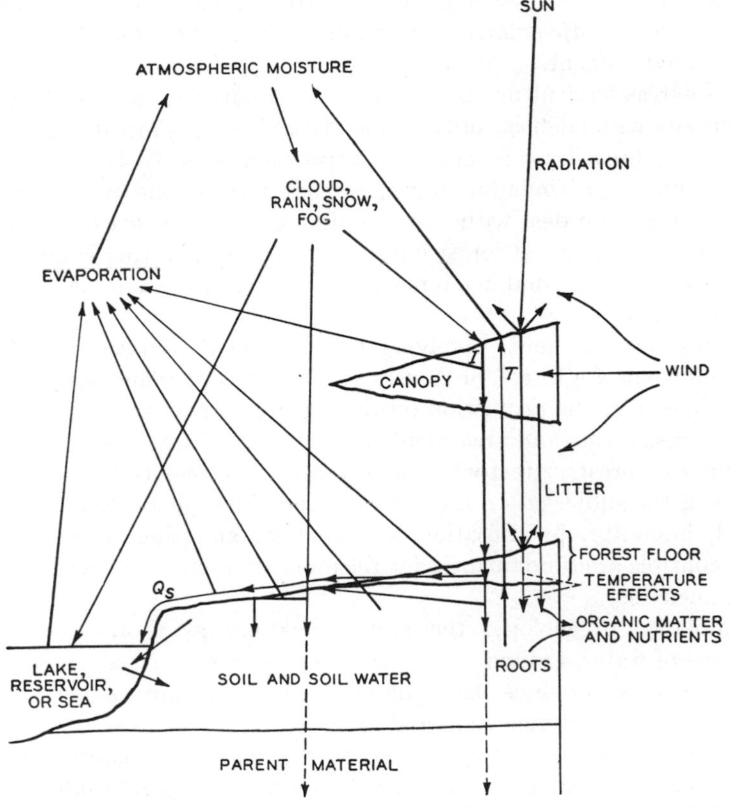

Fig. 1.—The subject matter of forest influences, including the factors, their relations, and the water cycle.

ecology and plant physiology; hydrology and hydraulics; meteorology and climatology; soil science; physiography; and geology. As the study of the effects of the forest vegetation on climate, water, and soil, forest influences tends to give considerable unity to an otherwise loosely related group of subjects. These relations of forest influences to other parts of forestry and to several other sciences have already been partially suggested by the definitions in Sec. 1.

Within the field of forestry itself, silvics, ecology, and forest physiology cover much the same ground but from different points of view. The reactions of ecology are to a large extent identical with influences. Otherwise, the viewpoint in influences represents the reverse orientation. Instead of being primarily concerned with the effects of habitat on the plants, the influences are the effects of the plants on the habitat. Successional trends, plant geography, transpiration, and the growing season are all important considerations in the influences of the forest on its environment.

Insofar as silviculture is concerned with cuttings, which affect the composition and density of the forest, these in turn affect its influences. Similarly, fires, by changing the vegetation, modify the influences. The problems of inflammability, rate of spread, and fire danger in forest protection deal with the same factors and are closely related to the climatic aspects of forest influences. Forest planting is one of the principal ways of making influences effective, for example, on bare eroding land.

Forest management, involving the treatment and control of the development of a forest, obviously also modifies its influences. In the specific case of the protection forest, the management requires special objectives, plans, and treatment based upon its influences.

Outside forestry, factors from several other sciences are essential parts of the subject. In meteorology and climatology, temperatures, wind, humidity, precipitation, evaporation, and their interrelations and controls are fundamental for the understanding of the same factors in influences.

Similarly, in geology, the rock formations as they affect soils, seepage of water, ground water, and normal erosion are all important. Physiography provides basic material on topography, slopes, elevations, erosion, and mountain and valley forms.

In soil science, inasmuch as vegetation is one of the major elements in the development of the soil profile, the relation to the influence of forest on soil is evident. Almost all the soil factors—physical, chemical, and biological—are involved in these relations. Soil moisture, structure, temperature, organic matter, and nutrients may be mentioned specifically.

In the engineering field, hydrology and hydraulics cover the subject of water in all its phases but without particular reference to the vegetation. Thus precipitation, evaporation, soil water, runoff, stream flow, and floods and their measurement indicate the close relation between these subjects and forest influences.

Forest influences is directly concerned with all these elements from other fields of knowledge. It attempts to analyze the effects of forest vegetation of different characteristics through and upon them, and to explore the possibilities of planning the management of forests to utilize their influences to obtain the greatest benefits. A different viewpoint is emphasized, but to a certain extent all these elements tend to be integrated in the subject of forest influences. At the same time an understanding of these related subjects adds appreciably to one's effectiveness in forest influences or watershed management.

4. Significance.—The significance of the study of forest influences is derived chiefly from the importance of water and soil and their conservation. It has additional incidental values in relation to other parts of forestry by contributing to the understanding of certain aspects of succession, natural reproduction, effects of cutting, fire danger, and the management of forest areas for their watershed values.

The importance of the conservation of water is obvious, particularly in semiarid regions where people live, and where it becomes the most essential natural resource for agriculture and for life itself. The control of floods and erosion and the maintenance of soil productivity are not far behind the conservation of water in their importance. They may even come first in humid regions where water is abundant. In either case the forested and other wild lands are the sources of most of the water. Forestry involves the management of these lands for whatever products and uses may be needed in the interests of human welfare. The forester is prepared to be and usually is the manager of the lands from which the waters flow.

Many examples could be cited of applications by the Forest Service in the national forests, by the National Park Service, by the Soil Conservation Service, in some of the schools, and by many water companies. The Prairie Shelterbelt Project and the flood-control surveys of the U.S. Department of Agriculture are examples of recent developments. Thus the rehabilitation of eroded lands by forest planting, the establishment of shelterbelts in nonforested regions, the management of forest lands for their watershed values, and research in the methods and results to be sought and obtained in such management are now active parts of forestry.

CHAPTER II

THE HISTORY OF THE DEVELOPMENT OF FOREST INFLUENCES

The development of the ideas of forest influences has been a long and interesting one, probably as long as the history of forestry itself.

5. Early Beginnings.—As early as 1215, Louis VI of France promulgated an ordinance under the title "The decree of waters and forests," which indicates the recognition even in those days of the close relation between waters and forests. As a protection against avalanches, the first "ban" or protection forest in Switzerland was reserved by a community in 1342. Between 1535 and 1777, 322 such forests were proclaimed.

The earliest expression of the opinion that forests promote precipitation is attributed to Christopher Columbus in the history published by his son in 1571. He ascribed the heavy rainfall in the West Indies to the great forests and trees and compared it with the conditions in the Azores and Canary Islands, which were formerly forested and well watered but "now that the many woods and trees that covered them have been felled, there are not produced so many clouds and rains as before."

The silting of the lagoons near Venice as a result of deforestation in mountain headwaters of the Po and other rivers was an early cause of recognition of forest influences. According to the Italian engineer, Lombardini, writing in about 1840, the volume of material transported to the sea annually by the Po between 1200 and 1600 was only 0.4 the amount since that date. The difference of 25 million cubic meters he attributed chiefly to the clearing of the mountains.

Interesting details are contained in a letter written in 1608 to the government of Venice and published in 1934 (298), from a translation of which the following excerpts are quoted:

> Hence, to correct these evils, it is necessary to remove the causes which, for the most part, had their origin a hundred years ago in the silting we now see. Other causes I would name are the great and frequent inundations and the huge quantities of refuse and mud which the mountain torrents and rivers are carrying and depositing at the present day in the lagoon, something unknown to antiquity. For then, both mountains and valleys were full of trees and vast forest, as a re-

sult of which in those times the rains, falling upon these woods, were soon dispersed, and all the water descending directly was almost wholly absorbed by the dead leaves and by the ground itself; that small portion which ran hither and thither through the forest, held back by the trunks and roots of the trees, turned into vapor, consuming itself almost completely in the ditches and gullies. Likewise, in those times, the snows lying in the shadow of the woods were but gradually liquefied, losing themselves in the soil in such a manner that but few of the waters due to rains and snows found their way from the mountains to the rivers. Wherefore, these did not create floods, but flowed leisurely within their own beds, doing no harm of any importance; and as most of the river banks were thickly covered with brush and willow thickets, the little mud or silt that the waters carried was nearly all deposited in that shrubbery. But at the present time, as the mountains of this exalted Dominion have been ruined and despoiled of their clothing, the rains, finding nothing to restrain them, and the snows remaining exposed to the sun, fall precipitately, straight down upon the lower levels. Laden with great quantities of residual matter, they swell the turbulent brooks and rivers to such an extent that each year, by the force of this impact, they break the dikes, lay waste the fields, destroy edifices, country homes, sometimes even entire hamlets (as we in our times have often seen) and, in their impetuous downward course, sweep all this material, the rich and heavy as well as the light, into the sea. . . . I am wondering if the fires that during this century have many times each year been set in your Serene Highness's mountains, are not the real and main causes of these evils, inasmuch as some people set fire to the undergrowth, where the woods have been cut, in order to convert the latter into fields and pastures. Others burn up the bushes and dry grass to enlarge the area of grazing land and to obtain a quicker and more tender crop of grass. Thus each year we find nearly all the mountains of your Serene Highness's realm burnt over several times.

All these benefits your Serene Highness's subjects will enjoy when once the ditches are made and the fire danger is removed, and within a short space of time one shall see the most ugly parts of the mountains all reclothed with young trees. These, holding back, as of old, the waters of the rains and snows, will cause the torrents and rivers of your Serene Highness's entire domain to flow slowly and limited to their own river beds, without carrying such masses of ashes and other material as they do at present and deposit in the lagoon.

The torrents in the Alps afforded the most striking examples of the relations of forest and lack of forest with water for many years in Austria and France as in Italy. In France alone an early survey enumerated 1,462 dangerous watersheds with an area of one million acres. The local governments in these countries passed decrees against clearing of the forests as early as the sixteenth century. In Austria, Denmark, France, and Germany several publications appeared on the subject between 1778 and 1826, including the work of von Zallinger in 1778; Viborg, 1789; Bremontier, 1798; Karsten, 1801; von Arretin, 1808; and of Duile and Hagen in 1826.

Reforestation of the mountains as a means of control of torrents began in Japan in 1683.

Early recognition of the menace of shifting sand following denudation is indicated by a decree of 1539 in Denmark, which imposed a fine for destroying certain species of sand plants on the coast of Jutland. The fixation of sand dunes by planting vegetation began in Japan with a local attempt in 1616 and more widely and successfully in 1763. In 1712 a report was published in Germany on the means used in Flanders and Holland to stabilize the dunes. Similar work was undertaken in Denmark in 1730, followed in 1770 to 1800 by substantial beginnings in Danzig, Prussia, and France.

The sand dunes of the west coast of France provide the classical example of the effects of deforestation and the stabilization resulting from reforestation. Originally, these dunes were forested, and their movement started only after the destruction of the forests. As early as 1717, small-scale attempts at reforestation were made by private individuals whose property was threatened. Later the Government undertook the stabilization, and in 1786 Bremontier made his notable report and obtained the appropriation to ascertain the possibilities of fixation of the dunes. Successful methods were worked out experimentally by 1793, and in 1799 the Commission of Dunes was created to carry on the work systematically on a large scale. The reforestation and fixation of 200,000 acres of moving sand was completed in 1865.

The sands and marshes of the Landes behind the dunes provided an almost equally difficult problem. Their reclamation by reforestation was begun in 1837, and between 1850 and 1892, 1,750,000 acres of pine forest, chiefly maritime pine, had been established.

In 1801, F. A. Rauch in France published his two volumes, "Harmonie Hydro-Végétale et Météorologique," indicating the recognition of the relationships between hydrology, vegetation, and climate.

In 1830, Hundeshagen in Germany studied the effects of the widespread practice of removing forest litter.

In Denmark in 1837, Bergsoe noted that the felling of the woods on the Atlantic Coast of Jutland had exposed the land to the strong sea winds, which had exerted a sensibly deteriorating effect on the climate of that peninsula. The sand dunes were stationary before the destruction of the forests to the east of them. The felling of the trees removed the resistance to the lower currents of the westerly winds, and the sands thereupon buried large areas of fertile soil.

In 1841, Surell published his excellent book, "Étude sur les Torrents des Hautes Alpes." This was the beginning of the long struggle in France, which has resulted in the stabilization of many of the destructive torrents and avalanches. Blanqui, in 1843, and several other writers in France and in the other countries bordering the Alps, wrote eloquently on the results of deforestation and the necessity for the control of torrents from then until 1860, when the French law gave the Forest Department the mission of controlling the torrents.

A. von Humbolt in "Ansichten der Natur" in 1849 remarked, "How foolish do men appear, destroying the forest cover without regard to consequences, for thereby they rob themselves of wood and water." In 1853, Becquerel first published in French his memoir upon forests and their climatic influence, which in 1869 was translated into English in the annual report of the Smithsonian Institution and republished in 1877 in Hough's "Report upon Forestry" (174). It is interesting to see in the following questions how clearly Becquerel formulated important problems of forest influences at that time.

1. What is the part that forests play as a shelter against the winds or as a means of retarding the evaporation of rain water?
2. What influences do the forests exert through the absorption of their roots or the evaporation of their leaves in modifying the hygrometrical conditions of the surrounding atmosphere?
3. How do they modify the temperatures of a country?
4. Do the forests exercise an influence upon the amount of water falling and upon the distribution of rains through the year, as well as upon the regulation of running waters and springs?
5. In what manner do they intervene in the preservation of mountains and slopes?
6. What is the nature of the influence that they may be able to exercise upon the public health?

Also in 1853, Belgrand wrote (44) that one need not expect from afforestation any effect on the regime of streams. However, the recorded reductions in the water level in wells and streams were attributed by von Wex in 1873 to the deforestation in the basin of the Danube. On the other hand, Valles, another engineer, studying the floods of the Seine, came to quite opposite conclusions. This is probably the beginning of the controversy over the effect of forests on stream flow and floods.

In 1873 appeared Ebermayer's account of the forest-meteorological observations taken at a series of stations in Bavaria from 1866 to 1871.

6. United States.—The history of forest influences in the United States did not lag far behind that of the European countries.

The movement of and damage by shifting sand resulted in measures for stabilization as early as 1739. In that year, the citizens of Truro, Massachusetts, passed an act providing a penalty for grazing of the meadows near the beach. This was necessary because cutting of the woods and grazing just back from the beach had resulted in shifting sands burying meadows near the shore. Later in the same year a similar act was passed for Plum Island near Ipswich, Massachusetts, which, in its preamble, recognized fires as well as cutting of the trees as a cause of the movement of sand. Penalties were provided both for grazing and for setting fires. Similar laws were passed in the subsequent colonial period for several localities where shifting sand became a menace along the Atlantic Coast.

In 1799, Noah Webster presented a paper before the Connecticut Academy of Arts and Sciences on the supposed change in the winter temperatures, in which he said,

> From a careful comparison of these facts, it appears that the weather in modern winters in the United States is more inconstant than when the earth was covered with woods at the first settlement of Europeans in the country; that the warm weather of autumn extends further into the winter and spring encroaches upon the summer; that the wind being more variable, snow is less permanent, and perhaps the same remark may be applicable to the ice of the rivers. These effects seem to result necessarily from the greater quantity of heat accumulated in the earth in summer since the ground has been cleared of wood and exposed to the rays of the sun, and to the greater depth of frost in the earth in winter by the exposure of its uncovered surface to the cold atmosphere.

This is apparently the first published reference to the subject in the United States.

In 1849, a report of the U.S. Patent Office discussed the destruction of forests and the resulting influence on water flows. In 1858, R. U. Piper, in "Trees of America," mentioned the effect of woods in retarding the melting of snow, and quoted W. C. Bryant as follows. "Streams are drying up and from the same cause, the destruction of our forests, our summers are growing drier and our winters colder." Five years later, in 1863, the first edition of "Man and Nature," by Marsh, appeared, which was later revised under the title "The Earth as Modified by Human Action" (239). This interesting book brought together in a chapter of 240 pages under the heading "The Woods," an excellent summary of the European opinions and findings on forest influences up to that time, with special emphasis on the torrents in the Alps.

In 1865, a paper entitled *Forests, their influence, uses and reproduction*, in the *Transactions* of the New York State Agricultural Society, by Watson, gave the observations and conclusions from northern New York as to forest influences on climate, winds, water systems, streams, freshets, dew, evaporation, and soil. In 1867, Lapham, Knapp, and Crocker published their "Report on disastrous effects of destruction of forests in Wisconsin."

A resolution was introduced in Congress in 1872 calling for an inquiry as to whether a certain percentage of public lands sold must be planted with trees or a certain percentage of existing forests preserved for the purpose of preventing or remedying drought. No action was taken.

The Agricultural Appropriation Act of 1876 contained a rider authorizing the expenditure of $2,000 to prosecute investigations to ascertain the influence of forests upon climate and the measures that have been successfully applied in foreign countries, or that may be deemed applicable in this country for the preservation and planting of forests. Dr. R. B. Hough was appointed and prepared a report upon forestry, which appeared in 1877 and included a section of 162 pages on "The connection between forests and climate" and "reboisement" (reforestation). This was a surprisingly good compilation of evidence of the beneficial effects of forest cover and conversely of the deleterious effects of its removal on climate, water, and soil (174).

Also in 1876, a bill was introduced in Congress "for the preservation of the forests of the national domain adjacent to the sources of the navigable waters and other streams of the United States," and to determine what should be reserved in order to prevent such rivers becoming "scant of water." The bill failed of passage, but it was the first attempt to create national forests and was based on the conception of their value for the regulation of stream flow.

About the same time a writer in the *Virginia Enterprise* of Nevada remarked, "It will be but a very short time before we shall be able to observe the effect that stripping the fine forests from the sides and summit of the Sierras will have on the climate of this state and California."

The acquisition of municipal forests for the protection of watersheds was authorized by the State of Massachusetts in 1882.

The California State Board of Forestry, first established in 1885, during its temporary existence prepared a report entitled *A presenta-*

tion of results of torrent action and reduced springs consequent upon forests fires in southern California.

The Act of Mar. 3, 1891, authorized the President to "reserve any part of the public lands wholly or in part covered with timber or undergrowth, whether of commercial value or not, as public reservations." Within two years, 17,500,000 acres had been withdrawn including the present Angeles and part of the San Bernardino National Forest, evidently in recognition of their value for watershed protection. The Act of June 4, 1897, provided for the protection and administration and clarified the purpose of the forest reserves. It contained the clause " . . . no public reservation shall be established except to improve and protect the forest within the reservation or for the purpose of securing favorable conditions of water flows and to furnish a continuous supply of timber."

The Federal Division of Forestry in 1901, in cooperation with the Geological Survey, began an examination of 9,600,000 acres of forest lands in the southern Appalachian Mountains to ascertain their suitability for forest reserves as a step toward the protection of the watersheds and navigation in the lower reaches of the streams. This was a forerunner of the Weeks' Law of 1911.

In 1902, the *Forest Reserve Manual* gave the object of the Forest Reserves to maintain forests on the lands for two principal reasons:

1. To furnish timber.
2. To regulate the flow of the water. This they do
   - *a.* By shading the ground and snow and affording protection against the melting and drying action of the sun.
   - *b.* By acting as windbreaks and thus protecting the ground and snow against the drying action of the wind.
   - *c.* By protecting the earth from washing away and thus maintaining a "storage layer" into which rain and snow water soak and are stored for the dry seasons, when snow and rain are wanting.
   - *d.* By keeping the soil more pervious, so that water soaks in more readily and more of it is thereby prevented from running off in time of rain or when the snow is melting.

From this it follows that the more extensive the forest and the better its condition, the better it will serve the purpose. It is evident that an open park, an old "burn," an extensive "slash," or an open stand of scattering trees or chaparral does not serve this purpose of the forest as well as a close, thrifty stand of young timber.

The period from 1877 to 1912 might be called the "period of propaganda" as far as forest influences in the United States are concerned. An enthusiastic effort was being made to conserve forest and other natural resources that were being depleted. A flood of writings stressed the evil effects of deforestation on climate because that carried a strong popular appeal. It may be noted also that these effects were least likely to be disproved. Relatively little scientific evidence on the subject was produced during the period in the United States.

The propaganda led to exaggerated claims for the influence of forests and inevitably also to reaction, conflict, and opposition. The opposition came from authoritative sources. Chittenden, of the U.S. Army Corps of Engineers, published in 1909 his lengthy article "Forests and reservoirs in their relation to stream flow," which is still cited by those who wish to minimize the influence of forests. Moore, Chief of the Weather Bureau, issued his pamphlet in 1910, "A report on the influence of forests on climate and on floods" (259), which likewise aimed to show that forests were an insignificant factor. Finally, Smith, Chief of the Geological Survey, stated, "What man does with forests will have little effect on erosion."

The recognition of the inadequate basis for conclusions in this controversial discussion led first to a lull in the writings on forest influences from about 1912 until 1925 and second to the initiation of attempts to obtain quantitative scientific data to provide convincing answers to some of the debated questions. The forest experiment stations of the U.S. Forest Service, established between 1908 and 1912, early initiated the collection of forest meteorological data in the Southwest, in the Rocky Mountains, in the Pacific Northwest, and in California. The forest and stream flow experiment at Wagon Wheel Gap, Colorado, begun in 1909, was one of the first projects of the Colorado station.

In the East the evidence of the value of forests for the protection of stream flow was sufficiently established that the Act of Mar. 1, 1911, known as the Weeks' Law, provided "for the protection of the watersheds of navigable streams and to appoint a commission for the acquisition of lands for the purpose of conserving the navigability of navigable rivers." This was the basis for the acquisition of lands for national forests in the East until the passage of the Clarke–McNary Act of 1924.

In 1912, Zon summarized the evidence for the forests under the title "Forests and water in the light of scientific investigation" (381).

7. Recent Developments.—The recent revival of interest in forest influences may be ascribed chiefly to four things. The first was the high value of water in the West, which resulted from the large demand

and inadequate supply; second, the flood of 1927 in the Mississippi River and the inadequacy of the then-existing protective measures; third, the serious erosion and ruin of cleared lands in the Appalachian and Piedmont regions; and, fourth, the disastrous dust storm of May, 1934, which originated in the southern Great Plains.

Several results developed from the growing concern for the conservation of soil and water. The ten erosion experiment stations were provided for in 1930. The Soil Erosion Service (later renamed Soil Conservation Service) was established in 1933. The Prairie States Forestry (shelterbelt) Project was launched in 1934. In the Soil Conservation Act of 1935, Congress "recognized that the wastage of soil and moisture resources on farm, grazing, and forest lands of the Nation, resulting from soil erosion, is a menace to the national welfare and that it is hereby declared to be the policy of Congress to provide permanently for the control and prevention of soil erosion and thereby to preserve natural resources, control floods, prevent impairment of reservoirs, maintain the navigability of rivers and harbors, protect public health and public lands, and relieve unemployment."

Still more recently, the Omnibus Flood Control Act of 1936, with amendments, provided that investigations with respect to watersheds and measures for runoff and water-flow retardation and soil-erosion prevention shall be under the jurisdiction of and prosecuted by the U.S. Department of Agriculture. Both surveys and action programs, when the estimated benefits are in excess of the costs, have grown out of this Act. Coordinated plans are made jointly by the bureaus of Forest Service, Soil Conservation Service, and Agricultural Economics. The forest and range experiment stations of the Forest Service have taken part in the flood-control surveys. They have also been accumulating large amounts of quantitative data as a much-needed foundation for intelligent watershed management. Engineers, soil scientists, and others in the Soil Conservation Service and in the Hydrology Section of the American Geophysical Union are simultaneously contributing valuable materials for that foundation.

8. Historical Review: Summary.—The sources of interest in the subject of forest influences have arisen from observations

1. That changes in climate, and particularly in precipitation and the resulting flow of springs, were associated with the destruction of the forests.

2. That damage to property by shifting sands was associated with deforestation.

3. That damage to life and property by torrents and avalanches was associated with the cutting of forests.

In the United States the historical development has been quite similar to that in other countries.

1. The earliest interest resulted from the necessity of preventing damage by shifting sand.

2. Supposed effects of deforestation on climate were early sources of concern.

3. The third stage emphasized the unfavorable effect of timber cutting upon the regulation of stream flow and the maintenance of navigation.

4. The propaganda stage followed when many of the alleged benefits of forest influences were put forward in the movement to "sell" forestry.

5. The recent and present stage of scientific inquiry and support for the effects of forests on climate, soil, and water, which began in Europe about 1870 and has been continued, at least intermittently, in Germany, Austria, France, Russia, Sweden, Switzerland, Japan, and China, began in the United States about 1908 with the initiation of the collection of forest meteorological data at the forest experiment stations. More recently the emphasis has been focused on erosion, soil stabilization, and flood control.

CHAPTER III

CLIMATIC, SOIL, PHYSIOGRAPHIC, AND FOREST REGIONS AND THEIR RELATION TO FOREST INFLUENCES

The character and intensity of forest influences may logically be expected to vary between different regions characterized by distinctive vegetation, climate, soil, and physiography. Recognizing the existence of such differences, it is evidently unsafe to generalize conclusions derived in one region by assuming their applicability to another region without supporting evidence. This caution is particularly necessary in a country as large and diverse as the United States. It deserves even greater emphasis when applications from foreign countries to the United States are concerned.

9. Regional Subdivisions and Their Significance.—Regional subdivisions of the country are available from various sources, based upon vegetation, climate, soil, and physiography. For the most part, the regional classifications have been made independently. If corresponding regions are thus distinguished, that correspondence means that the factors of vegetation, climate, soil, and physiography reinforce one another. It is then logical to expect that such regions will be different in respect to forest influences.

10. Climatic Regions.—Climate may be considered first because it is only locally subject to modification by vegetation, soil, or physiography and is, therefore, more nearly an independent variable for purposes of comparison. The elements of climate that are most significant in forest influences are water, solar radiation, temperature, and wind. Of these, water in its various forms is much the most important. Theoretically, therefore, moisture should provide the best basis among the climatic factors for the differentiation of regions corresponding to differences in forest influences.

There are two outstanding moisture factors: one, precipitation as the source of water, and the other evaporation as it affects water losses from plant and soil. Incidentally, evaporation represents an integration of the effects of temperature, insolation, atmospheric humidity, and wind movement. A combination of precipitation and evaporation, therefore, should provide a useful basis for a classification of climate with the likelihood of correlation with forest influences therefore, the soil regions should be related to regions based on climate, physiography, or vegetation.

The system of soil classification commonly used in the United States distinguishes numerous small subdivisions represented by the soil series, of which there are now more than 2,000 in the United States, each with from one to five texture classes or soil types as subdivisions. These soil series and types are much too detailed for a

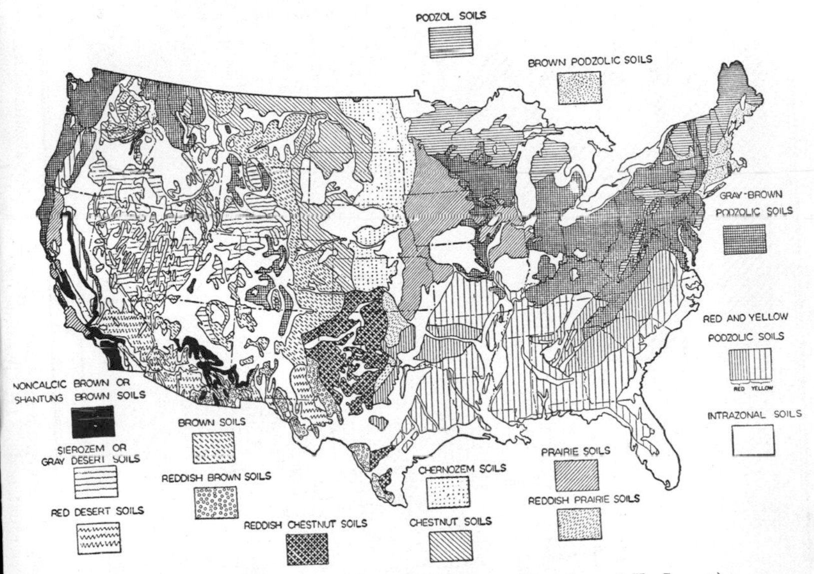

Fig. 4.—Soil regions of the United States. (*Courtesy of H. Jenny.*)

regional classification. They have been combined in zonal groups, however, and the resulting map of the soil regions of the United States, Fig. 4, from Jenny (188) and others (349) affords a reasonably satisfactory basis for a comparison with the regional classifications based on features other than soil. The 16 groups or soil regions are characterized only by name. Descriptions are available in Jenny's (188) or any recent work on soils. The map serves to separate soil regions that are somewhat homogeneous and reasonably distinct from one another. Unfortunately, the map, like most soil surveys, is in parts unsatisfactory for forest purposes, especially in the West, because most of the rough and mountainous areas are included in a single category that actually includes soils of diverse characteristics. Obviously, the shallow rocky soils of mountainous areas are variable and difficult to map, but, notwithstanding that fact, some basis

because precipitation and evaporation are two of the important factors in the water cycle.

Precipitation, which may be represented by P, is the original source of the water supply. Evaporation E is one part of the water that is returned to the atmosphere and not available for human use. The remainder, $P-E$, represents the part of the water contributing to runoff, stream flow, floods, and erosion.

The relation between precipitation and evaporation has been developed as a ratio of P/E by Transeau (346) and later by Livingston and Shreve (223). The form of the ratio that gave the most satisfactory relation to the distribution of vegetation resulted when P was taken as the normal precipitation for the average length of the frostless season plus the preceding 30 days. The evaporation factor was based on Russell's data for the one year 1887–1888 and for a rather small number of stations. The evaporation for this purpose was measured from a free water surface as providing a uniform surface of known area and constant composition. When lines each representing a certain value of the P/E ratio are drawn on a map as in Fig. 2, climatic regions that form a familiar pattern are delineated, the southern pine region at one extreme with a ratio of over 0.90 and the piñon–juniper and desert scrub at the other with a ratio of less than 0.20.

11. Physiographic Regions or Provinces.—The provinces that have been distinguished on the basis of physiography are broad subdivisions of the land in each of which there is some uniformity of topographic expression. That implies also a certain degree of uniformity in the geological structure, in physiographic process, and in the stage of development of the physiographic cycle. At the same time, there are variations in geology and topography within each province that would be expected in regional subdivisions as broad as are these provinces. For the same reason their boundaries on the ground are not always so distinct as they appear in Fig. 3.

The physiographic provinces are useful because they reflect the influences of geological formation, degrees of slope, differences in elevation, and stages of erosional development. Physiographic differences of this character would be expected to show relations to various elements in forest influences including factors of climate, runoff, stream flow, and erosion.

12. Soil Regions.—Soils are the product of climate and vegetation acting on a geological parent material whose influence becomes progressively less apparent as the soil matures. On theoretical grounds,

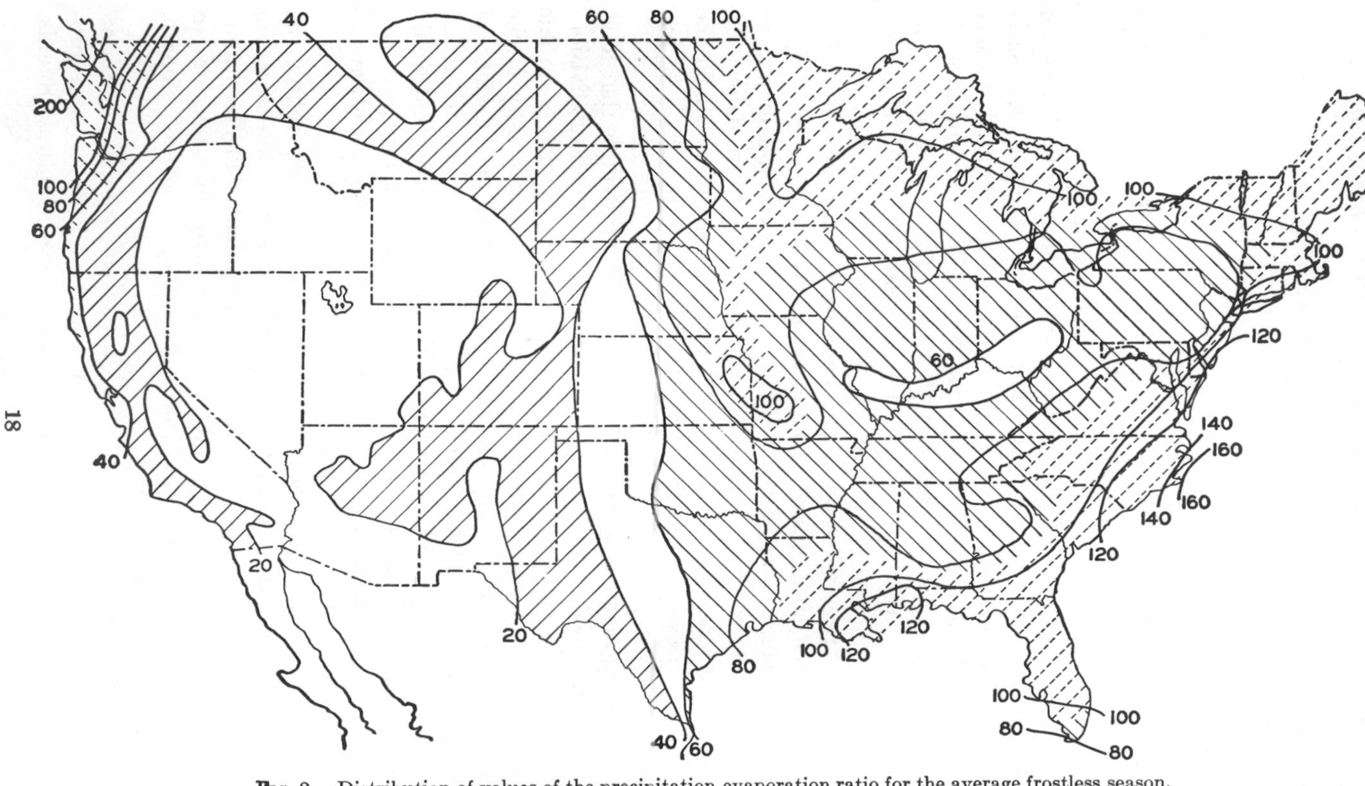

Fig. 2.—Distribution of values of the precipitation-evaporation ratio for the average frostless season. [*Courtesy of Livingston and Shreve, Carnegie Institution of Washington* (223).]

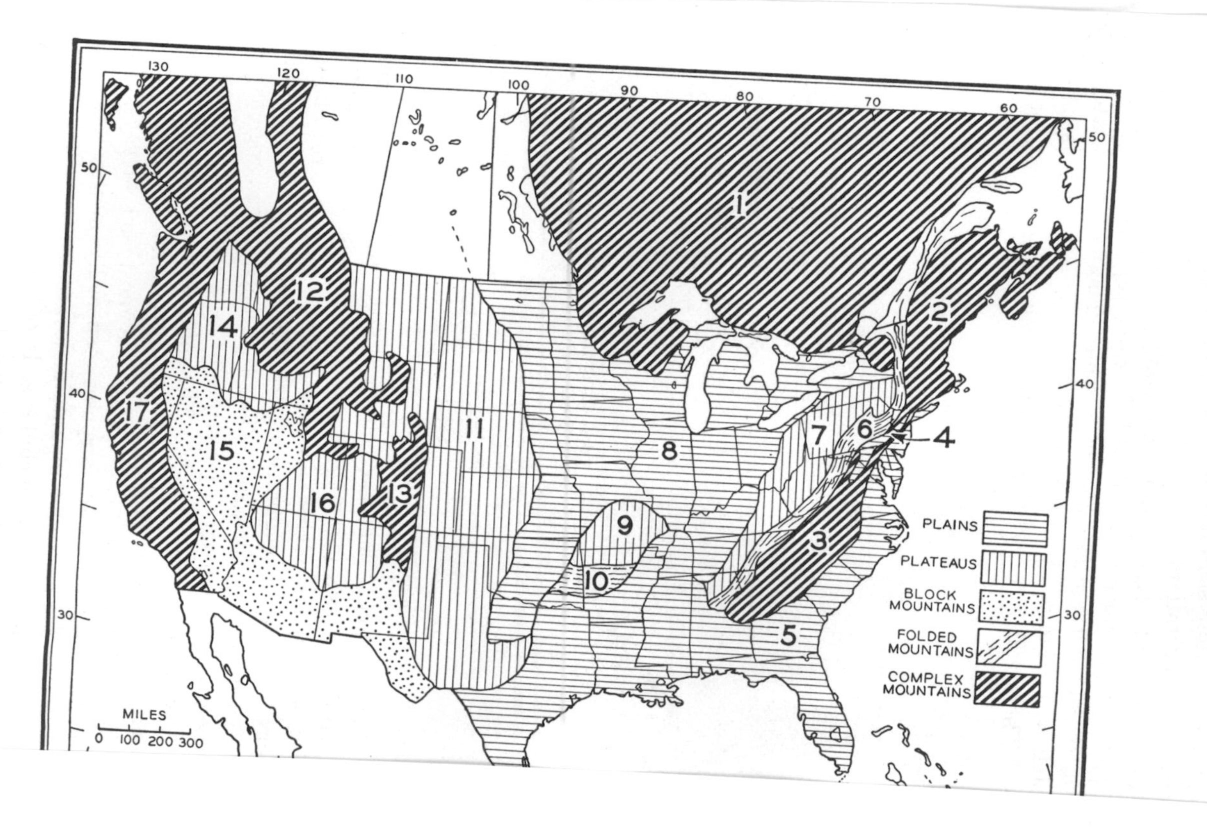

because precipitation and evaporation are two of the important factors in the water cycle.

Precipitation, which may be represented by P, is the original source of the water supply. Evaporation E is one part of the water that is returned to the atmosphere and not available for human use. The remainder, $P-E$, represents the part of the water contributing to runoff, stream flow, floods, and erosion.

The relation between precipitation and evaporation has been developed as a ratio of P/E by Transeau (346) and later by Livingston and Shreve (223). The form of the ratio that gave the most satisfactory relation to the distribution of vegetation resulted when P was taken as the normal precipitation for the average length of the frostless season plus the preceding 30 days. The evaporation factor was based on Russell's data for the one year 1887–1888 and for a rather small number of stations. The evaporation for this purpose was measured from a free water surface as providing a uniform surface of known area and constant composition. When lines each representing a certain value of the P/E ratio are drawn on a map as in Fig. 2, climatic regions that form a familiar pattern are delineated, the southern pine region at one extreme with a ratio of over 0.90 and the piñon–juniper and desert scrub at the other with a ratio of less than 0.20.

11. Physiographic Regions or Provinces.—The provinces that have been distinguished on the basis of physiography are broad subdivisions of the land in each of which there is some uniformity of topographic expression. That implies also a certain degree of uniformity in the geological structure, in physiographic process, and in the stage of development of the physiographic cycle. At the same time, there are variations in geology and topography within each province that would be expected in regional subdivisions as broad as are these provinces. For the same reason their boundaries on the ground are not always so distinct as they appear in Fig. 3.

The physiographic provinces are useful because they reflect the influences of geological formation, degrees of slope, differences in elevation, and stages of erosional development. Physiographic differences of this character would be expected to show relations to various elements in forest influences including factors of climate, runoff, stream flow, and erosion.

12. Soil Regions.—Soils are the product of climate and vegetation acting on a geological parent material whose influence becomes progressively less apparent as the soil matures. On theoretical grounds,

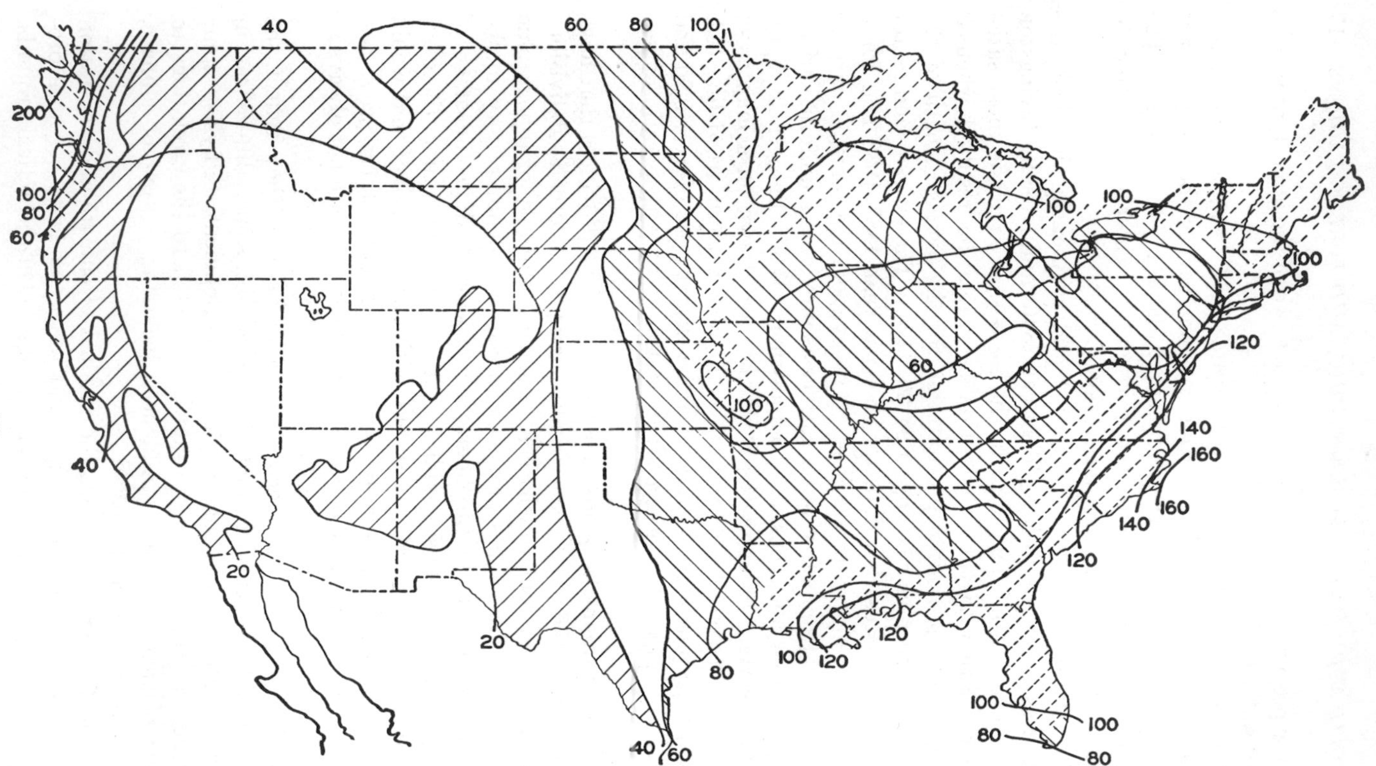

FIG. 2.—Distribution of values of the precipitation-evaporation ratio for the average frostless season. [*Courtesy of Livingston and Shreve, Carnegie Institution of Washington* (223).]

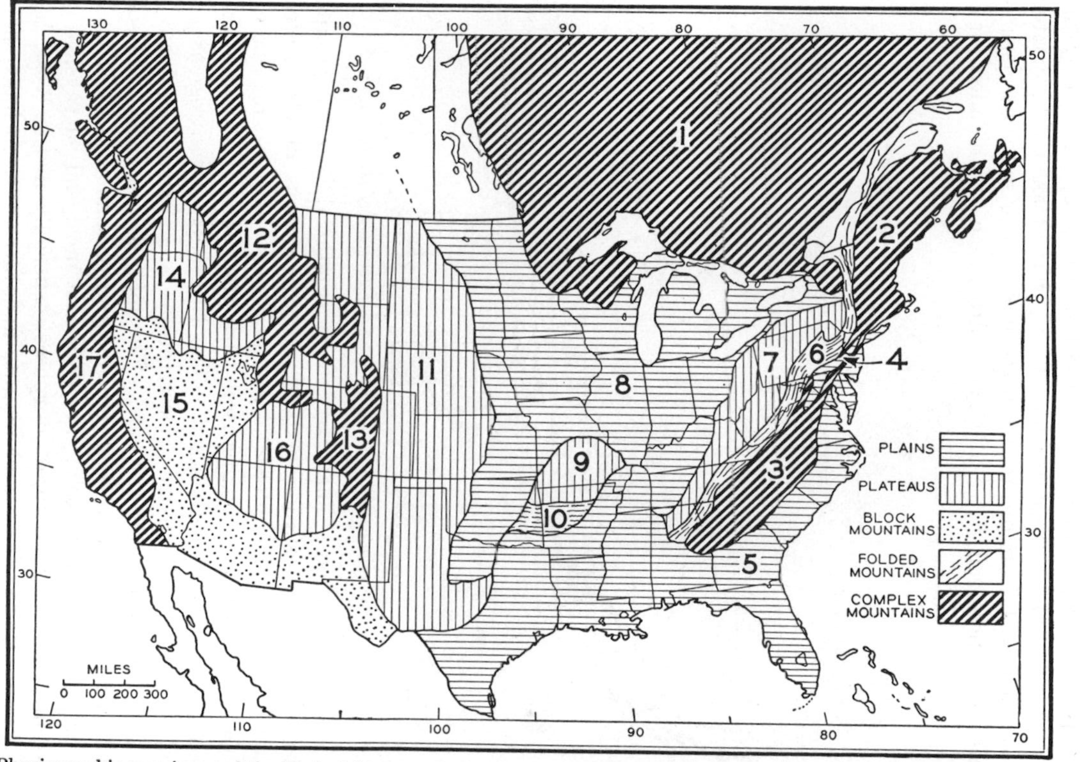

FIG. 3.—Physiographic provinces of the United States. 1. Laurentian upland. 2, 6 and 7. Appalachian highlands. 3 and 4. Piedmont. 5. Atlantic plain. 8. Central lowland. 9 and 10. Interior highlands. 11. Great plains. 12 and 13. Rocky Mountain system. 14. Columbia plateaus. 15. Basin and range province. 16. Colorado plateaus. 17. Pacific ranges. (*From "Atlas of American Geology" through the courtesy of A. K. Lobeck and the Geographical Press, Columbia University.*)

therefore, the soil regions should be related to regions based on climate, physiography, or vegetation.

The system of soil classification commonly used in the United States distinguishes numerous small subdivisions represented by the soil series, of which there are now more than 2,000 in the United States, each with from one to five texture classes or soil types as subdivisions. These soil series and types are much too detailed for a

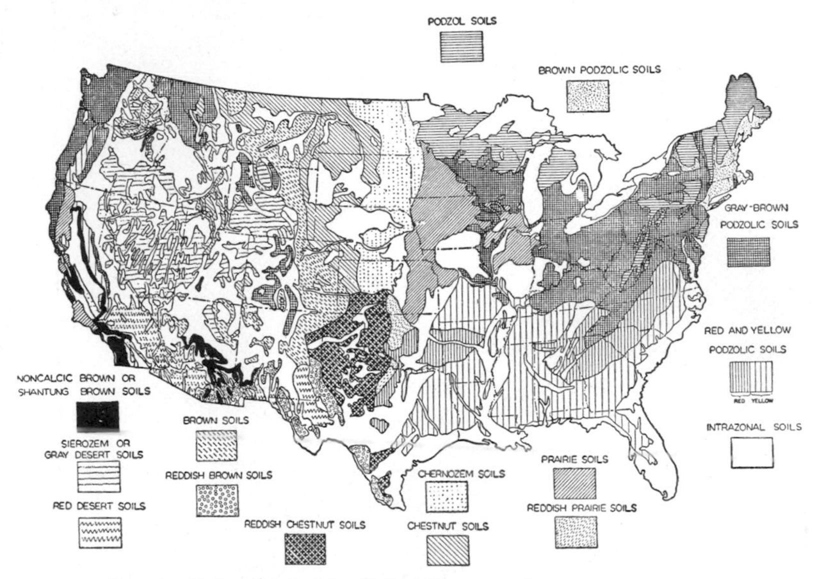

FIG. 4.—Soil regions of the United States. (*Courtesy of H. Jenny.*)

regional classification. They have been combined in zonal groups, however, and the resulting map of the soil regions of the United States, Fig. 4, from Jenny (188) and others (349) affords a reasonably satisfactory basis for a comparison with the regional classifications based on features other than soil. The 16 groups or soil regions are characterized only by name. Descriptions are available in Jenny's (188) or any recent work on soils. The map serves to separate soil regions that are somewhat homogeneous and reasonably distinct from one another. Unfortunately, the map, like most soil surveys, is in parts unsatisfactory for forest purposes, especially in the West, because most of the rough and mountainous areas are included in a single category that actually includes soils of diverse characteristics. Obviously, the shallow rocky soils of mountainous areas are variable and difficult to map, but, notwithstanding that fact, some basis for the subdivision of the soils in these large areas is badly needed.

When allowance is made for the consequent lack of differentiation in the West and the inclusion in the intrazonal category of such diverse soils as the sand hills of Nebraska and the peat swamps of northern Minnesota, the soil regions are quite similar to those of climate and physiography.

13. Forest and Vegetation Regions.—A map of the distribution of natural vegetation in the United States has been prepared by the U.S. Forest Service (351) on the basis of one by Schantz and Zon (313). It is reproduced as Fig. 5. The regions are designated by the predominant or characteristic species and form the pattern for the country as a whole, familiar to all foresters. Because vegetation is the resultant of climate, physiography, and soil, the vegetation regions would be expected to resemble those on the other bases.

14. Hydrological Regions.—Another approach to a regional classification, more directly related to forest influences, may be made by using as a basis the differences between precipitation and stream flow as measured at Weather Bureau and stream-gaging stations. These differences represent the part of the precipitation that does not reach the streams. From the viewpoint of human use, they are losses of water, and in large part they are caused by or related to the vegetation on the drainage basins. When these losses are entered on a map and lines of equal annual loss drawn, the different ranges of loss between lines form a pattern of regions, again somewhat similar to those previously outlined. Such a map and the resulting regions are shown in Fig. 6 (202). The losses vary from more than 30 in. annually in the southern pine and northwest coast fir and redwood to less than 10 in. in the sagebrush and desert-shrub regions.

15. Correspondence between Regions.—The differentiation of the regions on the different bases is clear. As indicated in Table 1, one vegetation region is distinct in all five respects, two in four, three in three, and thirteen in two of the regional classifications. Except for the forest types, the beech-birch-maple-hemlock is not differentiated from the white–red–jack pine and the Pacific Douglas fir is not distinct from the redwood. In both instances the types overlap and mixtures occur frequently. Minor differences between them in soil and climate are not reflected in the broad regional classes that have been used. Some of the other cases of lack of differentiation are attributable to differences in the degree of subdivision in the regional classifications.

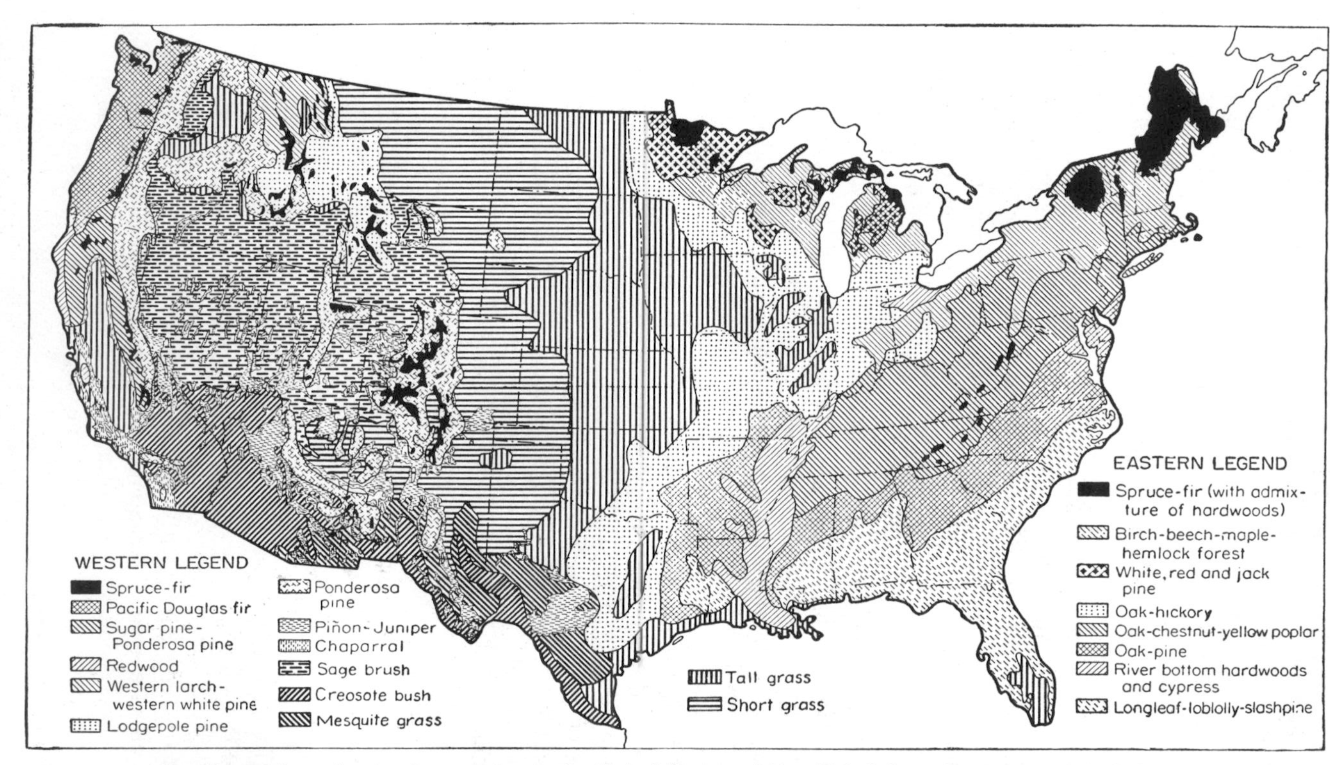

FIG. 5.—Regions of natural vegetation in the United States. (*After United States Department of Agriculture.*)

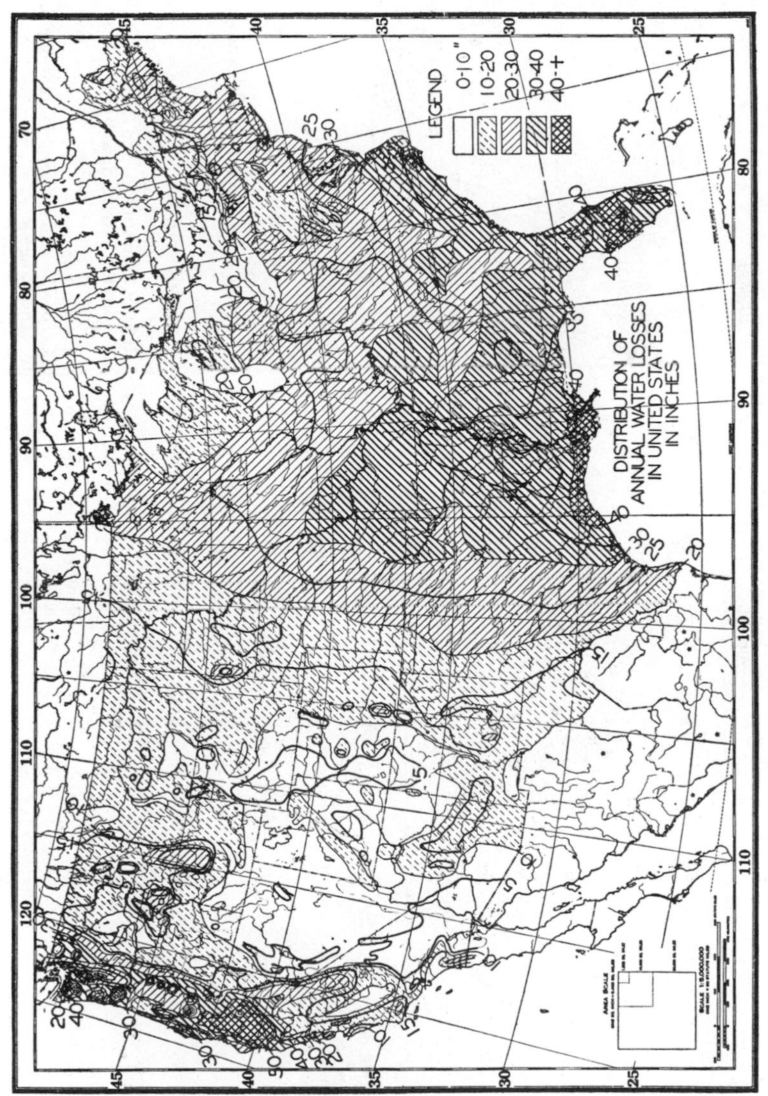

FIG. 6.—Distribution of annual water losses in the United States.

TABLE 1.—COMPARISON OF REGIONS ACCORDING TO VEGETATION, HYDROLOGY, CLIMATE, SOIL, AND PHYSIOGRAPHY

Vegetation	Annual water losses, in.	P/E	Soil groups	Physiographic provinces
Eastern regions				
Longleaf–loblolly–slash pine	30–40	0.9–1.2	Yellow podsolic	Atlantic plain
River-bottom hardwoods and cypress	30–40	0.9–1.2	Alluvial intrazonal	Alluvial valley
Oak–pine	25–35	0.7–0.9	Red podsolic	Piedmont
Oak–chestnut–yellow poplar	20–30	0.6–0.9	Gray–brown podsolic	Appalachian highlands
Oak–hickory	20–30	0.6–0.9	Gray–brown podsolic	Central lowland and interior highlands
Tall grass	20–30	0.6–0.9	Prairie	Central lowland
Birch–beech–maple–hemlock	15–20	0.7–1.0	Podsol and gray–brown podsolic	Appalachian highlands and central lowland
White–red–jack pine	15–20	0.8–1.1	Podsol and gray–brown podsolic	Appalachian highlands, central lowland, and Laurentian upland
Spruce–fir	10–20	1.0+	Podsol	Laurentian upland and Appalachian highlands
Western regions				
Pacific Douglas fir	25–60	0.5–2.0	Gray–brown podsolic	Pacific border
Redwood	25–55	0.4–0.6	Gray–brown podsolic	Pacific border
Sugar and ponderosa pine	15–40	0.2–0.3	Gray–brown podsolic	Sierra mountain
Western larch–western white pine	15–20	0.2–0.3	Gray–brown podsolic	Rocky mountain
Spruce–fir	10–20	0.2–0.3	Gray–brown podsolic	Rocky mountain
Ponderosa pine	10–20	0.15–0.25	Gray–brown podsolic and red	Colorado and Columbia plateaus
Short grass	10–20	0.3–0.6	Brown and chestnut	Great plains
Bunch grass	10–20	0.15–0.2	Chestnut	Columbia plateau
Mesquite grass	10–20	0.15–0.2	Noncalcic brown and reddish brown	Basin and range
Lodgepole pine	10–15	0.15–0.2	Rough and mountainous	Rocky mountain
Piñon–juniper	5–15	0.15–0.25	Rough and mountainous	Colorado plateau, basin and range, and Rocky mountain
Chaparral	5–15	0.2–0.4	Noncalcic brown and prairie	Pacific border
Sagebrush	5–10	0.1–0.2	Gray desert	Colorado and Columbia plateau and basin and range
Desert shrub (creosote bush)	4–10	0.1–0.2	Red desert	Basin and range

On the whole, there is a considerable degree of correspondence between the five systems of classification. This tends to indicate their significance and the probability that forest influences will be different in the different regions, and that conclusions from data of one region should not be applied in another without corroborative evidence.

16. Regions of the United States: Summary.—The subdivisions of the United States on the basis of the independent classifications of hydrology, climate (P/E ratio), physiography, soil, and forest show considerable correspondence. All five systems tend to reinforce one another; that is, in any one region the climate, hydrology, physiography, soil, and vegetation tend to act together in differentiating that region from any other region.

These supplementing regional distinctions indicate the strong probability that the forest influences in the different regions will vary in character and intensity.

Conclusions as to influences derived from one region should not be applied to another without verifying their applicability. Similarly, even greater caution is required in applying European conclusions as to forest influences to conditions in the United States.

CHAPTER IV

VEGETATIONAL FACTORS

The differences and the changes in composition, density, size, age, and rate of growth of forest and brush cover that are usually recognized in forestry may be expected to be reflected in local differences in their effects on climate, water, and soil. Differences in vegetation develop in response to environmental and historical factors, but, as the vegetation grows, it reacts on the site with changing intensity. Such differences and their reactions, therefore, will require consideration in the study of influences. Not only are there differences between trees and stands but also between parts of a given tree or stand. For example, the influences exerted by the crowns are distinct from those exerted by the roots. Stems, branches, flowers, and fruit are much less important than the foliage and roots.

17. Measures of Foliage and Their Relations.—The amount of foliage, whether expressed in weight or area or volume, and its relation to the area of ground occupied by a tree or stand may be expected to be related in a significant way to influences such as transpiration and interception of radiation or precipitation. Existing knowledge of foliage relations is far behind that for growth and yield of stems. It would be most useful to be able to convert data from yield tables into terms of foliage with assurance. For the same species it has been shown that greater current growth of stems is associated with greater amounts of foliage. Qualitatively, the relations of foliage to species, age, site, and density are likely to correspond to differences in wood increment.

Various expressions of these relations have been used in individual studies. Working with Scotch pine in Sweden, Tiren (343) found that needle area was directly proportional to dry weight and therefore that the ratio was approximately constant at 15 to 17.5 sq m per kg. Although this ratio varies with species, it is nearly constant for any one species as has been demonstrated for both forest and fruit trees (79). Thus needle area and dry weight of needles could be used interchangeably in determining the correlation with some other variable such as interception or transpiration. Leaf weights can be determined much more quickly and simply than areas, and if the factor of area per unit of weight is known, areas can be derived from the weights. For example, Tiren's 105-year-old stand of Scotch pine had 2.0 metric tons per acre or 4,900 kg per hectare dry weight of needles. Using a factor of 16 sq m per kg, the needle area was 78,500 sq m per hectare or 7.85 hectares per hectare, or 7.85 acres per acre. Oven-dry weight is preferable to green or air-dry weight because the latter involve undetermined differences in moisture content.

The similar relation between leaf surface A and volume V of needle fascicles of loblolly pine can be expressed by the equation,

$$A = 70.9V + 1.68$$

where A is in square centimeters and V is in milliliters, according to Kozlowski and Schumacher (208). For eastern white pine the equation was

$$A = 50V + 0.6$$

Both relations were sufficiently well defined as to be useful for estimating needle areas from their volumes. Volumes of leaves may be determined quite easily and with sufficient accuracy by immersion.

The factor of leaf area per unit of ground area is a useful expression, first, because it enables one to derive leaf areas directly for an acre or any area of ground surface, and, second, it is a dimensionless ratio and has the same numerical value regardless of the unit of area used. It may be expressed in either of two related forms, one based on the projectional area of the crown or crowns of the trees and the other on the total ground area of the stand. Obviously, the difference involves the area of openings between the crowns and is commonly expressed as crown density or the ratio of crown-projectional area to total ground area. Thus, by means of the crown density, which may be estimated to the nearest tenth, the ratio of leaf area to total ground area may be converted to the ratio of leaf area to crown-projectional area or vice versa. For example, an old stand of manzanita (*Arctostaphylos glandulosa*) on the San Dimas Experimental Forest had a ratio of leaf area to crown-projectional area of 2.8 and had a crown density of 0.87. The ratio of leaf area to total ground area was, therefore, 2.4. In general, the ratio of leaf area to ground area divided by crown density gives the ratio of leaf area to crown-projectional area and because crown density never exceeds 1.0, the latter always is greater than or at most equal to the former ratio.

18. Composition.—The differences in species between regions and in different sites or types within a region are obvious from the most casual observation. Such differences, upon analysis, include distinct features that have varying effects on the habitat. Variation in density between crowns of individual trees of different species is one factor. The difference between the density of the crowns of Digger pine and white fir is an extreme example among the conifers and between paper birch and sugar maple among deciduous species. In general, intolerant species have less dense crowns that the tolerant. The evergreens differ from the deciduous species in the degree and duration of the interception of radiation, precipitation, and wind movement, notably between winter and summer. These differences are found both in individual trees and in stands. In part they consist in different amounts of foliage characteristic of different species, as shown in Table 2.

TABLE 2.—LEAF AREA PER UNIT OF CROWN-PROJECTIONAL AREA AND PER KILOGRAM OF DRY WEIGHT IN WELL-STOCKED STANDS

Species	Age	National forest	Elevation, ft	Leaf area/ crown-projectional area	Sq m/kg
Ceanothus crassifolius....	17	Angeles N. F.	3,000	1.7	5.3
Photinia arbutifolia......	17	Angeles N. F.	3,000	3.4	9.6
Arctostaphylos glandulosa.	17	Angeles N. F.	3,000	9.2	7.1
Arctostaphylos glauca....	50	Angeles N. F.	5,000	4.2	7.9
Canyon live oak........	50	Angeles N. F.	5,000	8.5	9.7
Ponderosa pine.........	28	Stanislaus N. F.	5,000	4.2	8.8
White fir...............	27	Stanislaus N. F.	5,000	5.4	10.7

Leaf area includes total surface area of both sides of flat leaves. Evidently, there are marked differences between species in the area ratio even when they are growing in dense stands of the same age on similar sites, as for example between 1.7 and 9.2 for ceanothus and manzanita, respectively. Likewise, the ratios of area to dry weight of leaves vary between extremes of from 5.3 to 10.7 sq m per kg for the ceanothus and white fir of Table 2. Other species including the pine and oak are intermediate. The values for California black and Gambel oaks are 8.9 and 10.7, respectively (306). This ratio reflects the thickness and density of the leaves, and hence the relations between species are not the same as those for the ratio of leaf area to crown-projectional area. For the same species different individual trees show only a narrow variation from these averages, so that the ratios for a given species may be considered approximately constant. The differences in these foliage relations result in variations in light, heat, and precipitation reaching the ground and in differences in the amount of accumulation of leaf litter and its rate of decomposition.

Certain chemical and textural features of the foliage characteristic of different species may also be important. For example, the gums and resins of coniferous foliage affect rate of decomposition. The cutin and leathery texture of certain leaves, notably of the broad sclerophylls, are in marked contrast to the thin, soft leaves of birch and maple. The tannins of the oaks afford another example. Most of the species of eucalyptus introduced in the United States form little or no forest floor, and the question may be raised whether these species have the beneficial effect on the soil that is attributed to most other genera. The foregoing differences are reflected in the microclimate, in the accumulation of litter, and in the decomposition of the forest floor.

Differences in species composition are associated necessarily with differences in root habits and root systems. The depth, spread, and amount of root material and of absorbing root surface per unit volume of soil are likely to vary not only with the species but also with the density of the stand, the light conditions, and the available moisture and nutrients in the soil. However, Coile (93) has shown that the amount of roots less than 0.3 in. in diameter in the upper 4.5 in. of soil of shortleaf pine and of red gum–yellow poplar types on different soil types is nearly the same, that is, 1.7 metric tons per acre. This leads to the suggestion of the concept of "root capacity" to mean that "after fully stocked stands have reached a certain age, the number or weight of small roots in unit volume of the surface soil reaches a near constant." Species with deeply penetrating roots like the hickories would almost certainly have more roots in the deeper layers of the soil than shallow-rooted species like beech. Differences in penetrability also affect the amount of roots in the upper and underlying layers.

The relation of top to root varies with species and is unquestionably important as affecting the transpirational use of soil moisture. For example, on the same Georgeville soil in North Carolina, Duncan (113) reports data for several species of 1-year seedlings from which the following top/root ratios are derived: red gum, 2.3; eastern redcedar, 0.92; northern red oak, 0.45; and post oak, 0.24. The wide variation between species is evident. Mixed stands are quite different and usually more effective than pure stands in occupying and utilizing space, both in soil and air.

19. Density of Stands.—Density, within a climatic region as a result of the spacing and size of the trees, varies first with species. Those more tolerant of shade or more economical in their use of water and nutrients will stand more closely spaced and have a correspondingly greater density than the less tolerant. Density also changes with age, increasing from the seedling stages at least until the competition between individuals prevents further increase by the elimination of some individuals in the process of natural thinning. The density in even-aged well-stocked stands reaches its maximum at or before maturity and thereafter again decreases. Mixtures of species by the stratification of crowns and roots ordinarily show greater densities than do pure stands. In general, the intensity of reaction or of forest influences increases with the density of the stand.

Density is expressed in several ways that do not necessarily give corresponding comparisons. Those commonly used include number of stems per acre, basal area per acre, stand-density index, projectional area of crowns or crown density, percentage of light or radiation reaching the ground, and dry weight of plant or of foliage and leaf area. Amount of root material per unit of area or of soil volume would be informative if more data were available like those for jack pine in Vermont by Adams (2), who shows the kilograms of dry matter per acre below ground to be 2,870 with 10,140 trees per acre, 1,576 with 2,722, 1,657 with 1,210, and 1,236 with 680 trees per acre. The relation is not a simple one when density is expressed as number of trees per acre or as basal area or as stand-density index.

Density varies also with regional differences in climate and site. Moist regions have dense vegetation where it has not been disturbed, and arid regions have sparse and scattered vegetation. This is probably the result of the limiting effect of deficient soil moisture. The differences between species and regions in density may be seen in Table 3 by a comparison of the stand-density indices, basal areas, and number of trees per acre of well-stocked stands. The species are arranged in descending order of the stand-density index because that measure of density is independent of age and site. Red fir, redwood, and red gum lead with densities of over 600. The moderately tolerant species of relatively humid climates form a second group including white fir, white pine, Sitka spruce–western hemlock, and red spruce with indices of 440 to 600. The southern, ponderosa, jack, and lodgepole pines with Douglas fir, aspen, and cottonwood form a group of intolerant species with densities of from 250 to 400. The oak and northern hardwoods with indices of 230 and 210 are at the end of the list doubtless in part because the stands available for yield plots represented less than "normal" stocking.

Basal area and number of trees per acre reflect to some extent the differences in sites and ages of culmination of the periodic annual

TABLE 3.—DENSITIES AND PERIODIC ANNUAL INCREMENT BY SPECIES FOR AVERAGE SITES AT AGES OF CULMINATION

(*From yield-table data*)

Species	Average site index at 50 years	Periodic annual increment per acre at culmination		Average d.b.h., in.	Trees per acre	Basal area per acre, sq ft	Stand-density index
		Age, years	Volume, cu ft				
Red fir	45	110	273	14.3	360	437	680
Redwood	110	30	380	10.4	628	372	630
Red gum	100	15	177	3.3	3,700	123	620
White fir	70	55	270	10.0	585	317	585
White pine (Wis.)	60	35	195	5.0	1,600	195	520
Sitka spruce–w. hemlock	89	35	260	6.1	1,130	216	500
S. white cedar	50	20	140	1.8	7,400	140	500
W. white pine	60	60	167	5.8	1,190	221	470
Red spruce	39	45	165	4.1	1,800	162	440
Shortleaf pine	70	25	146	4.5	1,480	158	400
Slash pine	81	20	130	4.9	1,090	148	330
Loblolly pine	91	25	160	7.0	540	144	310
Lodgepole pine	38	50	69	3.8	1,490	118	310
Douglas fir (NW)	114	25	240	5.4	800	122	300
Ponderosa pine	51	25	95	3.1	1,900	100	300
Jack pine	53	25	110	3.4	1,680	108	300
Quaking aspen	60	45	77	5.2	810	116	290
Longleaf pine	71	20	100	3.8	1,150	93	250
E. cottonwood	140*	14	450	8.5	320	126	250
Virginia pine	46	15	58	2.5	2,240	68	250
Oak (Central states)	60	30	46	4.0	965	84	230
N. hardwoods (Lake states)	Medium	50	31	6.7	390	90	210

* May be better than average site.

increment, which differ widely with species. Those species which grow on good sites have high basal areas per acre relative to those on poor sites. Those of which the growth culminates late tend to have small numbers of trees per acre as compared with those of early culmination. The rank order by basal area is not very different from that of stand-density index. The foregoing measures represent densities of the stands according to the stems, but they are not necessarily correlated closely with amounts of foliage.

However, in stands of the same species and age growing on a reasonably homogeneous site, the relation of foliage to density of stand may be well defined. In six stands of canyon live oak 50 years old at about 5,000 ft elevation on the San Dimas Experimental Forest, the dry weight of leaves in metric tons per acre W was a linear function of the basal area per acre A, which could be represented by the equation,

$$W = 0.014A + 1.8$$

Similarly, in six stands of *Ceanothus crassifolius,* 17 years old at about 3,000 ft in the same locality, the dry weight in metric tons of leaves per acre when plotted over percentage crown density or coverage c, defined a straight line of which the equation was (203)

$$W = 1.48c - 0.06$$

These two examples based on limited data are only suggestive and require supplementing before the possibilities of generalization and application can be evaluated.

The top/root ratio is probably concerned in the matter of stand density. With ample moisture large tops can develop with small roots. On the other hand, if moisture is deficient, large roots are required to support small tops. Hence in dry regions the apparently open densities may consist of small tops above ground with large roots completely occupying the soil between the plants. In a specific case, however, the matter is not so simple, as shown by data of top/root ratio for 14-year-old jack pine in Vermont for plantations of different spacings (2). The ratios were 4.8 for 2- by 2-ft spacing, 7.0 for 4 by 4, 4.5 for 6 by 6, and 5.6 for 8 by 8 ft.

20. Size and Age.—The age and size of the trees in a stand have certain effects in forest influences in addition to those which have been mentioned as associated with density. A few large old trees with high crowns may exert a quite different influence on wind velocities and temperatures within the stand, for example, than many small trees with low crowns, although the densities might be the same. Similarly, trees of different crown classes (and sizes) have quite different amounts of foliage. Thus in a 23-year-old plantation of white pine in Vermont, Adams (3) reports the following dry weights of needles per sample crop tree of each crown class and corresponding diameter breast high (d.b.h.):

Crown Class	D.b.h., in.	Weight, kg.
Dominant	5 –6.3	5.4
Codominant	4.3–4.8	3.5
Intermediate	2.1–4.3	2.0

The distribution among crown classes for stands as units is indicated by Burger (60) for Norway spruce in Switzerland as follows:

Age, years	Total dry weight, metric tons/acre	Per cent of weight by crown classes			
		Dominant	Codominant	Intermediate	Suppressed
35	6.3	54	37	5	4
45	6.9	51	38	7	4

Trends of development with age are usually so well defined that the actual relations to density or size may be overlooked. Thus the green weight of white pine needles as a percentage of the total mass of the trees decreased with age, according to Burger (63) as follows:

Age, years	3	20	30	40	60	70
Needle weight, %	70	8	5	4	3	2.5

The needles constituted from 7 to 9 per cent of the total dry matter above ground in 23-year-old white pine and 9 to 11 per cent in 21-year-old Scotch pine, according to Adams (3).

Data for leaf weight W in relation to diameter D for individual trees of several species, when plotted on double logarithmic paper, form linear trends in the range of from 1 to 30 in. in diameter at 4½ ft above the ground level. The general equation expressing the relation is

$$\log W = b \log D - a$$

where b is the slope and a is the Y intercept in logarithmic units. Values of b and a for a variety of species, sites, and ages are given in Table 4 (199).

Although the variation in the constants indicates small differences in the slopes and positions of the trend lines between species and sites, the trend for any one species and site is sufficiently well defined that leaf weight could be predicted from the diameter with only a small error. In fact, for white pine and Douglas fir no large error would result from using an identical equation for both species. The changes in the constants as site index increases are in opposite directions for the white pine and jack pine so that the indicated relation to sites is not

TABLE 4.—SUMMARY OF VALUES OF THE SLOPES b AND THE Y INTERCEPTS a IN THE RELATION BETWEEN LEAF WEIGHT IN KILOGRAMS PER TREE W AND D.B.H., IN INCHES D EXPRESSED BY THE REGRESSION EQUATION, $\log W = b \log D - a$, FOR DIFFERENT SPECIES, LOCALITIES, SITES, AND AGES

Slope b	log of Y int. a	Species	Location	Elevation, ft	Site index, ft	Age, years
			Leaves oven dry			
1.15	−0.46	Ponderosa pine	California	5,000	20–39	24–69
1.67	−0.73	Ponderosa pine	California	4,000–5,000	20–76	5–69
1.47	−0.53	White pine	Vermont	....	90	24
2.09	−1.22	White pine*	Switzerland	1,300	72–67	21–70
2.32	−1.15	White pine	Vermont	....	47	28
1.96	−0.91	Douglas fir	Switzerland	....	80–110	20–45
2.56	−0.94	Red pine	Vermont	....	43	28
2.66	−1.36	Canyon live oak	California	5,000	....	50
2.87	−1.58	Jack pine	Minnesota	....	45	37
3.15	−1.16	Jack pine	Vermont	....	54	14
			Leaves air dry			
3.3	−2.2	Scotch pine	Switzerland	1,300	57†	32
3.0	−1.7	Scotch pine	Switzerland	1,300	43†	32
3.0	−1.6	Scotch pine	Switzerland	6,300	20†	32
2.1	−1.1	Scotch pine	Switzerland	3,500	29†	32
3.4	−2.0	Norway spruce§	Switzerland	1,500	69‡	40
2.8	−1.2	Norway spruce§	Switzerland	5,280	33‡	40
2.7	−1.5	Norway spruce¶	Switzerland	1,500	60‡	40
1.8	−0.7	Norway spruce¶	Switzerland	5,280	31‡	40
			Leaves green			
2.1	−0.8	White pine*	Switzerland	1,300	72–87	21–70
2.2	−0.5	Norway spruce	Switzerland	....	45‡	35
2.0	−0.4	Norway spruce	Switzerland	....	Average	all
2.1	−0.5	Silver fir	Switzerland	....	Average	all
1.7	−0.5	European beech	Switzerland	....	Average	all
1.51	−0.42	European beech	Switzerland	....	....	80

* Same stand and data except for degree of dryness.
† Dominant height at 32 years.
‡ Dominant height at 40 years.
§ From seed originating at 1,600 ft elevation.
¶ From seed originating at 6,100 ft elevation.

consistent. In even-aged stands the relations hold at least to ages of 70 years and in all-aged stands of tolerant species, to diameters of 30 in. The conversion from green to dry weight involves only a change in the value of the Y intercept, which for the white pine decreased from −0.8 to −1.2. For a given species and site, the same equation applies to trees of different crown classes and to stands of different density.

In all these equations for diverse species, sites, and regions the constants for the slopes of the trend lines vary only from 1.1 to 3.4, and, for the logarithms of the Y intercepts, from −0.4 to −1.6, ranges that are surprisingly narrow. Such equations with appropriate and perhaps estimated constants offer the possibility of predicting leaf weights from diameters and by summation, the totals for stands from stand tables.

This close relation between leaf weight and diameter is not wholly consistent with the logical expectation that amounts of foliage should be most closely correlated with current increment of stems. The latter culminates at early ages while diameters continue to increase with age. The two relations might be reconcilable up to ages not too far above the culmination of growth. None of the even-aged stands for which equations have been presented was over 70 years old. Although most of the available data do not indicate flattening of the trends with increasing age, the possibility is suggested as a caution in applying, without checking, a leaf weight–diameter equation to older stands.

Similar linear trends were found between dry weight of leaves and crown diameters of shrubby species in the chaparral of Los Angeles County when both measures were expressed in logarithms (203). For dry weight W in kilograms and crown diameters D in feet, the regression equations were for *Arctostaphylos glandulosa,*

$$\log W = 2.13 \log D - 1.46$$

for *Ceanothus crassifolius,*

$$\log W = 2.40 \log D - 2.01$$

These two equations applied to the winter and spring seasons before the new foliage had developed. By including the leaves for the current season the equation for *Ceanothus* became

$$\log W = 2.27 \log D - 1.79$$

The change in the constants with the seasonal change in amount of foliage is not large but neither is it negligible. Seasonal changes of somewhat similar magnitude would doubtless be found for other evergreen species. It is interesting that the numerical values of the constants for these shrubs when the crown diameters are expressed in feet correspond so closely to those for trees when the stem diameters are expressed in inches.

TABLE 5.—FOLIAGE AND RELATED DATA FOR PONDEROSA PINE AND WHITE FIR STANDS AT 5,500 FT ELEVATION ON THE STANISLAUS AND FOR CANYON LIVE OAK AT 5,000 FT ON THE ANGELES NATIONAL FOREST, CALIFORNIA

	Ponderosa pine				White fir		Canyon live oak						
Age, years	24	28	65	69	27	38	55	55	55	55	55	55	55
Ave. d.b.h., in.	1.3	4.5	8.2	5.9	0.4	1.0	2.7	3.2	3.9	4.0	4.2	4.3	4.4
Basal area per acre, sq ft	28	181	443	343	56	53	135	230	185	220	184	161	203
Trees per acre	3,070	1,633	1,206	1,826	8,740	8,380	3,395	4,000	2,640	2,520	1,920	1,616	1,920
Stand-density index	130	450	800	750	50	210	410	740	590	600	480	410	520
Site index	20	30	39	30	20	20	18	25	25	26	29	31	27
Dry-leaf weight of average sample tree, kg	0.40	1.82	4.56	2.60	0.26	0.32	0.63	1.40	1.28	1.62	1.75	2.2	2.3
Dry-leaf weight, metric tons per acre	1.3*	2.5*	4.4*	4.8*	2.3	2.7	2.6*	5.0*	4.4*	5.0*	4.4*	4.2*	4.8*
Needle area per kg of dry weight, sq m per kg	11.0	8.8	24.1	25.5	10.7	11.8	9.5	9.6	9.5	10.2	9.9	10.1	9.2
Needle area per unit of crown-projectional area	3.1	4.2	19.6	14.4	5.4	5.9	7.3	9.2	6.6	8.5	11.6	8.1	8.6

* Computed from stand table by equation relating leaf weight to d.b.h.

The trend of leaf weight with age in well-stocked stands should be similar to that for current annual growth, as previously suggested. When the regression equation for ponderosa pine

$$\log W = 1.67 \log D - 0.73$$

is applied to the stand tables for that species for site index 60 (199), the following figures are obtained:

Age, years	20	40	60	80
Dry-leaf weight, metric tons per acre	1.8	3.4	3.0	2.8

The culmination of the trend of leaf weight between 20 and 40 years corresponds to that for wood increment. In general, it is probably true that the maximum amounts of foliage in fully stocked even-aged stands are found between the ages of 15 and 60 years, depending on the species.

For well-stocked stands the ratio of leaf area to ground area changes with age as does the current growth. Thus white pine in Switzerland had ratios of 14.4 at 21 years, 16.6 at 53 years, and 15.9 at 70 years. The maximum at or before 53 years doubtless represents about the age of the culmination of the current annual increment. Similarly, Scotch pine in Sweden gave 7.1 at 35 years, 10.2 at 55, and 6.6 at 105 years (343). Four Scotch pines of different sizes from the same site and the 150- to 160-year age class in Germany (110) had ratios of 6.6 to 9.3 that did not vary systematically with size of trees. Manzanita (*Arctostaphylos glandulosa*) in the Angeles National Forest had a ratio of leaf area to crown-projectional area of 9.2 at 17 years and 2.9 at 45 years of age. However, the difference cannot be ascribed wholly to age because the former stand was at 3,000 and the latter at 5,000 ft elevation, so that the sites were not comparable. Some figures for ponderosa pine, white fir, and canyon live oak are given in Table 5 along with related data.

The maximum dry weight of foliage in this table for a well-stocked stand of ponderosa pine 69 years old is 4.8 metric tons per acre. Since this stand is well above the age of culmination of the periodic increment, it seems likely that the amount of foliage on older stands is usually less than this. For dense willow saplings 10 ft high Lee (249) reports 1.1 metric tons per acre dry weight of leaves.

21. Site Quality and Rate of Growth.—The fact that greater growth is found on better sites has a corollary—that the greater growth results from a larger area and volume of foliage, and probably also from a larger volume of roots. These differences in site quality and growth, therefore, are likely to be reflected in greater influences or reactions on the site factors.

Suggestions of the effects of differences in site quality on foliage may be seen in Table 5. The 65- and 69-year ponderosa pine stands differ only slightly in stand-density index, but the former has site index 39 and the latter 30. Associated with the better site are higher dry-leaf weight per average tree and higher needle area per unit of crown-projectional area. Similarly, in the two canyon live oak stands of the same age and stand-density index of 410, the one with site index 18 in comparison with that of 31 shows lower dry-leaf weight per tree and per acre and lower ratio of leaf area to crown-projectional area. As more data become available it may be possible to generalize these relations.

The marked differences in amount of top and of root and in the ratio of top to root resulting from growth on favorable as contrasted with unfavorable soils are shown in Table 6 from Burger's work in Switzerland (64) with pine, spruce, and fir.

The much greater growth of both tops and roots and for all three species on the chalk–limestone representing a good site is evident. The ratios are greater for pine and fir on the good site, but the spruce shows a higher ratio on unfavorable limey sand.

Of the relations between leaves and wood volume or increment, only a few indications are available. For white pine in Switzerland, Burger (63) found that a needle area of 1 hectare per hectare or between 1,000 and 1,100 kg of green needles were required to produce 1 cu m of wood. However, the relations between weight of leaves and total weight of wood vary with age and with site conditions. The latter may be illustrated by Burns' figures in Table 7 for bald cypress seedlings in the greenhouse (69). It seems unlikely that leaf weights or areas can be predicted from volumes of stems or branches without considering other variables.

The maximum periodic annual increment in the development of a stand presumably is laid on in the year when the amount and efficiency of the leaves are greatest. At least the age of maximum foliage in the life of a stand may be assumed to be the same as that of the culmination of the current increment. Furthermore, the volume of increment should be a reasonably reliable index of the amount of foliage. Comparisons of foliage between species on this basis evidently will be affected by differences in efficiency of the leaves in producing stem wood. This source of discrepancy may be allowed for, and the species allocated to groups according to increment and presumably also to maximum amount of foliage per acre. Cottonwood, redwood, red fir, white fir, Sitka spruce–western hemlock, and Douglas fir would

TABLE 6.—WEIGHTS AND RATIOS OF TOPS AND ROOTS OF 6-YEAR SEEDLINGS ON SOILS OF GOOD AND POOR SITES

Soil formation	Species		
	Scotch pine	Norway spruce	Silver fir
	Weight of tops, g		
Chalk–limestone	3,575	2,926	159
Limey sand	658	366	80
	Weight of roots, g		
Chalk–limestone	668	609	100
Limey sand	185	55	70
	Top/root ratio		
Chalk–limestone	5.4	4.8	1.6
Limey sand	3.6	6.6	1.1

constitute a group with large amounts of foliage. The white pines, red gum, red spruce, southern white cedar, and loblolly pine form an intermediate group. The seven less tolerant pines, aspen, and oaks

TABLE 7.—PERCENTAGES OF DRY WEIGHT IN DIFFERENT PARTS OF BALD CYPRESS SEEDLINGS UNDER DIFFERENT SITE CONDITIONS

Site	Part of plant				
	Leaves	Branches	Stems	Main roots	Fibrous roots
Water table above soil surface	10	13	47	25	5
Water table at soil surface	4	7	33	55	1
Drained soil	7	14	26	53	0

of the Central states make up the group with light foliage. This grouping, with a few exceptions, is quite similar to that on the basis of stand density. Where data for amounts of foliage are lacking, the density and periodic-increment figures of Table 3 give an indication of the probable relative intensities of the influences of well-stocked stands of different species on their respective average sites.

The relation between air-dry weight W of needles in kilograms and 10-year periodic increment i in cubic feet is linear for the four Scotch pines of different sizes, 150 to 160 years old, from the same site, according to the data of Dengler (110) and can be expressed by the equation

$$W = 37.8i - 1.2$$

For red pine, 28 years old in Vermont (70), the equation is

$$W = 30.1i + 0.2$$

In this and the three following equations, i represents 5-year periodic increment instead of 10. For jack pine in Minnesota (308) the equation is

$$W = 23.1i - 0.1$$

For Douglas fir plantations in Switzerland (62) it is

$$W = 15.5i - 0.5$$

For white pine, 28 years old in Vermont (3) it is

$$W = 13.8i + 0.1$$

Thus for five different species the regression coefficients vary only between 13.8 and 30.1 (if it is assumed that the 5-year periodic increment for Scotch pine would be about half that of 10 years), and the Y intercepts are not far from 0. On the other hand, the relation of dry weight of needles to current annual increment of stems may not always be linear as indicated by the eight sample trees from a 47-year stand of Norway spruce (73), which show a decelerating trend, while the white pines from stands of different ages and sites (3) yield an accelerating trend. A general statement of the relation must await further evidence, including variations associated with differences in age and site.

The fact that growth in basal area and in volume are closely related suggests that leaf weight would also be a linear function of basal area. This proves to be true for an 80-year stand of European beech (61) when oven-dry weight W of leaves in kilograms is expressed as a function of annual growth g in basal area in square feet per tree by the equation

$$W = 256.8g + 2.1$$

The relation of needle area to periodic increment is also linear for the Scotch pine, although the points show a slightly wider dispersion about the trend line. Needle area was also found by Tiren working with

Scotch pine in Sweden (343) to be a linear function of five-year periodic annual increment. The two were proportional at each age but did not give the same straight line at different ages, although the slopes were not widely divergent.

22. Cuttings and Logging, Insect, and Disease Damage.—Forest cuttings vary in kind and degree from a light thinning to a clear cutting that removes everything. In general, the effects of cuttings are (*a*) to reduce the density of the stand, (*b*) to change the proportion of sizes, or (*c*) to alter the proportion of species. Either one or more of these three results may be obtained in the same operation. These changes in the upper canopy react in turn upon the understory of woody and herbaceous species and usually tend to augment its size and density and to modify its composition, frequently by the increase of deciduous species of soft, easily decomposing foliage.

The roots of the trees cut and of the vegetation destroyed incidental to the cutting, if the species do not sucker or sprout, decay and form channels favorable to the movement of soil moisture and for the root development of subsequent vegetation. Decayed taproots of shortleaf pine in a field in North Carolina left large channels more than 3 ft deep. A certain amount of the slash, including limbs and foliage of the trees cut, is left on the ground to rot. This material partially shades certain portions of the area and contributes organic matter to the soil, part as rapidly decomposing foliage and part as the more bulky, less rapidly decomposing woody materials.

The logging operations following cutting result in the thinning or clearing of vegetation or denuding, or even channeling parts of the surface of the ground in the process of skidding. On slopes, if the skidding is straight downhill so that the skidding trails have steep gradients, maximum velocity of the surface runoff waters which concentrate in the trails is assured. The proportion of area cleared and denuded in logging tends to increase with the power and speed used in logging, although it rarely exceeds 25 per cent. Trucks and tractors cause much less denudation than donkey and cable methods.

Insects and diseases cause openings in the canopy that affect the influences in ways quite similar to the effects of certain cuttings. The resulting changes are less sudden, however, and the disturbance less serious, particularly at the ground surface or in the soil.

23. Fires.—Forest fires kill all or a part of the living trees, shrubs, and other vegetation in their path. The smaller, more inflammable parts may be wholly consumed, while the larger woody parts remain in a more or less charred condition. Fires also destroy part or all of the forest floor. Several degrees of destruction may be distinguished. Patches of vegetation and forest floor may be left unburned although surrounded by burned area. Patches that are scorched only on the surface may be left. Part or all of the area may be left with only charred remains of organic matter on the surface. Finally, part or all of the area may be burned completely down to the mineral soil.

In any case, fires leave a certain amount of ash and small blackened pieces on the surface. These materials are light and blow or float readily. The nitrogenous constituents of the organic matter are volatilized wherever the forest floor has been reduced to ash, and other mineral constituents, notably potassium, phosphorus, calcium, and magnesium are released from the organic complexes and, probably with some loss, are left in more soluble form. Thus they are more easily translocated by water and are more readily available for plant nutrition. There are also left after a fire the charred woody remnants, which are highly resistant to decay.

The intensity of a burn varies with the local climate and the season, with the degree of slope, with the exposure and topographic location, and more generally between different regions. The results of fires in forest influences are likely to vary accordingly. The heating effects of fires have been measured by means of thermocouples. In leaf litter in the Appalachian region, Korstian (207) found maximum temperatures of 300 to 1000°F within 1½ to 3 minutes after the beginning of burning with a subsequent reduction to 300° or less in 8 minutes after the start of the fire. In a slash fire in Douglas fir in the Pacific northwest, Hofmann (155) recorded the following temperatures at different locations with respect to the surface of the ground.

Location	Temperature, °F
30 in. above ground	850
¾ in. under duff	120
Under 1 in. of mineral soil	75
1¼ in. under surface of rotten log	65
In mineral soil under 1½ in. of duff	60

In nine different areas of chaparral in northern California, Craddock and Sampson (305) recorded maximum temperatures during fires at 1.5 in. below the surface of the mineral soil of 630, 395, 240, 235, 230, 215, 160, and two below 100°F. The downward heating effect of an intense forest fire, even a short distance below the surface of the mineral soil or other noninflammable material, may or may not be sufficient to cause the alteration of part of the colloidal and organic material.

The character of the burn affects the remaining vegetation, the forest floor, and the succession of vegetation that may follow. In all three ways it also affects forest influences. The more severe the burn the greater is the change from the previous condition and the greater the change in influences. Fires, in destroying a dense forest, are sources of great changes in influences, and consequently they have been the causes of such serious disturbances in matters of runoff, floods, and erosion that they have been instrumental in emphasizing the importance of forest influences.

24. Natural Succession.—Cutting or fire always results in the development of new vegetation, which may often take the form of a new secondary succession. The new community differs from the old in age, size, and density and may differ in composition. The greater the disturbance, the greater is the change in the succeeding vegetation. After a hot fire that destroys everything, the succession may start back with the pioneer herbaceous plants like fireweed, or with the pioneer tree species like lodgepole pine, jack pine, or aspen. After a light thinning the only change may be in the density and proportion of sizes. There are all intermediate degrees of cutting or fire damage.

When the climax or near-climax stages are destroyed or severely disturbed they tend to be replaced by widely different vegetation, whereas pioneer stages when destroyed tend to reestablish themselves. For example, the climax sugar maple–beech is replaced first by fireweed and raspberries, which in turn are followed by aspen and birch before the climax is reestablished. On the other hand, lodgepole pine destroyed by fire is immediately replaced by a new stand of lodgepole.

As the succession progresses toward the climax the forest tends to become more tolerant and in better adjustment with the site. It more completely utilizes the growing space because of the greater variety of species and the stratification of their tops and roots. In the earlier stages in the succession the species are less tolerant, less well adjusted to the site, and less effective in utilizing the growing space. Ordinarily, the number of species is limited, and they are usually of about the same size.

As the site is less favorable the succession becomes less complex. In the chaparral, for example, on poor sites the same species may represent both the pioneer and climax stages. On good sites in the east, on the other hand, there is a complex succession from herbs through pine to oak and finally to sugar maple and its associated climax species.

Inasmuch as each disturbance by cutting or fire results in a change in the vegetation, and any change in vegetation is expressed in differences in size, composition, or density, corresponding differences in influences may be expected with successional changes of any kind or with retrogressions caused by cutting or fire.

25. Vegetational Factors: Summary.—The differences in forest or other vegetation that may be expected to be reflected in forest influences include the following:

1. Composition, species, crowns, and roots.
2. Density of individual trees and stands.
3. Size and age.
4. Site quality and rate of growth.
5. Cuttings and logging, insect, and disease damage.
    a. Kind and degree.
    b. Slash.
6. Fires.
    a. Effect on vegetation.
    b. Effect on the forest floor and soil.
7. Natural succession.

CHAPTER V

DETERMINATION OF INFLUENCES

Vegetation is almost infinitely variable, and it is reasonable to expect that its influences will be equally variable. The vegetation reacts on an environment that is also extremely complex and variable. One of the greatest obstacles to scientific development in the field of forest influences has been the difficulty of obtaining and applying conclusions from these complex groups of interacting variables.

26. Methods.—One method of approach is the experimental in which, as nearly as possible, all except one factor are held constant, and that one is varied under known and controlled conditions. The method can never be applied rigorously in natural habitats and in its accepted form is limited to laboratory use. It can be extended from the laboratory by partially controlled experimentation with the same material in the field, using phytometers, including some with cores of natural soil as an intermediate step between the laboratory and the natural site. The constancy or control of the factors in such a series becomes progressively less with the approach to natural field conditions. In other words, scientific experimentation requires that all the conditions of the experiment be exactly known and hence exactly comparable. This requirement can never be attained in natural sites because two or more such sites are never exactly known or comparable. For that reason, comparisons between different areas, however similar they may appear, will always lack the requirements of true experimentation. If the dissimilarity of different areas is overcome by using the same area over a period of time, partly before and partly after artificial modification of the vegetation or some other factor, the variation of the climatic and other factors with time again obviates the possibility of complying with the conditions of experimentation. In most of the problems of forest influences, laboratory experiments cannot be carried out on a scale sufficiently extensive to cover the range of field conditions or to be applied directly and satisfactorily to natural sites. The application of results from laboratory experiments becomes particularly unsatisfactory in the frequent instances when it is desired to make quantitative applications to a large area such as a whole drainage basin. Laboratory experimentation is desirable and essential for the solution of those questions which can be answered by that method, but the application and generalization of conclusions from laboratory determinations require a different method of approach.

Another method is the biometric analysis of the variation found under natural conditions. Such analysis requires a sufficient number of random samples representative of the variations to be evaluated. The results of the analysis of the data provided by these samples can then be interpreted in terms of probability for any desired degree of precision. In this way the relations between the important factors in natural environments can be made quantitative. Similarly, the applicability to any area from which an adequate number of representative and random samples has been collected, can be tested.

Both the experimental and biometric methods have their uses in forest influences, although unfortunately they have been little used. Depending on the nature of the problem, either may be used independently, or the former may be used to develop and test an hypothesis, and the latter to determine its degree of applicability under field conditions. Unfortunately, it is still necessary in forest influences to use data and to attempt to obtain indications from material that has not been determined or tested by either method and is correspondingly unsatisfactory.

Several methods have been used in making comparisons to evaluate the effects of vegetation. The first and simplest is the comparison of two areas with different cover, usually forest versus open. For a valid comparison, they must be far enough apart not to influence each other, which means several times the height of the trees, but at that distance there is no assurance that factors other than the cover are identical between the two areas.

Second, on the same area, records may be taken in and below the crowns for comparison with those above the crowns by instruments located on towers. This comparison, however, is questionable because the difference in elevation may cause differences that are not attributable to the vegetation.

Third, a comparison may be made on the same area by obtaining records in the forest and then removing the forest and continuing the same records for a subsequent period. In this case, however, the climatic factors vary widely with time, and a long series, probably 20 years or more, is necessary to average out that variation. Considering the irregular climatic cycles, that is perhaps impossible.

Fourth, by combining one and three, two areas may be compared, both forested and as nearly identical as possible, for a preliminary period to establish the relations between them either as ratios or differences. Then by removing the forest from one of the areas and comparing the ratios or differences between them after deforestation with those before, comparisons may be obtained that have some degree of control and should be reliable.

Fifth, it would also be possible to combine two and three, obtaining relations before and after cutting as a control. This has not been used but would be possible and desirable.

The first four methods have been used, the fourth, which is undoubtedly the best, less commonly than the first and third. It will be necessary to use data collected in any of these ways to establish relations with climatic and other factors, but the weaknesses in the methods by which comparisons are established should be recognized.

Mention should also be made of the rational method, much used in engineering, in which functional mathematical relations are formulated on the basis of physical laws. Often empirical constants must also be determined. Their applicability is then tested by comparisons of computed and observed values. Evaporation and runoff as functions of climatic and channel characteristics have been analyzed fruitfully by this method. It is to be recommended whenever there is a sufficient physical basis for the formulation of the functional relations. Unfortunately, this is rarely the case with the complexes of natural vegetation and soil.

Whatever the method, the attempt will be made to give quantitative expression to factors and relations. The influences on individual climatic factors will be taken up in a sequence from independent to more or less dependent so that interrelations will be clear from what has gone before. In such a sequence, solar radiation takes first place.

CHAPTER VI

SOLAR RADIATION AND LIGHT

The sun is the source of the radiant energy and light reaching the surface of the earth and, at times of clear sky, consists of 80 to 90 per cent of direct sunlight and 10 to 20 per cent of sky light. In forest influences the radiation as a source of energy and heat is a more important aspect than the illumination. In general, the daily solar radiation at sea level on a horizontal surface with clear sky at 40° north latitude varies with the season of the year from 100 to 600 cal per sq cm.

27. Variations Not Associated with Vegetation.—The actual amount of radiation at a station varies with latitude, season of year, altitude, degree, and aspect of slope and condition of atmosphere. The considerable variation with season and slope may be exemplified by the figures from Wagon Wheel Gap, Colorado, in Table 8 (37). The differences between south and north exposures are much greater in winter than in summer.

TABLE 8.—TOTAL DAILY GRAM CALORIES PER SQUARE CENTIMETER ON DIFFERENT ASPECTS AND SLOPES AT 9,500 FT ALTITUDE

Station	*A*–2	*B*–2	*A*–1	*B*–1
Exposure	SE	SE	NW	NE
Slope, deg	34	30	31	37
Date:				
Dec. 21	379	402	0	0
Jan. 21	....	....	11	
Feb. 21	....	....	105	46
Mar. 21	588	596	259	181
June 21	643	641	608	555

28. Influences of Forest.—For local comparisons between forest and open or between different types, densities, or sizes of vegetation, it is usually convenient to take the sun's radiation at noon on a clear day as a standard and express the radiation under different conditions of cover relative to that standard as 100 per cent. The thermopile, an instrument that is equally sensitive to all wave lengths of light, is probably the most satisfactory instrument for the measurement of the effects in forest influences. It measures the radiant energy rather than the illumination, thus including the heating effects of the infrared rays. The duration of sunlight, the length of day, and the alternation of clear and cloudy periods with the resulting variable total amounts of radiation for any given period are all important factors that are modified by the forest canopy.

Data on radiation as it is affected by the cover are available from several sources. The following percentages of area shaded in the chaparral of the Angeles National Forest in southern California based on averages of 21 plots in each of the two watersheds are reported by Plummer (289).

Watershed	Elevation, ft	Aspect			
		N	S	E	W
		Percentage shaded			
Pasadena	2,000–3,000	84	57	63	61
Pasadena	3,000–5,000	66	70	75	68
Santa Ana	2,000–3,000	64	39	51	49
Santa Ana	3,000–5,000	60	52	59	58

In general, the north aspects receive less and the south aspects more insolation than the others, with exceptions at the higher elevations above Pasadena. The differences between exposures and elevations are doubtless partly to be attributed to differences in the species of chaparral associated with aspect and altitude.

The light intensity in percentage of full sunlight measured with a radiometer by Shirley (317) in a red pine forest in northern Minnesota was (*a*) for old-growth pine 17, (*b*) for dense pine reproduction from 0.9 to 8, (*c*) for hazelnut and other brush from 0.7 to 16 per cent. It is interesting that the dense deciduous undergrowth, such as the hazelnut, may reduce the radiation reaching the ground to a greater extent than the forest or any other form of vegetation for which records are available.

In white pine in New Hampshire, Li (220) and Smith (326), using the difference between black and white atmometers, computed the light intensity reaching the forest floor on the basis of 100 per cent in a denuded area. Certain relations to basal area and age are suggested by the following figures:

Basal area per acre, sq ft	0	110	127	155	316
Age, years	...	30	50	55	Uneven
Radiation at ground, %	100	6	22	14	6

CHAPTER VII

AIR TEMPERATURE

The evidence in the last chapter that a forest canopy may reduce the solar radiation to less than 1 per cent of that in the open leads logically to the expectation that, because the sun is the source of the heat, the daytime temperatures where part of the sun's radiant energy is intercepted by the trees will be lower than those in the open. Moreover, it is common experience that it is cooler in the shade than in the sun. But how much cooler? The difference is obvious qualitatively but not quantitatively. And is the difference the same close to the ground surface as it is a few feet above? If not, are the vertical distributions of temperature in the forest and in the open distinct and definable? These questions require numerical answers for which some figures are available.

Similarly, a forest canopy will act as a blanket in reducing the effectiveness of outgoing radiation from the surface. Thus under a forest an increase in the minimum temperature may be expected.

30. Physical Relations.—As a result of the differences in the duration and intensity of solar radiation there are both seasonal and diurnal changes in air temperature. Temperatures also vary with the passage of cyclonic disturbances and with the altitude and latitude, season, slope, exposure, and wind movement. These various modifying factors tend to obscure variations caused by the forest in intercepting parts of both incoming and outgoing radiation.

The means and ranges of temperature are the averages and the differences, respectively, between maxima and minima. Hence, if a forest canopy reduces the maximum and increases the minimum, the range will be reduced and by an amount greater than the change in either one. The effect on the mean depends upon the balance between the opposing changes in maximum and minimum. If the maximum is lowered more than the minimum is raised, the mean will be reduced or vice versa. If the changes in maximum and minimum are equal, also being in opposite directions, there will be no change in the mean. More generally, the influence of a forest canopy is likely to be obscured by figures for means as compared with maxima or minima. These relations can be seen in the graphs of diurnal trends of temperature in Fig. 7. In the usual case as at Berkeley, California, the maximum is lowered more than the minimum is raised, and consequently the mean is reduced. In the extreme example from northern Arizona at an elevation of 7,250 ft, the night minimum in the forest was increased by 30°F while the noon maximum was not lowered, and thus the mean was greatly increased. In both instances the daily range in the forest was lower than that in the open.

Maximum temperatures are modified by the forest also through its effect in minimizing the heating of the ground surface and thus reducing

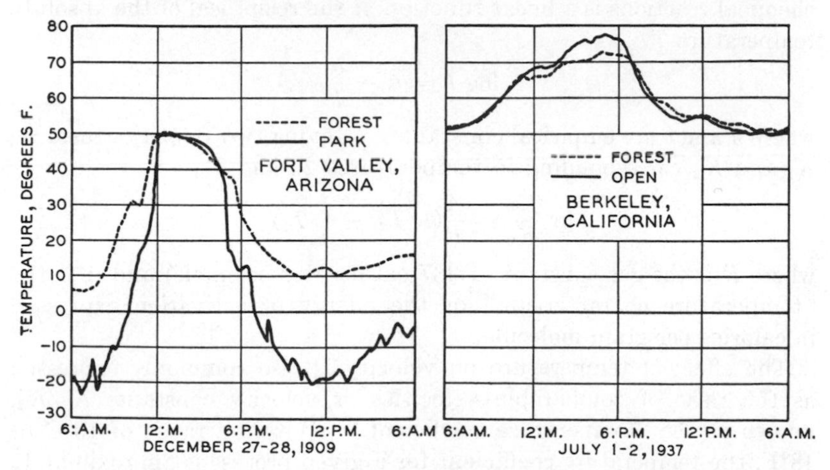

FIG. 7.—Comparison of diurnal trends of air temperature in forest and open. (*From data of United States Forest Service.*)

the resulting convectional currents of warm air. This influence is closely related to and acts in the same direction as the direct solar radiation. Both convection and radiation are most effective close to the surface, and thus the influence of vegetation decreases upward from the ground.

Another factor that affects chiefly the night minimum temperatures is the flow of cold air from higher to lower elevations down canyon bottoms or into depressions. This tends to reduce the minima. However, it is not much affected by the forest canopy. Moreover, it tends to reduce the minima at night, whereas the influence of a forest canopy is to increase them.

Temperature influences the rate of most chemical, physical, and biological processes. The influence of the forest on temperature derives importance from its effect on the length of the growing season; on occurrence of frosts or excessively high temperatures; on soil temperatures, the viscosity of the soil solution, and the activity of microorganisms; on humidity and evaporation; on transpiration; on snow melting; and on the movement of soil water.

Molecular activity increases with rising temperatures, and consequently it is not surprising that the velocities of many physical, chemical, and biological processes increase with temperature. The relations between temperature within limited ranges and velocity show well-defined trends, which have been given mathematical expression. Thus the logarithm of the velocity constant k for many chemical reactions is a linear function of the reciprocal of the absolute temperature T.

$$\log k = a - \frac{b}{T}$$

where a and b are empirical constants. Having two velocity constants K_1 and K_2 corresponding to temperatures T_1 and T_2,

$$\ln \frac{K_1}{K_2} = \frac{A}{R}(1/T_2 - 1/T_1)$$

where R is the gas constant (1.987 cal per deg per mole) and A is the "temperature characteristic" or the energy of activation expressed in calories per gram molecule.

The effect of temperature on velocity is also commonly expressed as the ratio of comparable velocities or velocity constants K_1/K_2, known as the "temperature coefficient." Within a range of 10°C or 18°F, the temperature coefficient for a given process is approximately constant and is expressed by the symbol Q_{10}. Thus

$$Q_{10} = \frac{K_1}{K_2}$$

The relations between Q_{10} and A for different 10°C ranges of temperature have been determined, and their values for several processes summarized by Barrows (33). When $Q_{10} = 1$, $A = 0$. When $Q_{10} = 2$, $A = 11{,}800$. As his figures show, Q_{10} varies with temperature, and in many processes A does also. However, if log K plotted over $1/T$ gave a strictly linear trend, the temperature characteristic A would necessarily be constant and independent of temperature and thus preferable to Q_{10}.

Actual values for Q_{10} according to Barrows averaged, for pressure of water vapor at saturation from 1.98 for 32 to 50°F to 1.74 for 86 to 104°F; for evaporation from a free water surface, 1.95 for 32 to 50°F and 1.54 for 50 to 68°F; for rate of ice formation, 2.00 at −4°F. The corresponding A's were 11,200, 9,800; 10,200, 7,200; and 11,400.

Q_{10} for transpiration of potted conifer seedlings varied from 1.3 to 1.7 on the basis of Roeser's figures (296). Assuming a temperature coefficient of 2, Livingston (223) has applied it in his temperature efficiency index I which is expressed in degrees centigrade in the equation,

$$I = 2^{\frac{t-4.5}{10}}$$

or for the Fahrenheit scale,

$$I = 2^{\frac{t-40}{18}}$$

where t is temperature.

Starting a little above the surface, the decrease of temperature with increase in altitude in the absence of disturbing factors is at the rate of about 1°F for each 300 to 455 ft, depending on the topography.

31. Effect of Elevation of Thermometers.—Within the forest itself air temperatures also change with minor differences in elevation above the ground surface. Bates (34) gives the following differences resulting from an increase from 1 to 20 ft in elevation of the thermometers.

Type	Temperature difference	
	Mean annual	Mean growing season
Ponderosa pine	−2.6	−3.0
Engelmann spruce, northeast slope	+0.3	−1.0

By changing the heights of the thermometers from 5½ to 8 ft during June and July, Pearson (282) found the mean maximum temperature was reduced by 0.6°F and the mean minimum increased 2.1°F. A change from a height of 8 in. to 18 ft above the ground during the period from July to September in Vermont, based on temperature summations above 40°F gave the following results reported by Adams (1):

Type	Unthinned	Thinned
White pine	Warmer by 1.8%	Cooler by 9.2%
Scotch pine	....	Cooler by 14%

The foregoing figures illustrate several aspects of the influence of forests on air temperature. In the ponderosa pine the effect is doubtless due to the greater intensity of convection and radiation from the heated ground surface at the 1-ft as compared with the 20-ft level. In the Engelmann spruce the effect of cold-air drainage in a canyon during fall and winter reduced the night minima at the 1-ft level more than at 20 ft. In general an increase in distance above the ground within the stand results in a reduction of maximum and of mean temperatures and in an increase of minimum temperatures.

Because small changes in elevation of thermometer in relation to ground and crowns cause marked differences in temperature, thermometers installed to provide comparisons should be at the same level with respect to the surface.

32. Vertical Distribution of Temperature near the Surface.—Variations in temperature at different distances above the surface such as those already mentioned result from a combination of physical causes including the temperature of the surface, the intensity of convectional currents and turbulent mixing, and the density, absorption of radiation, and temperature of the vegetative canopy. In the simple case of bare ground in the open the law of vertical distribution may take either of two forms (338). In one, if the atmosphere is unstable (vertical movement prevalent) the temperature t is proportional to the logarithm of the height above the surface h, and

$$t = c \log h$$

The constant of proportionality c can be determined from several records of temperature at different heights. This relation is valid only for heights of less than 50 ft.

In the other, if the atmosphere is stable (vertical movement negligible) the function can be expressed as

$$t = t_0 + \Delta t_1 h^{\frac{1}{n}}$$

where t_0 is the temperature close to the surface, Δt_1 is the difference in temperature per unit of height to the $1/n$ power, and n is nearly constant, varying between 5 and 7. For example, on an open south exposure at noon in August, in Berkeley, California, this equation represented the Fahrenheit gradient closely in the first 3 ft when $\Delta t_1 = 12$ and $n = 5$. However, the conditions at the time were presumably unstable.

In a forest the vertical distribution of temperature would be expected to be different from that in the open because the canopy intercepts, absorbs, or reflects considerable parts of the radiation. Thus during the day the heating near the ground surface is less, but in the upper surface of the canopy a local maximum develops that is less intense however, than that at the ground in the open because there is less concentration of heating over the irregular exposure of the crowns than on a nearly plane ground surface. The resulting vertical distributions as sketched by Geiger (134) showed, in a dense forest at noon, the maximum at the upper surface of the canopy and the minimum close to the ground with an S curve between. In the morning the maximum was in the crowns. However, the total range between maximum and minimum was less than 2°F. The vertical distributions tend to approach those in the open as the crown density decreases.

33. Influence of Forest on Temperature.—Considerable evidence is available to show that a forest or other canopy of vegetation influences air temperatures significantly if not by large amounts.

In the Wagon Wheel Gap experiment in Colorado (37), the net changes in temperature attributable to the deforestation of watershed *B* resulted in increases in the maxima, minima, and means in every month of the year. The net changes are given in the following tabulation. The extremes in individual months are underlined.

Annual	Oct.	Nov.	Dec.	Jan.	Feb.	Mar.	Apr.	May	June	July	Aug.	Sept.
Mean maxima												
2.1	3.6	2.6	1.6	1.8	3.2	3.2	1.2	1.4	0.8	1.0	1.6	2.8
Mean minima												
0.7	0.5	0.5	0.4	0.1	0.4	0.6	0.8	1.0	1.1	1.0	1.1	0.9
Monthly means												
1.3	1.6	1.4	0.9	1.1	2.8	1.6	1.0	1.1	1.2	1.3	1.2	1.4

The increases in the maxima were greater than the increases in the minima except in June and July. The increases in the maxima were greatest in the months from September to March inclusive, probably owing to the loss of foliage of the young aspens on watershed *B* in contrast to the effect of the evergreen cover on *A*. The increases in the minima were greatest from May to August.

In northern Idaho during July and August, 1931, the average maximum air temperature in a clear-cut area of the western white pine type was 86.8°F (187). An adjacent uncut area showed an average maximum 6.4°F lower. Departures in mean temperatures for the period from July to October in white pine stands in southern New Hampshire were reported by Li (220) as follows:

Measure of temperature	Departures from open, °F	
	30-year stand	55-year stand
Maximum	−4.2	−4.0
Minimum	+3.0	+3.0
Difference in daily range	−7.1	−7.1

The greatest daily difference of 7° in the maximum occurred on bright sunny days, and the greatest daily difference in the minimum, of 6.5°, on still, clear nights.

In the San Bernardino Mountains, where records were kept by Munns[1] from 1912 to 1917, three stations at an elevation of about 6,000 ft gave the following comparisons of different kinds and degrees of cover. The open station was in a 30-acre flat continuously exposed to the sun during the day. The dense chaparral, containing three

Measure of temperature	Actual temperature in open, °F	Departures from open, °F	
		Chaparral	Forest
Absolute maximum	85	+7	+1
Absolute minimum	10	−1	+6
Absolute monthly range	56	+8	−5
Maximum daily range	39	+10	−2
Minimum daily range	8	+1	−2
Mean daily range	27	+8	−5
Mean monthly maximum	63	+5	−2
Mean monthly minimum	36	−2	+5
Mean annual	50	+8	−5
Mean for days with snow on ground	34	−2	−1
Mean for days without snow on ground	51	+1	+2
Number of days freezing or with frost	126	+12	−32

[1] The figures are derived from an unpublished manuscript in the files of the U.S. Forest Service, Studies in forest influences in California, by E. N. Munns, who has kindly permitted their use.

species of oak, two of ceanothus, and one of manzanita and chamise, was 5 to 8 ft high and located about 400 yd west of and 50 ft higher than the control station. The forest station was in a 2-acre patch of scrubby short-boled Jeffrey pine with 0.8 crown density and 20 trees per acre averaging 30 in. in diameter and 80 ft high. There was no understory of brush.

The temperature extremes in the chaparral and notably the higher maxima may be attributed to stagnation of air movement in the dense brush combined with the effect of lowering the secondary maximum of the canopy to a level where it intensifies the primary maximum close to the heated ground surface. The temperatures in the open are intermediate between those of the chaparral and the forest, resembling the forest more closely in the warm parts of the year and the chaparral in the cold parts. The forest cover increased the minima more than it reduced the maxima, thereby contributing to an increase in the mean annual. This is the same phenomenon noted in ponderosa pine at 7,000 ft elevation in Arizona and is apparently a characteristic of high altitudes in the southwest, where the clear sky and still air contribute to intense radiation at night. The chaparral exhibits an influence opposite to that of the forest, probably because of the canopy close to the ground, which makes its influence similar to that of young low-crowned forest stands and distinct from older taller stands, in which natural pruning and thinning have eliminated most of the lower branches.

The influence of different degrees of cutting in Colorado may be exemplified from the four plots in a north slope stand of Douglas fir, which had the following residual stands: one clear cut, one with 25 per cent basal area, one with 33 per cent basal area, and one uncut (34). These data indicate that, as the degree of cover becomes heavier from open to uncut, the mean maximum temperature in January is decreased by 3.1°, and in July by 2.2°F. The mean minimum in July and in October is increased by 2.5°. The mean daily range in January is increased by 4.5°, while in July it is decreased by 2.9°. The absolute minimum in January increased by 4.4° and in July by 1.1°F. The intermediate degrees of cutting showed temperature departures between those of open and uncut but not in proportion to the basal areas of the residual stands.

After a cutting that removed two-thirds of the stand of western white pine type (187), the mean maximum temperature in July was closer to that of the clear cut than of the uncut, while the mean minimum, on the contrary, more closely approached that of the uncut (Table 9).

Not only the degree of cutting but also the kind of cutting affects the influence of the residual forest on air temperature. In the Douglas fir just mentioned, the difference between a shelterwood and a selection cutting was sufficient to obscure the effect of a difference in basal area between the two partially cut plots. Similarly, in Europe, von Wrede and Schubert have concluded that temperatures after shelterwood cuttings have a wider range, thus being more like clear cuttings than after group-clear cuttings, which are more like the uncut stand. The maximum ranges of temperature in August and September were recorded in large openings.

TABLE 9.—INFLUENCES OF FOREST ON AIR TEMPERATURES

Forest type and location	Departures, forest from open, °F				Ref.
	Mean maximum		Mean minimum		
	Jan.	July	July	Jan.	
Beech climax, N. Y.	.....	−8.3	+4.5	.....	(258)
W. white pine climax, Idaho	.....	−6.0	+7.0	.....	(187)
Hemlock, N. Y.	.....	−4.9	+3.9	.....	(258)
White pine, N. H.	.....	−4.8	+3.2	.....	(220)
W. white pine, two-thirds cut, Idaho	.....	−2.3	+5.2	.....	(187)
Douglas fir, 9,000 ft, Colo.	−3.1	−2.2	+2.5	+2.1	(34)
Jeffrey pine, 6,000 ft, S. Calif.	+0.7	−1.4	+6.7	+3.6	Munns
Jack pine, 25 yr, Neb.	−0.1	−1.3	−2.8	+1.1	(275)
Aspen–Engelmann spruce, 9,000 ft, Colo.	−1.8	−1.0	−1.0	−0.1	(37)
Douglas fir, three-quarters cut, 9,000 ft, Colo.	−1.0	−0.9	+1.1	+1.2	(34)
Ponderosa pine, 7,250 ft, Ariz.	−1.0	0	+5.0	+6.0	(282)
Chaparral, 6,000 ft, S. Calif.	+1.4	+5.0	−2.1	−1.6	Munns

The departures of forest from open for a variety of types, sites, ages, and regions are summarized in Table 9, arranged in ascending order of the mean maxima in July. In general the sequence of types in this table is from the tolerant, dense, climax types at the top to the intolerant, light-crowned, pioneer types at the bottom. The table supplements the specific examples that have been cited in support of the statements in the following summary:

34. Air Temperatures: Summary.—1. Slight changes in the elevation of the thermometers in relation to ground surface and crowns cause marked differences in temperatures.

2. Forest cover in general reduces the maximum temperatures throughout the year. The reduction in monthly maxima is more pronounced in July, when the departure from the open may be as much as −8°F, and −3° in January. Exceptions are found at 6,000 ft in the San Bernardino Mountain Jeffrey pine in January with +0.7°, and in chaparral, +1.4° in January and +5.0° in July.

3. Forest cover in general raises minima throughout the year, the monthly minima by as much as 6.0° in January and by 6.7° in July. These high values occur in the Southwest at high elevations where the night radiation in the open is intense. Exceptions occur in the chaparral and to a less extent in the spruce–aspen forest in Colorado.

4. Forest cover reduces the range of temperature—annual, monthly, or diurnal—the mean monthly by as much as 12.8° in July (under beech–maple climax in New York) and by 5° in January. An exception is found in the chaparral, which increases the monthly range by 7° in July.

5. Forest cover usually reduces the mean temperature because the maximum is lowered more than the minimum is raised. Exceptions occur in southern California and Arizona, where the minimum is increased more than the maximum is lowered, and therefore the mean is raised.

6. Cuttings produce effects on air temperatures intermediate between those of open and uncut forest. The lighter the cutting the more the temperatures resemble the uncut forest and vice versa, with exceptions.

7. Temperature effects of forest vary in different regions and with the character and density of the cover. The reduction of the maxima in summer increases with crown density. Evergreens are more effective in winter in reducing maxima and raising minima. Chaparral may show influences opposite to those of the forest.

CHAPTER VIII

WIND

Wind movement and particularly its velocity is important in forest influences; first, as the cause of moving sands and other soils; second, in connection with the use of trees in belts or groves as windbreaks; and, third, as one of the factors in evaporation, transpiration, and the measurement of precipitation by means of rain gages. In association with the movement of air masses it may also be related to changes in temperature and humidity.

35. Velocities and Kinds in the United States.—The United States is in the belt of prevailing southwesterly winds, although the directions are subject to much local variation as are also the velocities, depending upon cyclonic movements, local topography, relative heating of land and water surfaces with diurnal changes in the sun's heat, elevation above the ground surface, or altitude in the mountains. The average hourly wind velocities in the United States at 100 ft above the ground level vary from 7 to 15 mph, being highest near the oceans, plains, and Great Lakes. The velocities in summer average 2 or 3 mph less than in winter. At low elevations on the average they are 2 or 3 mph greater at noon than at night. At high elevations, as for example on Pikes Peak in Colorado, this relationship is reversed, and the velocities are 3 to 7 mph greater at night than by day. The average velocities on Pikes Peak vary from 13 to 38 mph.

From the viewpoint of shelter or protection from possible damage the important winds in the United States belong to three categories: (*a*) the cyclonic storm winds of the Atlantic Coast and Great Lakes, (*b*) the anticyclonic westerlies of the plains and prairies, which are dry and hot in summer and dry and cold in winter, and (*c*) the Chinook and Santa Ana winds of somewhat local occurrence in the western mountain regions, which are ordinarily descending desiccating winds of autumn, winter, or early spring.

36. Vertical Distribution of Wind Velocity.—All natural surfaces are more or less rough and therefore exert frictional resistance to the movement of the air above them. The rougher the surface, the greater is the friction. Moreover, the effect of the frictional resistance decreases with the distance above the surface, so that wind movement increases with elevation above the ground or canopy surface.

The vertical gradient of wind velocity above a rough surface is susceptible of physical analysis, which can be illustrated by the data of Fons (124) for velocities at different heights above the ground for winds of 5, 10, and 15 mph at the highest level of 116 ft. When these three series are plotted on semilogarithmic paper with the heights on the logarithmic scale, they form straight lines that nearly converge to a common point at height k, 0.19 ft, and velocity v_t, 2.9 mph (Fig. 8).

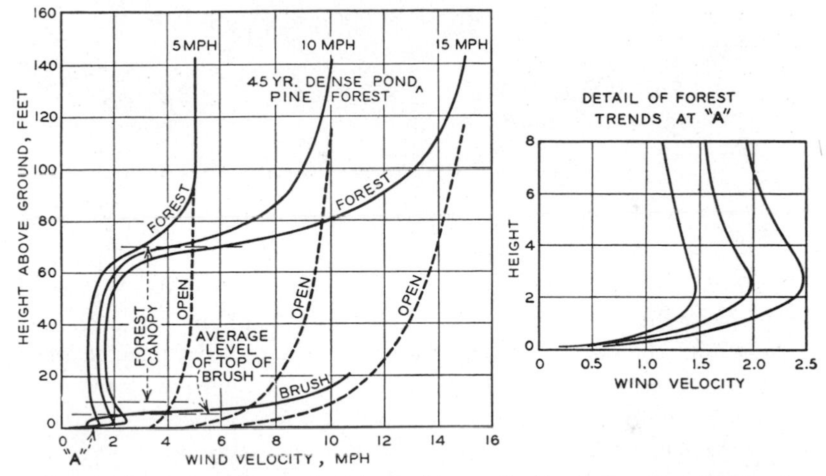

FIG. 8.—Vertical gradients of wind velocities on the Shasta Experimental Forest. [*After Fons* (124).]

Apparently the conditions are similar to those during sand blowing in the desert, for which Bagnold (24) has shown that the velocity v at any height h can be expressed as a function of the velocity gradient or friction velocity v^* by the equation,

$$v = 5.75v^* \log \frac{h}{k} + v_t$$

where

$$v^* = \frac{m}{5.75}$$

when m is the slope of the semilogarithmic trend.

Thus,

$$m = \frac{v_2 - v_1}{\log h_2 - \log h_1}$$

where subscript 2 represents a higher and 1, a lower level.

Hence,

$$v = m \log \frac{h}{k} + v_t$$

For any given condition of surface roughness, k and v_t will be constant regardless of the wind velocity at higher levels. For the three trends on the Shasta National Forest, $k = 0.19$ and $v_t = 2.9$, so that the equations become

$$\text{For 5 mph at 116 ft,} \quad v = 0.8 \log \frac{h}{0.19} + 2.9$$

$$\text{For 10 mph at 116 ft,} \quad v = 2.5 \log \frac{h}{0.19} + 2.9$$

$$\text{For 15 mph at 116 ft,} \quad v = 4.3 \log \frac{h}{0.19} + 2.9$$

These equations are of the general form for a logarithmic relation, namely, $v = m \log h + c$, suggested by Sverdrup (338) as applicable for high velocities and low atmospheric stability. He also suggested for stable conditions that the power law, $v = ch^{\frac{1}{n}}$, might better express the relation, where n is a constant for which empirical values between 4 and 7 have been derived. Fons' data (124) for the velocity distribution above the pine canopy form a linear trend on double-logarithmic paper that can be expressed for winds of 10 mph at 142 ft as

$$\log v = 0.21 \log h + 0.635$$

or

$$v = 4.3h^{0.21}$$

in which n has a value of 4.8.

Both forms of equation indicate that wind velocity increases with the height above the boundary surface. The similarity of form of these equations with those of the vertical distribution of temperature in Sec. 32 is noteworthy.

The wind velocity at a level H above the influence of the boundary was used by Fons (124) as the basis for predicting the velocity at any lower level h. He found that the relation could be expressed by an equation of the form

$$v_h = a + bv_H$$

where v is velocity in miles per hour and a and b are constants that, for the data in the open, had values for a between 1 and 1.5 and for b between 0.34 and 0.79. These constants would vary with the elevation and with the density of the vegetation. According to this equation the wind velocity at a lower level increases as the velocity above the influence of the boundary increases.

The relative velocity expressed as a percentage P_h becomes 100 v_h/v_H and

$$P_h = 100\left(\frac{a}{v_H} + b\right)$$

This may be interpreted that the velocity at the lower level relative to that above the influence of the boundary decreases as the latter increases. Figures illustrating the changes in absolute and relative velocities are given in Table 10.

TABLE 10.—ACTUAL AND RELATIVE WIND VELOCITIES AT DIFFERENT ELEVATIONS WITH RESPECT TO THE CANOPY OF PONDEROSA PINE 70 FT HIGH

Elevation, ft	Position relative to canopy	Velocities, mph			Relative velocities, %		
142	Above	5	10	15	100	100	100
30	In	1	1.3	1.6	20	13	11
2.5	Below	1.4	1.9	2.5	28	19	17

Not only do the curves of wind velocity in relation to height above the ground or canopy surface have characteristic trends, but similar trends obtain in reverse with distance downward from the canopy surface as the boundary. These relations are illustrated by the trends in Fig. 8.

Another expression for the relation of wind velocity to heights above the ground may be derived from Rohwer's data of the ratio of observed velocity v at different heights h to velocity close to the ground v_0 for several widely separated stations (297). The trend is essentially linear on double-logarithmic paper and may be expressed by the equation

$$\log \frac{v}{v_0} = 0.29 \log h + 0.06$$

The difficulty with this relation would be in the determination of the velocity close to the ground where the gradient is steepest and the velocities are most affected by the roughness of the surface. However, it is likely that the v_0 would approximately equal the v_t of the earlier formula and would be constant. In that case the formula reduces to the power law,

$$\log v = 0.29 \log h + K$$

where

$$K = \log v_t + 0.06$$

For these data, the value of n is 3.4, somewhat smaller than the lower limit of 4 suggested by Sverdrup.

37. Variation with Tree and Stand Differences.—A forest or other cover provides mechanical obstruction to wind movements and intercepts or deflects air currents with consequent effects on the force, direction, and velocity of the wind. The degree of influence of the cover varies with the height, length, and width of crown, and densities of individual crowns and of the stand. It therefore varies also with age, size, and species of tree.

TABLE 11.—RELATIVE WIND VELOCITIES IN PERCENTAGES OF THOSE ABOVE THE CROWNS FOR LEVELS IN AND BELOW THE CANOPIES OF PINE AND PINE WITH SPRUCE UNDERSTORY

Elevation, m	Wind class					
	Strong		Moderate		Light	
	Pine	P + Sp	Pine	P + Sp	Pine	P + Sp
13.7	75	46	83	48	77	47
1.1	50	17	58	17	62	27

When an air current reaches a forest stand, part is deflected upward with only a small change in velocity. Another part passes under the crowns with rapidly decreasing velocity, and a third part passes through and among the crowns with very low velocity. When the wind strikes an extensive forest there is a marked reduction in velocity close to the forest margin and only slight further reductions within the forest.

The vertical distribution of wind velocity in and above the forest has been suggested previously for ponderosa pine. Confirmatory and more detailed studies on microclimate in Germany by Geiger (135) give the wind velocities at 1.1 m above the ground, in the canopy at 13.7 m, and above the canopy at 16.9 m for pure Scotch pine stands and for Scotch pine with understory of Norway spruce. The data were classified according to wind velocity above the canopy in three groups: (*a*) strong winds, (*b*) moderate winds, and (*c*) light winds. The velocities for the three groups and for the two kinds of stand relative to the velocities above the crowns as 100, are shown in Table 11.

The figures suggest several conclusions. The wind velocity in the crowns is much less than above them and somewhat greater than below them. An understory reduces the wind velocity in the canopy and even more below it. The reduction of actual velocities in the canopy and below it is greater in strong than in light winds. The relative reduction varies little with velocity.

38. Effect of Shelterbelts.—Windbreaks or shelterbelts are usually narrow strips of trees and represent the simplest example of the influence of trees on wind. The chief interest is ordinarily in the effects to leeward of the trees, where velocities are reduced. This calming effect depends upon the mass of air calmed, which is determined by the height of the trees and upon the velocity in the calmed area, which is determined by the density of the stand and its foliage. The reduction

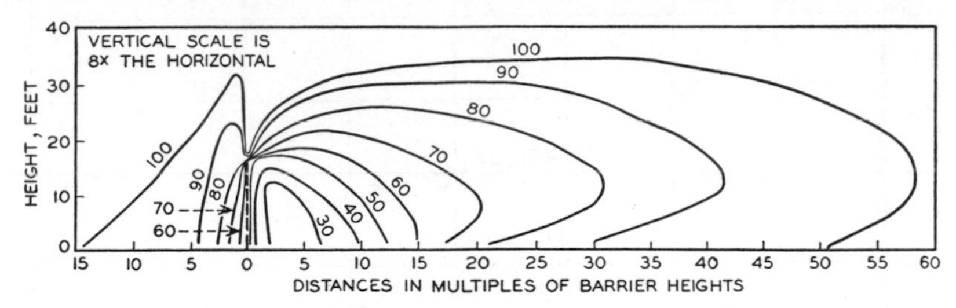

FIG. 9.—Distribution of wind velocities as percentages of velocities in the open at corresponding heights at different distances from a half-solid barrier. [*From Bates and Stoeckeler, United States Forest Service* (38).]

of wind velocity at different distances to leeward of barriers and belts of trees has been studied by Bates (36, 38). The distances to leeward were expressed as multiples of the height of the trees, a useful device facilitating application to belts of any height. The distribution of velocities relative to that in the open as 100 per cent at different distances and elevations to leeward and windward of a 16-ft wooden barrier constructed of 6-in. boards spaced 6 in. apart is shown in Fig. 9.

Such a barrier corresponds to a low-crowned shelterbelt. The reductions of velocity at 16 in. above the ground extend to fifty times the height of the barrier to leeward and ten times, to windward. The maximum effect is at three to five times the height to leeward, where the velocity is less than 30 per cent of that in the open. The 90 per cent line comes at thirty times the height. Beyond $15h$ the maximum reductions of velocity are found at heights of 8 to 12 ft instead of close to the ground level. For example, at 16 in. above the ground the relative velocity of 80 per cent is at $23h$, while at 12 ft it is at thirty-one times the height to leeward. Moreover, the influence of the barrier extends above its height so that, from 10 to $25h$ to leeward the velocities are 90 per cent of those in the open at 30 ft above the ground or almost twice the height of the barrier.

Several comparisons of the influences of actual windbreaks from Den Uyl's work in Indiana (111) are shown in Table 12. A dense belt of four rows of Norway spruce was more effective than two rows of spruce and pine, and these in turn were more effective than a light deciduous belt in winter. For these three, at 5 mph in the open, the relative

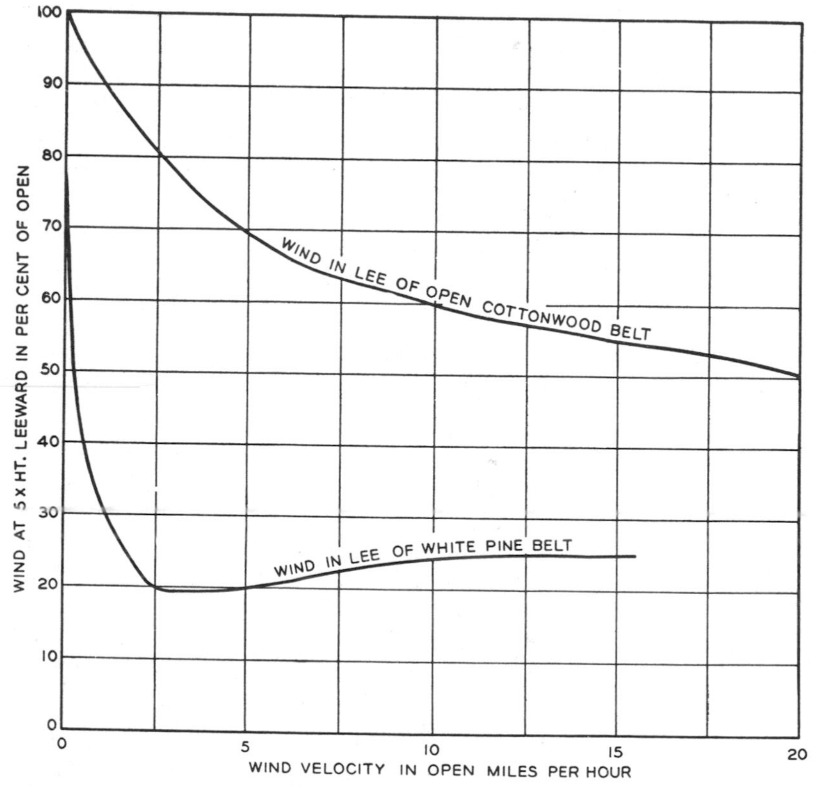

FIG. 10.—Reduction of wind velocity by shelterbelts. [*From Bates, United States Forest Service* (36).]

velocities were 4, 20, and 40 per cent and at 15 mph, 20, 27, and 53 per cent. The 4 per cent represents the maximum reduction in velocity so far published. A single row of willow reduces velocities more than fifteen times the height to leeward. Single rows of blue gum reduce velocities to one-third or one-half of those in the open.

Illustrating the effect of species with different crown characteristics, Bates (36) showed that at five times the height to leeward, an open cottonwood belt with high thin crowns reduced the velocity progressively as the velocity in the open increased up to 20 mph. At that point the velocity $5h$ to leeward amounted to 50 per cent of that in the open (Fig. 10). A white pine belt with lower denser crowns on the other hand showed maximum effect at velocities of 3 to 5 mph when the velocity $5h$ to leeward amounted to only 20 per cent of that in the open. In general the denser the belt of trees, the closer to the trees is the line where maximum reduction of wind velocity is found.

39. Effect of Forests.—The wind velocities in larger forest areas as compared with those in the open are also illustrated from several sources in Table 12. In the Wagon Wheel Gap experiment at elevations of about 9,000 ft in the Rocky Mountains, the wind movement after deforestation averaged 3.1 mph whereas before deforestation it averaged about 1 mph or three-tenths of the velocity in the open. A plantation of jack pine, 40 ft high, in the Nebraska sand hills reduced the average wind velocity at 23 ft above the ground to 28 per cent of that in the open (275). As the trees increased in height from about the 23-ft level of the tower in 1922 to 40 ft in 1939, the velocities decreased at an essentially uniform rate from 6.5 to 1 mph or from 85 to 13 per cent of the velocity in the open. Expressed in another way, as the trees grew 1 ft in height, the wind velocity at 23 ft among the crowns decreased by ⅓ mph or by 4.2 per cent of that in the open. Apparently in this stand the point of minimum velocity was lower than 17 ft below the tops, if it is assumed that the velocity increases again in the space below the living crowns.

At 8,000 ft on the east slope of the Rocky Mountains the wind velocities in a mature stand of Douglas fir on a north slope measured at 1 ft above the ground are reported by Bates (34) as follows:

Character of stand	Mean annual velocity, mph	Velocity, % of open	
		Mean annual	Growing season
Open	2.3	100	100
25 per cent left (shelterwood)	1.4	63	54
35 per cent left (selection)	2.0	89	75
Uncut	0.9	40	33

During the growing season when the actual wind velocities were reduced, relative reduction by the forest cover became greater. The stand left after the selection cutting, although 35 per cent was left, was less effective in reducing wind velocities than the 25 per cent residual stand left in the shelterwood system.

TABLE 12.—EFFECT OF FORESTS AND SHELTERBELTS ON WIND VELOCITIES

Forest		Miles per hour		Percentage relatives in forest, open = 100	Reference
		Open	Forest		
Western white pine, Idaho		1.4	0.2	12	(187)
Western white pine, half cut, Idaho		1.4	0.8	58	(187)
Jack pine plantation, Nebr.		7.7	2.1	28	(275)
Aspen, Colo		3.1	1.0	32	(37)
Douglas fir, Colo.		2.3	0.9	33	(34)
Douglas fir, three-quarters cut, Colo.		2.3	1.4	63	(34)
Ponderosa pine, Ariz.		5.3	2.7	51	(282)
Birch-maple-fir, cutover, N.Y.		...	...	80	(332)
Shelterbelts	Distance to leeward multiples of height h	Open	Shelter-belt	Percentage relatives in shelterbelt, open = 100	
Blue gum, S. Calif.	$3h$	21	7	34	(ms*)
Blue gum, S. Calif.	$6h$	21	10	49	(ms*)
Norway spruce, 4 rows, dense, Ind.	$2h$	5	0.2	4	(111)
Norway spruce, 4 rows, dense, Ind.	$2h$	{10, 15}	{2, 3}	20	(111)
Spruce and pine, 2 rows, medium dense, Ind.	$2h$	{5, 10}	{1, 2}	20	(111)
Spruce and pine, 2 rows, medium dense, Ind.	$2h$	15	4	27	(111)
Deciduous, in winter, density light, Ind.	$2h$	5	2	40	(111)
Deciduous, in winter, density light, Ind.	$2h$	10	5	50	(111)
Deciduous, in winter, density light, Ind.	$2h$	15	8	53	(111)
Willow, 1 row, medium dense, Ind.	$15h$	5	4	80	(111)
Willow, 1 row, medium dense, Ind.	$15h$	10	9	90	(111)
Willow, 1 row, medium dense, Ind.	$15h$	15	14	93	(111)

* From an unpublished manuscript, through the courtesy of Woodbridge Metcalf, Extension Forester, California Agricultural Experiment Station.

The rather open mixed forest of yellow birch, sugar maple, spruce, and balsam fir in the Adirondacks of New York was found by Stickel (332) to reduce the wind velocity measured at 2 P.M. by 20 per cent, based on records from April to October for 3 years. At 7,200 ft in the ponderosa pine area of northern Arizona, Pearson (282) found that the velocities in the forest averaged 51 per cent of those in the open park. The maximum difference was found in the middle of May, when the velocity in the park was 7.2 mph, as compared with 3.4 in the forest, and the minimum was in the middle of August, when the respective velocities were 3.5 and 2.1 mph. In the western white pine forest with understory of western redcedar and western hemlock in northern Idaho, Jemison (187) showed the effects of cutting on wind movement per day for July and August for the period 1931–1933, as follows:

Wind velocity	Clear cut	Half cut	Uncut
Miles per hour	1.4	0.8	0.2
Relatives (%)	100	58	12

The wind velocities are not directly proportional to the degrees of cutting.

The data that have been used in the foregoing examples and recapitulated in Table 12 also serve to support the following summary of conclusions:

40. Wind: Summary.—1. The degree of influence of forest cover on wind movement varies with the height, length, width, and density of individual crowns and with the density of the stand.

2. The velocity may be reduced even by a shelterbelt to only 20 per cent of the velocity in the open. In the forest, it is usually 20 to 60 per cent of that in the open.

3. Reduction of actual velocity by the forest is greater as the velocities in the open are greater. The relative reduction varies less with velocity.

4. Velocities in the forest are characteristically low, usually ranging from 1 to 2 mph on the average.

CHAPTER IX

ATMOSPHERIC MOISTURE

The amount of water vapor in the atmosphere has significance in relation to forest cover both directly as in the addition by transpiration and indirectly as a factor in evaporation, condensation, precipitation, and radiation.

41. Measures and Determination.—Different measures of atmospheric humidity are available and have been used to express the influences of forest cover. These measures are related but not identical, and the interpretations derived from any one may or may not be the same as those indicated by another. Commonly, the humidity, however expressed, is obtained from the readings of the wet- and dry-bulb thermometers of a psychrometer.

Atmospheric humidity depends first on the amount of water vapor in the atmosphere. This may be expressed as either absolute humidity or density, the mass of water vapor per unit volume of air as either grains per cubic foot or grams per cubic meter, or as the vapor pressure resulting from the presence of the given amount of water vapor, expressed in inches or centimeters of mercury. Differences in absolute humidity or vapor pressure are independent of temperature; that is, the mass of water vapor in unit volume of atmosphere, as expressed by a given value of either absolute humidity or vapor pressure, is the same whether the temperature is high or low. On the other hand, when the air is saturated with water vapor both the absolute humidity and vapor pressure increase with temperature. The values of vapor pressure at saturation in inches of mercury for different Fahrenheit temperatures are given in psychrometric tables (242) and for absolute humidities in grains per cubic foot by Meyer (251). The trend of saturated vapor pressure e_s over temperature t is nearly linear on semilogarithmic paper when vapor pressure is plotted on the logarithmic scale. It may be expressed by the empirical equation

$$e_s = 0.18 \times 10^{\frac{7.5t-240}{t+395}}$$

The constants in the exponent take different values if the air is over ice instead of water.

Absolute humidity at saturation H_a in grains per cubic foot may be computed from the vapor pressure in inches of mercury by the equation

$$H_a = \frac{0.622\rho e_s}{B_0[1 + 0.002(t-32)]} = \frac{11.746e_s}{1 + 0.002(t-32)}$$

where the constant 0.622 is the ratio of the molecular weight of water vapor, 18, to that of dry air, 28.9; ρ is the weight of a cubic foot of dry air in grains under standard conditions; B_0 is the barometric pressure at sea level in inches of mercury; and 0.002 is the coefficient of expansion of air per degree Fahrenheit.

Specific humidity H_q is another expression of atmospheric moisture that may be defined as the mass of water vapor in a unit mass of humid air. It may be derived approximately from the vapor pressure e and barometric pressure B as

$$H_q = 0.622\frac{e}{B}$$

Specific humidity is independent of the units in which e and B are expressed, provided only that they are the same.

If the air is not saturated, then the wet-bulb temperature in addition to the dry-bulb temperature is required to obtain any of the measures of humidity. The difference in temperature between dry and wet bulb $(t - t')$, referred to as wet-bulb depression, is sometimes used as a measure of humidity. In conjunction with the dry-bulb reading, it makes possible the determination of the other measures of humidity. Psychrometric tables give values of humidity corresponding to different values of the wet-bulb depression.

Vapor pressure e may be computed by the psychrometric formula,

$$e = e_s - 0.000367B(t - t')\,[1 + 0.00064(t' - 32)]$$

where e_s is the saturated vapor pressure at the temperature of the wet-bulb t'. It may be obtained more quickly from the psychrometric tables in which the actual vapor pressure is equal to the saturated vapor pressure at the temperature of the dew point. The tables give dew-point temperatures for different values of air temperature and wet-bulb depression.

The tables also give values of relative humidity H_r for different values of t and $(t - t')$. If the actual and saturated vapor pressures (or absolute humidities) are known,

$$H_r = \frac{100e}{e_s}$$

Similarly, saturation deficit H_d is

$$H_d = e_s - e$$

or

$$H_d = e_s \frac{(100 - H_r)}{100}$$

Under natural conditions, higher vapor pressures or absolute humidities are ordinarily associated with higher temperatures and vice versa. The curve of increasing vapor pressure at saturation with increasing temperature within the range of from 30 to 90°F indicates that, for a 1°F change in temperature, there is a change in vapor pressure in the same direction, amounting to about 0.02 in. Similarly, the values of actual vapor pressure plotted over temperature from the data presented by Stickel (332) indicate an increase of 0.011 in. for an increase of 1°F. This relationship is not linear, and consequently the foregoing figures represent only approximations within the temperature ranges usually encountered.

If it were desired to measure the amount of water vapor added to the atmosphere by transpiration from a forest, vapor pressure, absolute humidity, or specific humidity would be the most useful of the humidity measures. Actually, differences in absolute humidity between forested and open areas are small and tend to be minimized by the rapid diffusion of water vapor in the air and by wind movements, which further accelerate the mixing process.

Saturation deficit also increases with an increase in temperature at an accelerating rate. Within the usual temperature ranges this increase amounts to from 0.02 to 0.025 in. per °F. In this case a change of this magnitude for a unit change of temperature in the same direction does not indicate any change in the amount of water vapor in the atmosphere. On the other hand, a given saturation deficit at whatever temperature it may occur always represents the same pull on any free water with which the air may come in contact, provided the water is at the temperature of the dew point. This proviso is important because water is only exceptionally at the dew point temperature. If, for the saturated vapor pressure of the air, the vapor pressure at saturation corresponding to the temperature of the water is substituted, a vapor-pressure difference is obtained that should always give the same effect on evaporation irrespective of the temperature. This vapor-pressure difference should be the most useful climatic measure to indicate the evaporation from water, soil, or foliage, or from the stomatal openings.

Although relative humidity has been more commonly used in forestry than any of the other measures of humidity, it is probably the least satisfactory from the viewpoint of forest influences. Just as saturation deficit represented the difference between the actual and the saturation pressure, so relative humidity is the ratio between these two quantities ($100e/e_s$). A constant relative humidity, however, represents neither a constant amount of water vapor in the atmosphere nor a constant evaporative power. Relative humidity varies inversely with temperature in such a way that, for a 1°F change, there is a change of from 1 to 5 per cent in relative humidity in the opposite direction. Thus a change of this magnitude and direction in relation to temperature does not represent a difference in the content of atmospheric water vapor. Consequently, if the forest exerts an influence on temperature as it usually does, a change of relative humidity in the opposite direction is likely to reflect partly the temperature difference.

42. Vertical Distribution of Humidity.—Humidity varies with distance from a boundary surface as did temperature. From measurements over snow and ice in the open, Sverdrup (338) found that vapor pressure decreased with distance above the surface even when temperature increased. His data could be expressed by the equation

$$e = e_s - kh^{\frac{1}{n}}$$

where k is a constant less than 1, e_s is the saturated vapor pressure at the temperature of the snow or ice, and n had values between 5 and 6.

43. Influence of Forest on Vapor Pressure and Absolute Humidity. Actual records other than for relative humidity to indicate the magnitude of forest influences are rare. At 9,000 ft elevation in Colorado, Bates and Henry (37) found a net change of 0.006 in. or about 2 per cent in the mean annual vapor pressure as a result of deforestation. The vapor pressure was higher in the forested condition. This difference is very small but probably significant because the departure was in the same direction in 11 of the 12 months of the year and was not in the direction to reflect the influence of temperature. The temperature was lower in the forested condition, and hence the vapor pressure would tend to be lower also but actually was higher.

The following averages of absolute humidity based on observations at 4:30 P.M. in the western white pine type in northern Idaho during July and August in the years from 1931 to 1933 were recorded by Jemison (187).

Forest Cover	Absolute Humidity, Grains per Cubic Foot
Clear cut	2.82
Half cut	2.89
Uncut	3.34

These figures indicate a significant addition of moisture to the atmosphere by the forest, becoming progressively greater in the denser stand, because they obtained when the mean maximum and doubtless also the 4:30 P.M. air temperatures for the corresponding stands progressively decreased.

The extensive compilation of German and Swedish seasonal averages at the bases of trees and in the crowns for both deciduous and evergreen forests led to the conclusion (121) that there is "no distinct difference in absolute humidity between tree crown and base, or between glade and plain, or between woods and open." Except for the examples already cited, there is no good reason up to the present time to modify this conclusion. It requires as a corollary, to account for the known fact that forests give off into the air through the process of transpiration large amounts of water vapor, the recognition of the rapid diffusion, transportation, and mixing of that water vapor with the very large air masses of the general atmospheric circulation.

44. Influence of Forest on Relative Humidity.—The inverse variation of relative humidity with temperature not only diurnally but also in its vertical distribution above the ground level in the forest has been pointed out by Geiger (134). The maximum is close to the ground surface where water vapor from the saturated soil air is continually diffusing upward into the drier layer above the ground surface. This distribution close to the ground is indicated in figures from Europe by Stocker (333) from spruce high forest in mid-July at 10 A.M. The relative humidity at 6 cm among plants of *Oxalis* was 84 per cent; at 30 cm among *Myosotis*, 67 per cent; and at 100 cm in the open forest, 59 per cent. From the maximum near the ground, the relative humidity decreases at a decreasing rate upward, becomes stable from a distance considerably below the crowns up into them, and then decreases upward through the canopy and into the open air above. At noon just below the base of the canopy there may even be a secondary maximum corresponding to the secondary minimum in the vertical distribution of temperature. With rare exceptions, other relative-humidity records from different stands and localities are in the direction and of about the magnitudes that would be indicated by the inverse relation to temperature. A summary of the seasonal departures of forest from open in a variety of stands and localities arranged in descending sequence of the departures in summer, is given in Table 13.

As for temperature, the high departures, particularly in summer and fall, are associated with the dense, tolerant, climax types, and they become smaller or negative in the open, intolerant, pioneer types. Whether the differences in density are inherent in the species and types or are created by cuttings, the effects on relative humidity are in the direction opposite to the temperature change.

TABLE 13.—MEAN SEASONAL RELATIVE HUMIDITIES

Type	Departures, forest from open, %				Ref.
	Winter	Spring	Summer	Fall	
Western white pine, Idaho	.....	.....	+8.5	.....	(187)
Beech–oak, Germany	+2	+1	+7	+5	(121)
Scotch pine, Germany, base	+2	+4	+5	+4	(121)
Scotch pine, Germany, crowns	+1	+2	+4	+3	(121)
Scotch pine, Sweden	+1.5	+4.5	+4	+2	(121)
Ponderosa pine, 7,250 ft, Ariz.	−5	+1	+3	−1	(282)
Maple–birch–balsam fir, N.Y.	.....	+1.5	+1.8	+2.7	(332)
Western white pine, half cut, Idaho	.....	.....	+1.7	.....	(187)
Maple–hemlock, Wis.	.....	+1.4	.....	+1.8	(256)
Jack pine, Wis.	.....	−3.2	.....	+0.3	(256)
Jeffrey pine, 6,000 ft, S. Calif.	+4.0	0	−2.7	−0.3	(ms)
Chaparral, 6,000 ft, S. Calif.	+1.0	−7.0	−5.3	+1.7	(ms)

Because maximum temperatures, associated with minimum relative humidities, are the result of absorption of solar radiation, it is not surprising to find that Gast (133) established a high negative correlation between relative humidity H_r and solar radiation R in Massachusetts. The correlation coefficients were between −0.84 and −0.88. The regression equation as he gave it with solar radiation as the dependent variable was

$$R = 303.2 - 2.53H_r$$

where R is in gram calories per square centimeter.

The relative-humidity figures are unsatisfactory not only because they reflect differences in temperature and not necessarily differences in the moisture content of the atmosphere, but also because they are means of observations taken at arbitrary hours of the day. Daily maxima of relative humidity, which commonly approach 100 per cent during the night, are not at all distinctive, but the minima, which usually occur in the early afternoon would probably be much more instructive than the means that are reported. For this purpose it would be necessary to make humidity determinations near the time of the minimum in the afternoon or preferably to obtain them from hygrograph traces.

From such records for 12 days in September, Geiger (135) shows the diurnal trend of relative humidity under the canopy of a stand of Scotch pine in Germany at a height of ½ m above the ground and at 16 m just above the crowns of the trees. The trends for both heights are quite similar and show a maintained relative humidity above 95 per cent for the night until 7 A.M., then a uniform decrease to the minimum between 50 and 60 per cent about 3 P.M., followed by a progressive rise becoming flatter toward midnight. He also points out the interesting fact that the excess of the relative humidity at ½ m over that at 16 m is least and only about 1 per cent during the night and increases to a maximum that may sometimes reach as much as 25 per cent about 5 in the afternoon, 2 hours after the minimum relative humidity has been passed.

45. Influence of a Shelterbelt on Saturation Deficit.—Saturation deficit like relative humidity varies with temperature without any necessary corresponding variation in the actual moisture content of the atmosphere, but it has been little used in the study of forest influences. Figures for saturation deficit at different distances to leeward of a windbreak are given by Bates (36) as follows:

Multiple of Height to Leeward	Saturation Deficit, In. Hg
1	0.743
5	0.788
10	0.776
In open	0.697

The corresponding Fahrenheit temperatures were 85.1, 86.7, 86.9, and 84.9, so that the differences in saturation deficit reflect the differences in temperature rather than in moisture content of the air.

On the basis of the foregoing figures, although admittedly unsatisfactory, the following summary of indications, rather than conclusions, may be suggested.

46. Atmospheric Humidity: Summary.—1. Absolute humidities as expressed by vapor pressure may average as much as 0.015 in. higher in the forest, or 0.013 in. lower as compared with the open.

2. Other measures of atmospheric humidity vary with temperature, fferently with different measures. The influence of forest, if it is

significant at all, is small and usually such as might be due to difference in temperature. It may be reversed at different times of day or seasons of year.

3. Relative humidities tend to be slightly higher in the forest by amounts up to 11 per cent, which usually correspond to the lower air temperatures in the forest as compared with the open. Exceptionally, they may be lower by as much as 7 per cent, notably at high elevations in the Southwest. Relative humidities in thinned or open stands are intermediate between those of dense forest and of open areas. In dense stands, the relative humidities are highest close to the ground and lowest above the crowns with intermediate values in and below the canopy.

4. Humidity measures taken in the early afternoon and at night or as daily maxima and minima from hygrographs should show more consistent relations than means.

CHAPTER X

PRECIPITATION

Because precipitation is the source of water for infiltration into the soil, runoff, stream flow, and floods, and a chief cause of erosion, it is one of the most important elements in the study of forest influences.

For information on precipitation in general there is considerable literature in the fields of meteorology, climatology, and hydrology. The publications of the U.S. Weather Bureau contain much material on the subject, and references to precipitation as it relates to vegetation and other natural phenomena are widely distributed in forestry, botanical, geological, soil science, and engineering literature. A very small part of the literature, however, is concerned specifically with the influences that forest and other vegetation may exert on the precipitation and the resulting water on the earth's surface.

47. Forms and Causes of Precipitation.—The rain, snow, hail, and sleet that constitute precipitation are the principal sources of all the water on the earth's surface.

The capacity of the atmosphere to hold water is controlled by temperature and varies with temperature. If the temperature is reduced below the dew point, condensation and fog or precipitation result. This reduction in temperature and consequent condensation may be caused in three principal ways:

The first is by contact or radiational cooling, which often occurs on clear nights when the ground surface and the vegetation lose heat by radiation. The temperatures of the surface and of the adjacent air are also reduced, and to a greater degree as the wind movement is smaller. After the dew-point temperature is reached or passed, dew or frost is deposited at temperatures respectively above or below freezing.

The second form of cooling and condensation is by the mixture of air masses of unequal temperature. This process seldom produces more than a slight cloud or fog.

The third and most important cause of precipitation is by expansional or dynamic cooling resulting from vertical convection. This is the process that causes all considerable condensation. Three distinct cases may be recognized: One is the result of temperature gradients often caused by heating of the ground surface. This produces the convection or instability precipitation better known as the "thunderstorm" type. The second case results from converging winds caused by unequal heating of land and water masses or such as occur in the front half of cyclonic storms. This is commonly called the "cyclonic" type of precipitation. The third case is that of a forced rise of air masses either by the underrunning of cooler winds, as along the squall line of a storm, or by the flow of air over barriers of cold air

TABLE 14.—CHARACTERISTICS OF RAINDROPS AND PRECIPITATION OF DIFFERENT FORMS

Form	Intensity of rain, in./hr	Median diam of drops, mm	Velocity of fall, ft/sec
Fog	0.005	0.01	0.01
Mist	0.002	0.1	0.7
Drizzle	0.01	0.96	13.5
Light rain	0.04	1.24	15.7
Moderate rain	0.15	1.60	18.7
Heavy rain	0.60	2.05	22.0
Excessive rain	1.60	2.40	24.0
Cloudburst	4.00	2.85	25.9
		4.00	29.2
		6.00	30.5

or land elevations. The latter is known as the "orographic" type of precipitation. It will be helpful to distinguish the convective, cyclonic, and orographic forms of precipitation in considering the influence of forest cover.

The different forms of precipitation with quantitative expressions for some of their physical characteristics are given in Table 14 (178, 213).

Recent findings by Laws (215) indicate that the terminal velocities depend on the nature of the drop and not on the distance of fall. The relation between median diameter of drop in millimeters D and intensity of rain in inches per hour i is linear on double-logarithmic paper and can be expressed by the equation

$$D = 2.23i^{0.182}$$

48. The Measurement of Precipitation.—Precipitation varies widely with the character of the storm, with the topography, with elevation, and with seasons and cycles. Precipitation records also vary surprisingly within short distances. Actual records have shown a difference of 20 per cent in the catch of two rain gages only 15 ft apart. The Weather Bureau instructions for the standard installation of a rain gage provide that it be placed in the open so that no obstacle projects within the inverted conical surface having the top of the gage as its apex and a slope of 45 deg. Under forest conditions this would mean that, if the trees were 50 ft high, the gage with its top 3 ft above the ground level should be at least 47 ft from the nearest tree. Thus the only locations in the forest which meet the Weather Bureau standard are those in quite large openings. Moreover, this standard may not be sufficiently exacting. Recently Brooks (55) has stated that, for an unaffected catch, a gage should be at a distance of at least four times the height of an obstruction. To the extent that wind causes the reduced catch, the evidence as to the distance to leeward to which a shelterbelt reduces wind velocity is applicable and would indicate that a gage should not be within twenty times the height of an obstacle. On the other hand, records of rainfall for 2 years in rows of gages extending east from a north-south border of a pine plantation in Berkeley, California, indicated that the 40-ft trees did not effect the catch of rain to a distance of more than one-half their height. Within a dense forest, there are no openings that provide the required exposure for a gage.

Because precipitation is variable within short distances it is necessary to have several gages to obtain averages of precipitation for an area and to evaluate the reliability of those averages. This becomes particularly necessary in the forest, where the variation tends to be greatest. After records from several gages for a preliminary period are available, it is possible to determine how many are needed for any desired degree of accuracy in the mean value for the area by the relation between number of observations N, standard deviation SD, and the desired standard error of the mean SE.

$$N = \left(\frac{SD}{SE}\right)^2$$

The number of gages used can be reduced by changing them to new random locations frequently or after each storm.

Such an analysis for the steep broken topography of the San Dimas Experimental Forest, based on arbitrary standard errors of 10 per cent for storms of less than 0.5 in., 5 per cent for storms of 0.5 to 1 in., and 2.5 per cent for larger storms, showed, for drainage basins of 1.3 to 5.7 square miles, that from 4 to 24 gages per square mile would be required (370). This is many more than are available now or are likely to be available for a long time to come in most forest areas. A similar study over a much larger area in Ohio (90) led to the conclusion that a gage for every 37 square miles would not obviate underestimates of flood-producing rains on areas of over 200 square miles.

Variation in the catch of rain gages is caused in large part by the wind as it eddies across their orifices. Moreover the wind factor usually tends to reduce the catch of a rain gage below the true amount of precipitation. Various devices such as Nipher and Alter shields and the flaring margins around the tops of snow bins have been designed to minimize the error in the measurement of precipitation. Sinking a gage to its own depth in a pit has the same purpose. Some tests with rain gages with their tops 3 ft above the ground surface as compared with others 3 ft higher on the slope but in pits with their tops at the level of the ground surface, indicated that the latter collected about 7 per cent more precipitation than those on the standard support. This phenomenon of decreasing catch with increase in elevation associated with increasing wind velocities must be distinguished from the effect of increase in altitude on precipitation which is in the opposite direction.

Comparisons based on all storms of four seasons have been made on a 40 per cent slope at Berkeley, California, between gages of different types with and without shields. The slope-trough gage, 10 ft long and 4 in. wide placed up and down the slope close to the surface, gave the highest total catch of rainfall. If this gage is used as the standard or 100 per cent relative, the gage with a Nipher shield and its orifice parallel to the slope, after correction to the area of a horizontal orifice, caught 99 per cent. The 1 per cent difference was only such as might have resulted from chance. The average of two standard gages without shields but with orifices parallel to the slope was 98 per cent, a significant difference. A vertical recording gage 12 in. in diameter gave 88 per cent, and a vertical gage on an 8-ft tower with an Alter shield gave 84 per cent.

A more elaborate comparison of gages without shields but corrected for sloping orifices has been reported by Storey and Hamilton (336) from the San Dimas Experimental Forest. Their figures of average percentage deviations from the catch on a concrete slab 10 ft in diameter at the ground level and on the slope have been rearranged in Table 15 in ascending order of the departures of storms of more than 1 in. averaged for all exposures. With exceptions, this is also ascending order for the smaller storm classes and for the individual exposures. In general, either tilting a gage so that its orifice is parallel to the slope or lowering it closer to the surface reduces the departures. The trough gage with its long axis parallel to the slope although 3 ft above the ground, gave the lowest departures. It should be noted that the sliding cover by which projectional area was kept constant whatever the slope, was in effect a correction to equivalent horizontal exposure. The catch of the tilted gages was also corrected by multiplying the catch by the secant of the angle of slope. The standard gage gave the

TABLE 15.—PERCENTAGES BY WHICH CATCH IN RAIN GAGES OF DIFFERENT TYPES IS LESS THAN THAT ON A 10-FT DIAMETER SURFACE SLAB ON DIFFERENT EXPOSURES AND FOR DIFFERENT AMOUNTS OF RAIN

Type of gage	Class of storm			Class of storm								
	0–½ in.	½–1 in.	1+ in.	0–½ in.			½–1 in.			1 in.+		
	All exposures			Exposure								
				E	S	NW	E	S	NW	E	S	NW
Trough	6.4	3.4	1.6	5.2	7.6	....	3.2	3.7	....	1.4	1.9	....
Tilted, at surface	8.5	5.2	2.9	5.5	8.6	11.0	4.3	3.3	8.0	2.9	2.0	3.7
Tilted, 3 ft high	10.9	4.1	3.3	6.8	10.7	15.9	2.7	4.3	3.6	2.9	2.2	4.6
Standard, at surface	9.2	4.0	3.6	6.2	9.3	12.4	2.7	3.2	6.1	2.7	4.9	3.3
Tilted, 1 ft high	10.9	5.6	3.9	6.9	11.0	15.2	3.6	4.0	9.3	4.1	2.2	5.2
Stereo	10.9	4.8	4.0	10.0	10.3	12.8	4.3	4.7	5.6	4.4	3.6	3.9
Standard	9.2	5.4	4.4	4.8	10.0	12.8	2.4	5.9	7.7	2.5	6.4	4.4

greatest departure followed by the stereo gage in which the long axis of the gage was vertical but the orifice truncated parallel to the slope. With few exceptions the departures were greatest for the storms of less than ½ in. and lowest for those of over 1 in. Departures for the northwest exposure were decidedly higher than those for the east and south, a fact presumably related to the prevalence of south and east directions of the rain-bearing winds.

An estimate of the true precipitation, if it is assumed that the catch in a gage equipped with a Nipher shield, with its orifice parallel to the slope and corrected to the horizontal, approximates the true precipitation P, may be made from the linear relation between the catch of such a gage and that of a vertical gage without a shield P'. For 254 rains in six seasons on a 40 per cent south exposure at Berkeley, California, the relation was

$$P = 1.137P' + 0.002$$

The two gages were within 60 ft of each other on the same grassy slope. The unshielded vertical gage was a recording gage with an orifice 12 in. in diameter. The standard error of estimate of this regression is 10.3 per cent. In the mountains of southern California on the basis of 22 paired tilted and vertical gages without shields, the linear trend could be represented by

$$P = 1.16P'$$

although Storey and Wilm used an exponential equation as the best fit (337).

Standard gages on a south exposure in northern Idaho, compared with stereo gages sunk in pits so that the orifices were in the plane of the slope, caught 5, 10, and 20 per cent less rain at elevations of 2,700, 3,800, and 5,500 ft, respectively, according to Hayes (147). The increasing discrepancy as elevation increased was doubtless associated with the increasing wind velocities.

Comparisons of the catch by gages with and without devices to obviate the effects of wind, based on 3 years' records by the Soil Conservation Service at Coshocton, Ohio, have been reported by Riesbol (294). Relative to the catch of a standard Weather Bureau gage as 100 per cent, the other gages gave the following values based on the means:

Gage	Per Cent
Fergusson recording	99
Standard Weather Bureau	100
Adapted Nipher shield	101
Improved Nipher shield	103
Alter type, wooden-lath shield	105
In walled pit, 10 ft in diameter	106

Again it appears that the catch in an unshielded standard gage is likely to be 5 per cent too low. However, when the differences in the means, from which the foregoing relatives were derived, were tested statistically, it was found that none of them were significant at the 5 per cent level. If the computations could have been made on the data paired for each storm the results might have been different. After a thorough review, Brooks (55) concluded that unshielded gages in only moderate exposures catch 5 per cent less than the true precipitation and the deficit may be as much as 50 per cent.

The catch of precipitation decreased by 4 per cent with an increase of 1 mph in wind velocity, according to Schubert (312). If wind velocities in the forest are from ¼ to 2 mph less than in the open, the catch of rain gages in the forest would for this reason be greater by from 1 to 8 per cent. If allowance were made for this source of error the differences in the rainfall between forested and open areas would be decreased correspondingly. Similarly, according to the vertical distribution of velocity for winds of 10 mph above the influence of the surface, the top of a gage 3 ft above the ground will be subject to wind velocities of 5.5 mph and when close to the ground they would be about 3 mph. The difference of 2.5 mph would result in a catch of 10 per cent more at the ground level.

As the catch of precipitation by a rain gage is affected by the wind, it is also a function of the inclination of the rain or snow to the vertical r and of the slope of the ground surface in a direction toward the wind s. Horton (173) has shown that the ratio of the actual precipitation falling on the slope P to the measured catch in a gage with horizontal orifice P' can be expressed by the equation,

$$\frac{P}{P'} = \frac{\cos (r - s)}{\cos r \cos s}$$

This is transformable into the expression derived by Pers (284) and Fourcade (128)

$$\frac{P}{P'} = 1 + \tan s \tan r$$

If a slope of 30 per cent and wind velocity at the top of a standard gage of 5.5 mph with drops of a moderate rain falling at 18.7 ft per sec is assumed, the true rainfall, according to the formula, would be 13 per cent more than the catch. Close to the ground where the velocity would be 3 mph, the excess would be 7 per cent. The difference of 6 per cent corresponds closely to the differences previously mentioned between gages in standard exposures and those in pits with their orifices at ground level. For a slope of 30 deg and snow falling at an angle of 45 deg the ratio would be 1.58, or, in other words, the actual precipitation would be 158 per cent of what would be recorded in a gage.

When the aspect of the slope b is not directly toward the wind, Fourcade goes on to show that a correction can be made for the direction of the wind w by the use of the angle $b-w$, so that the formula becomes

$$\frac{P}{P'} = 1 + \tan s \tan r \cos (b-w)$$

Cosine $(b-w)$ becomes 0 when $(b-w)$ is 90 deg, so that the inclination of the rain requires no correction of the gage catch when the wind is blowing at right angles to the aspect of the slope. If the direction of the wind is downslope instead of upslope, the sign of the tangent term becomes negative, and the catch of a gage exceeds the actual precipitation. If there is no wind so that the rain falls vertically, r becomes 0 and the catch equals the precipitation. On level ground the slope becomes 0, and again the two are equal. In this case it is evident that the formula does not correct for the effect of the angle of rainfall or for eddies over the top of a gage, which reduce the catch on level as on sloping ground.

For most purposes in forestry the desired measurement of precipitation is that which falls on the land surface whether level or sloping. The error associated with the slope of the ground can be eliminated at least in part in two ways. One is by the use of "slope-trough" gages exposed parallel to the slope, usually with the long axis up- and downhill. Secondly, standard gages placed so that the planes of their orifices are parallel to the slope serve the same purpose. In both these devices a correction must be made if it is desired to know the precipitation falling on a horizontal area. If it is assumed that rain falls vertically the correction consists in multiplying the catch by the secant of the angle of slope. A third method is to place the gages with their long axes vertical and then cut the tops so that the planes of the truncated orifices are parallel to the slope. In this case it is not possible to correct the catch to that of a horizontal area.

The effect of different ways of installing gages is illustrated by 22 gages with standard exposure on 100 acres of the San Dimas Experimental Forest in Los Angeles County giving a total annual catch 13 per cent lower than the paired gages tilted so that the orifices were parallel to the ground surface. A similar comparison between two single gages for 305 storms on a 40 per cent slope at Berkeley, California, gave a difference of 10 per cent that was significant. Both these gages had Nipher shields, and the catch in the tilted one was corrected to a horizontal exposure before making the comparison.

The effects of variations in wind direction can be evaluated by mounting the gage on a swivel with a suitable wind vane to maintain its orientation with respect to the direction of the wind and with separate collectors for the different quadrants. If the device has both horizontal and vertical orifices, the latter always toward the wind, the ratio of horizontal to vertical catch gives the tangent r of the inclination of the precipitation to the vertical in the formula

$$\frac{P}{P'} = 1 + \tan s \tan r$$

If the device has multiple vertical orifices of fixed orientation, the factor cos $(b-w)$ must also be evaluated (127).

After analyzing records from a vectopluviometer, Hamilton (143) has derived relations between wind velocity, inclination of fall of rain, and slope that can be used to estimate the error in the catch of a gage of standard exposure. First, the angle of inclination of the rain with the vertical r can be predicted from the wind velocity v_w, the velocity of fall of the raindrops v_r, and the slope s, by the equation

$$\tan r = \frac{v_w}{v_r - v_w \tan s}$$

The velocity of fall may be obtained approximately from the rain intensity from Table 14. For Hamilton's data the velocity of fall varied only between 13 and 16 ft per sec. The slope factor, tan s, constitutes a correction for updraft in upslope winds.

Having the angle of inclination of the rain r, in degrees, the percentage difference between the true rainfall P and the catch in a standard gage P' could be represented as

$$100 \frac{P-P'}{P} = 0.6r - 2.0$$

For the measurement of the intensity of rainfall, tipping-bucket recording gages are extensively used. For locations where the gages cannot be visited except at long intervals, the use of an oil film or, in freezing weather, of both the oil and a saturated layer of calcium chloride have been tried.

Precipitation records of the U.S. Weather Bureau in cities are usually at stations on the tops of buildings where the wind effect may be large. Moreover, they are often more or less remote from forested areas. For that reason they may not be applicable to surrounding forest areas, and it is always desirable to obtain supplementary precipitation records in or immediately adjacent to a forest area as a basis for adjusting the values from the nearest Weather Bureau station.

49. Determination of Areal Precipitation.—The necessity of using several gages and perhaps a large number to obtain average figures for precipitation in forested areas involves also their distribution. In general, ridge tops and canyon bottoms are least likely to be representative and should be avoided. Theoretically, each area of distinct aspect and slope, which is called a "facet of slope," and each area with distinct vegetative cover should be represented by at least one gage.

If marked differences in elevation characterize an area they must be considered in the distribution of gages or the estimation of areal precipitation because, in general, precipitation tends to increase with altitude. The steeper the slope, the more rapid is the increase. This relation holds up to a certain height, which in the Sierra Nevada seems to be between 5,000 and 6,000 ft, above which precipitation again decreases. To what extent this apparent decrease at higher altitudes may be ascribed to the deficient gage catch of snows driven by high winds has not been determined, although the water equivalent of the snow on the ground on Apr. 1 recorded in the snow surveys at the altitudes above 8,000 ft is greater than the figures purporting to give the total annual precipitation at those elevations.

For the usual relation below 6,000 ft, Mead (243) proposed an equation of the form,

$$P_h = P_1 + K\left(\frac{A-a}{100}\right)$$

where P_h and P_1 are the precipitation at higher A and lower a altitudes and K is a constant that in different localities may have values from 0.2 to 0.8. For the west slope of the central Sierra Nevada the annual precipitation and median altitude of the principal zones of vegetation are as follows:

Vegetation	Altitude, ft	Precipitation, in.
Grassland	400	9
Woodland	2,000	22
Ponderosa pine	3,700	40
Mixed conifer	5,300	45
Fir	7,500	35
Subalpine	10,000	35

If precipitation is plotted over altitude up to 5,300 ft, the trend is roughly linear and the slope, K in the formula, becomes 0.72. In the White Mountains of New Hampshire it is 0.7. In southern California, K is between 0.3 and 0.4, and in the Gila River basin in Arizona it is 0.2. The effect of altitude is too great to permit the determination of the influence of the forest by comparing precipitation at forest and open stations which are at different elevations.

If the increase of precipitation with altitude were generally consistent, records of a few gages at different altitudes would make possible the drawing of lines of equal precipitation, called "isohyets," on a contour map, because the isohyets would parallel the contours. Actually, however, many unaccountable variations of precipitation with local topography interfere with the drawing of isohyets parallel to the contours. The diversity of isohyetal maps of the same small drainage basins in different storms in the San Gabriel Mountains of Los Angeles County are illustrated by Wilm, Nelson, and Storey (370). For the same reason, Clyde (91) found no correlation between precipitation above 8,000 and that below 5,000 ft in Utah and concluded that valley precipitation is a poor index of precipitation higher in the mountains.

An example of a 20 per cent difference in recorded seasonal rainfall is contained in the following figures by Bauer (40) for the four cardinal exposures at the same elevation in the Santa Monica Mountains.

Exposure	N	E	S	W
Rainfall, in.	26.3	27.8	22.5	22.5

Usually there are a few rain gages in or near the drainage basin for which an average areal precipitation is needed. The problem is then to combine the records of those gages in a way to give an average representative of the whole basin. Four methods have been used to determine areal precipitation.

Method 1.—The simplest and quickest is the arithmetic mean by which the sum of the n individual gage measurements divided by n, gives areal precipitation.

Method 2.—A network of triangles covering the whole area is formed by joining adjacent gaging stations by straight lines. The center of gravity of each triangle is located. The precipitation on each triangular area will then be ⅓ of the sum of the precipitation at the three gages at the apices of the triangle. For any triangle with one or two apices outside the watershed, the center of gravity of the area within the watershed is located, and lines are drawn from each apex of the triangle through this point to the opposite side. Then the weighting to be assigned to each gage record will be the ratio of the length of the part of the diagonal between the center of gravity and the opposite side to the total length of the diagonal. The sum of the three weighting ratios should be equal to 1 as a check. Marginal areas not within any triangle are allocated to the nearest triangle. The sum of the products of the center-of-gravity precipitations thus computed, multiplied by the ratio of area of triangle plus contiguous marginal areas to area of watershed, gives areal precipitation.

Method 3.—The area is divided into blocks such that the dividing line between each two gages is perpendicular to the line joining them at its mid-point. The areas of these blocks are determined by geometry, planimeter, or by counting squares, and the areal precipitation for the whole area calculated by weighting the precipitation of each gage by the ratio of the area of the block that it represents to the total watershed area.

Method 4.—An isohyetal map of the watershed is prepared, using precipitation records as primary guide points and drawing the isohyets to parallel the contours as nearly as possible. The area of each isohyetal zone is determined by planimeter. The areal precipitation for the watershed is obtained by weighting the median precipitation of each zone by the ratio of the area of that zone to the total watershed area.

The decision as to which of the four methods to use must be made according to circumstances and largely on logical grounds. If an area has a considerable number of well-distributed gages, all four methods will give closely similar results. The isohyetal method, Method 4, weights both areas and elevations and is therefore logically to be preferred. However, in areas without marked differences in elevation, or with few gages or where precipitation records do not show an increase with altitude, it may be difficult or fruitless to draw an isohyetal map.

Methods 2 and 3, triangles and blocks, give weights to areas but not directly to elevation. When it is necessary to use precipitation records from stations some distance outside the particular area being studied, the triangle method provides a means of weighting such records. On the other hand, the blocks obviate the marginal areas that in Method 2 must be assigned to the nearest triangles.

The simple average gives weight correctly to neither elevation nor area and has only its simplicity to recommend it.

The differences in areal precipitation for individual storms on areas of 1 to 6 sq mi each with eight or more gages, computed as simple averages and by the isohyetal method for subdivisions of the San Dimas Experimental Forest, were usually less than 5 per cent and did not exceed 13 per cent (370). A statistically significant difference between the two methods was obtained on only one out of ten watersheds. This is an example of correspondence of results by different methods when numerous gages are distributed over an area.

50. Precipitation Records and the Estimation of Extremes.—The effectiveness of rainfall is often determined by the manner of its occurrence. Thus intensities and rates for hours, days, or storms and the maxima or frequency of extremes at a station are often more useful than the means in relation to runoff, floods, and erosion. These rates are obtainable from recording gages and are available in published form by 5-minute and longer periods from many of the standard Weather Bureau stations. Records in the United States show maximum rates of up to 0.8 in. in 5 minutes, 1.9 in. in 1 hour, and over 9 in. in 24 hours.

Maximum and minimum precipitation are important in the planning of flood control and water supply. Existing records are of too short duration to give precise indications of what extremes, greater or smaller than any in the records, may occur. For this purpose predictions in terms of probability must suffice until long-term weather forecasting is more advanced. However, attempts have been made to express the relation of rainfall intensity i in inches per hour to its duration T in minutes and the frequency with which it will be reached or exceeded once in F years, based on existing data. The resulting empirical formulas have the form,

$$i = \frac{KF^x}{(T + b)^n}$$

where the factors K, x, b, and n depend upon conditions associated with geographic location. At different places and by different workers they have been found to have the following range of values: K, 10 to 60; x, 0.15 to 0.3; b, 0 to 12; n, 0.4 to 0.9.

If the duration of rain T is less than 60 minutes the values of K and n are decidedly smaller than for durations of 60 to 1,440 minutes according to Bernard (354). He has prepared maps for the eastern United States giving lines of equal magnitude for K, x, and n so that the appropriate values can be read from them for any desired locality. In his equations the factor b is 0. For most purposes the expected intensity may be read with sufficient accuracy directly from Yarnell's charts (379) for different combinations of duration and frequency. Because of the wide spacing of intensity-recording gages, which, for that reason, rarely catch the maximum rainfall in a small area of any given storm and also because the records cover insufficiently long periods, it is almost certain that either the formula or the charts lead to underestimates of the maximum rains that may be expected over small areas with given frequencies.

Like the extremes the precipitation in any year or other period usually varies from the average of the years of record at a station. This average is known as the normal precipitation and is computed as the arithmetic mean of the series, although good reasons have been advanced for substituting the median as a normal. For some purposes it is helpful to express the precipitation for any year as a percentage relative using the normal as a base. Such a relative is called the "index of seasonal wetness." Thus for the Santa Ana River above Riverside, California, for the 50 years from 1871 to 1921, the index of seasonal wetness reached a maximum of 229 and a minimum of 47 per cent (74).

An obstacle that frequently has to be surmounted in using precipitation records is the doubtful reliability of a normal derived from a short period of record. A better estimate of the normal can be obtained by utilizing the data from the nearest station with a long period of record. To explain by an example, the 22-year record from 1899 to 1921 for Blue Canyon in Placer County, California, gave a normal of 66.17 in. annually. Summit had a 50-year record from 1871 to 1921 with a normal of 46.38 in. The average at Summit for the period 1899–1921 was 47.02 in. Then the probable long-term normal for Blue Canyon would be 66.17 (46.38/47.02) or 65.27 in.

Another frequent difficulty arises when records for one or more months or other period are missing from a series. It then becomes essential to make the best estimate of the missing figure to be inserted. Again for this purpose it is assumed that the precipitation at the station is related to that at neighboring stations. If there are enough stations in the vicinity an isohyetal map may be made and the missing record interpolated in the zone in which the station is located. This method may also be used to estimate the precipitation at a station where no record exists.

The second method is one of arithmetic interpolation that can be illustrated by an example. The record for October, 1909, at Yosemite was not obtained. The total for the 11 other months was 38.84 in. The records of Summerdale and Crocker's were complete for that season.

		Inches
Summerdale, 12 months		49.93
Summerdale, omitting October		47.38
Crocker's, 12 months		56.39
Crocker's, omitting October		52.48
Crocker's, $^{11}/_{12}$ months		0.931
Summerdale, $^{11}/_{12}$ months		0.948
Average ratio		0.940
Yosemite, 11 months		38.84
Yosemite, 12 months	38.84/0.940 =	41.32
Yosemite, October		2.48

Thus the missing record at Yosemite is estimated to be 2.48 in. The same method is used when the records missing are for not more than

6 months. For longer periods it is better to interpolate for the total seasonal precipitation by the use of percentages of normal at neighboring stations.

The rainfall or water year in most parts of the United States does not correspond to the calendar year because Jan. 1 comes during a period of accumulation of precipitation and soil moisture. The water year should start with the stream-flow year, when water in the ground, on the surface, and in snow storage is reduced to a minimum. August 1, Sept. 1, Oct. 1, and Nov. 1 have been used as initial dates of the water year. Probably Oct. 1 is the most satisfactory in the West because that represents approximately the end of the dry season. Other advantages of a water year beginning Oct. 1 are in the case of seasonal divisions of the year, for which Jan. 1 is not a dividing point, and to ensure that most or all of the precipitation of a year or period will appear as stream flow during the same year or period.

51. Effect of Forest on Precipitation.—The actual influence of the forest on precipitation involves two separate considerations: First, the effect on the condensation of water vapor, and hence on precipitation, in the air above and in the forest or for certain distances beyond its borders, and, second, the effect on the amount of precipitation that, intercepted by the crowns of the trees, never reaches the ground.

The first of these two divisions, the effect of forest on rainfall, was an early and prolific source of controversy with violent generalizations on both sides and no means of adequate proof. One school in the argument maintained that forests augmented precipitation and rendered the climate more humid. The other contended that forests can have no appreciable effect on the water content of the great masses of air whose circulation results in precipitation, and that more abundant and frequent rains result in development of vigorous forests which are the effect and not the cause of the heavier rains. Zon was one of the ardent advocates of the positive effect of the forests and summarized the conclusions in the following words (381):

> Forests increase both the abundance and frequency of local precipitation over the areas they occupy, the excess of precipitation, as compared with that over adjoining unforested areas, amounting in some cases to more than 25 per cent. The influence of mountains upon precipitation is increased by the presence of forests.
>
> Forests in broad continental valleys enrich with moisture the prevailing air currents that pass over them and thus enable larger quantities of moisture to penetrate into the interior of the continent. The destruction of such forests

. . . affects the climate, not necessarily of the locality where the forests are destroyed but of the drier regions into which the air currents flow.

On the other side, Moore, former head of the U.S. Weather Bureau, concluded (259) that the cutting of the forests has had no effect on droughts, that forestation has little or no effect on precipitation, and that there is no evidence in the period of record of an increase or decrease in precipitation. Not only is satisfactory proof for a decision between these conflicting viewpoints extremely difficult, if not impossible, to obtain, but also the broad generalizations are unsafe where many conditions vary widely. It is therefore essential to limit consideration to a single geographic climatic region in attempting to analyze the influence of the forest.

The subject has more recently been reviewed by Nicholson, and his analysis has been evaluated by Lowdermilk (273). Nicholson concluded that in certain regions and under certain meteorological conditions the rainfall over forested areas may be appreciably increased by the presence of the forest. The two general sets of conditions that he cites are, first, mountain forests where cloud drip may increase precipitation, and, second, those regions where convectional rainfall is a large component of the precipitation and where the "possibility and quantity of such rain is greatly increased by the presence of forests." The latter suggestion depends upon the hypothesis that a forest is more effective than other forms of vegetation in promoting those conditions of temperature and moisture in the atmosphere which contribute to precipitation. It has not yet been proved that this is the case, and with respect to temperature it may well be argued that a forest in reducing the temperatures at the ground surface may actually tend to decrease the frequency and intensity of convectional showers. The magnitude of the contribution by cloud or fog drip will be considered in a separate section.

A previous analysis of the influence of forests on rainfall by Brooks (54), which Nicholson criticized, deserves special attention, although it is not favorable to the claims of those who support the contention that forests increase rainfall. Brooks distinguishes the three types of precipitation previously outlined. In the case of orographical precipitation, the forest adds to the effective height of the land. If this increase in height averages 10 m, a corresponding increase in rainfall has been found to vary from 0.8 per cent in Silesia to 1.2 per cent in West Prussia. There is also the effect of friction, which is obviously greater over the irregular surface of the forest canopy and therefore tends to check the velocity of the wind. Hence if the same volume of air moves more slowly, it must occupy a greater depth, and therefore there must be some ascent. This is estimated as a possible cause of a 2 per cent increase in rainfall. Hence the total increase due to the orographical effect of the forest is unlikely to exceed 3 per cent of the precipitation in temperate countries. The forests that would exercise the greatest effect would be those on the crests of hills or on the higher windward slopes.

The effect of forest on cyclonic precipitation is dismissed as negligible and impossible of proof.

Convectional precipitation is caused by heating of a surface layer of air adjacent to the heated surface of the ground. The air over a forest would be likely to be cooler than that over bare ground on clear summer afternoons. This would tend to decrease rather than increase the frequency of convectional showers. The magnitude of this effect has not been determined.

The earlier investigations by the historical method based on the alleged change in precipitation with progressive deforestation were reanalyzed with the result that Blanford's examples from India and Walter's data for Mauritius were found to be explicable on the basis of changes in atmospheric pressure and circulation, and thus not exclusively attributable to changes by deforestation.

The best example by the comparative method is Hamberg's Swedish investigation, and this was reanalyzed. Twenty-four forest stations had an average altitude of 77.3 m and an average forested area of 58 per cent. Thirty-two open stations had a mean altitude of 66.8 m and an average forested area of 17 per cent. The average rainfall of the forest stations exceeded that of the open by 8.6 per cent, which is ascribed in part to three causes: (*a*) the greater height of the ground in the forest accounting for 1.9 per cent, (*b*) the more sheltered position of the gages in the forest and correspondingly less wind movement accounting for 3.0 per cent, and (*c*) the height of the gages above the ground accounting for 0.5 per cent. These three corrections total 5.4 per cent leaving, as the effect of the 41 per cent difference in forest cover, 3.2 per cent of the rainfall. This 3.2 per cent had a probable error of 2.3 per cent, so that with odds of 1 : 1 the true increase caused by the forest would be between 0.9 and 5.5 per cent. In reanalysis by partial correlation, Brooks found that the regression coefficient for the percentage change in rainfall associated with a 10 per cent change in forested area, when the effect of height was eliminated, became 1.02. This may be interpreted by the statement that the increase of 41 per cent in forested area would be associated with an increase in the rainfall from May to October of 4.2 per cent. This was compared with Hamberg's figure, after correction for the greater height of the ground in the forest, of 6.7 per cent. When the sum of Hamberg's second and third corrections, or 3.5 per cent, was deducted from 4.2, there remained 0.7 per cent as the residual influence of the forest. An increase of this amount might be expected as a result of the increase in the effective height of the ground due to the presence of the forest. Brooks concluded that the chief influence of the forest on rainfall resulted from its shelter, which reduces the velocity of the wind, and hence increases the catch of the rain gages rather than causes an actual increase in the rainfall. If the rainfall measured in the open is corrected to the same wind velocities as in the forest, the increases attributed to the forest are in most cases reduced to 1½ to 2 per cent.

The frequently cited findings from the 33-year record in a large forested area near Nancy in France, where the rainfall in a forest clearing averaged 30.2 in.; at the forest edge, 27.3 in.; and in a field at a distance of 10 to 12 km from the forest, 25.6 in.; cannot be considered conclusive because of the distance between stations and of a 100-m difference in elevation.

Recently evidence by Holzman (157) has made it clear that precipitation does not depend upon the amount of water vapor in the atmosphere but on the conditions favorable to its release, among which changes in temperature associated with changes in elevation and the mixing of air masses of different characteristics are important. Although forests add much water to the atmosphere by transpiration from the foliage, the contention that more water will therefore be condensed as precipitation either more frequently or in greater amounts does not necessarily follow.

Notwithstanding the convincing evidence from the science of meteorology that forests can have no appreciable effect on precipitation, one striking example, apparently to the contrary, comes from the Copper Basin in eastern Tennessee (181). In that area 11 sq m have been denuded by smelter fumes, producing a waste of gullied, bare soil much like that which resulted from the same cause at Kennett, California. In this fume-denuded waste and in the adjacent forested area, weather stations established by the Appalachian Forest Experiment Station recorded totals of 2 years' precipitation as 95 and 115 in. respectively, an excess of 20 in. or 21 per cent that might be attributed to the forest. The effect of differences in shelter from wind movement on the catch of the gages was not involved, because the same relation obtained for storms in which there was no wind. The example is cited not as proof that forests increase rainfall but as an unquestionable difference between the two conditions, which, with further study, will probably be found to be associated with the differing temperatures and convectional air currents rather than directly with the water vapor produced by the forest.

To the original question in its usual meaning, answer may be made that according to present evidence forests do not increase rainfall appreciably.

52. Effect of Forests on Rainfall: Summary.—1. Forest cover has no appreciable effect on cyclonic precipitation.

2. The orographical effect of forest may increase local rainfall by not over 3 per cent in temperate climates, because the effective level of the ground is increased by the height of the trees and the friction of the wind with the tree surface is greater than with open ground.

3. The catch of rain in forest clearings is usually greater than in neighboring open sites by amounts up to 10 per cent. This excess is caused mainly by protection of the gage from air currents in the forest.

4. When this protection is allowed for, the effect of forests is to increase the rainfall in clearings by about 1 per cent compared with similar situations in the open.

CHAPTER XI

INTERCEPTION AND STEM FLOW

Although forests do not noticeably affect general rainfall, their influence is not at all negligible in the amount and distribution of precipitation that reaches the ground locally. Beneath the crowns of the trees a certain amount of precipitation is intercepted and does not reach the ground. On the other hand, in the openings between the trees or in the lee of a shelterbelt there may be either more or less precipitation at the ground than would be recorded if the trees were not there. The effect of the forest on the amount of precipitation reaching the ground, therefore, is exerted through the interception of a portion of the precipitation by the crowns where it is subject to evaporation. Canopies of forest or brush may cover the ground more or less densely up to nine-tenths of the projectional area of the crowns. When rain or snow falls many drops or flakes are retained on the foliage or branches. The first drops wet the leaf and twig surfaces, after which many drops accumulate and run together to form larger drops that fall from upper to lower leaves or run down twigs and stems and thus reach the ground. Gradually, the whole upper surface of the foliage is wetted to its saturation or storage capacity, and thereafter part drops to the ground and part evaporates into the air. Trees have a large surface area of foliage, and considerable water is required to saturate it. At the beginning of a shower all the rain may be retained on the foliage, and if the rain stops soon enough no water reaches the ground. This is the familiar observation under shade trees at the beginning of a shower. If the rain continues, drops begin to fall from the crowns. When the rain stops, the dripping from the trees continues for a short time.

53. Measurement.—The process has been demonstrated experimentally by Grah and Wilson (139) as illustrated in Fig. 11. Plants of Monterey pine and *Baccharis pilularis* of approximately the same size were subjected to a fine spray of 0.7 in. per hour, while hung on a balance so that weights could be read every minute. The curves represent the cumulative amounts of water retained on the surfaces of the plants with increasing time. They rise steeply at first, then more slowly as dripping increases until, after about 15 minutes, a maximum or saturation point is reached when the curves no longer rise, because the losses by dripping and evaporation equal the rates of application. For the pine with a crown width of 9.5 in., this amounts to 0.053 in. depth on the projectional crown area and for the *Baccharis* with crown width of 9.0 in., 0.033 in. depth. The oven-dry weight of the pine was 58.8 g, including 41.6 g of leaves. Thus the interception storage capacity of 61 g of water exceeds the dry weight of the whole

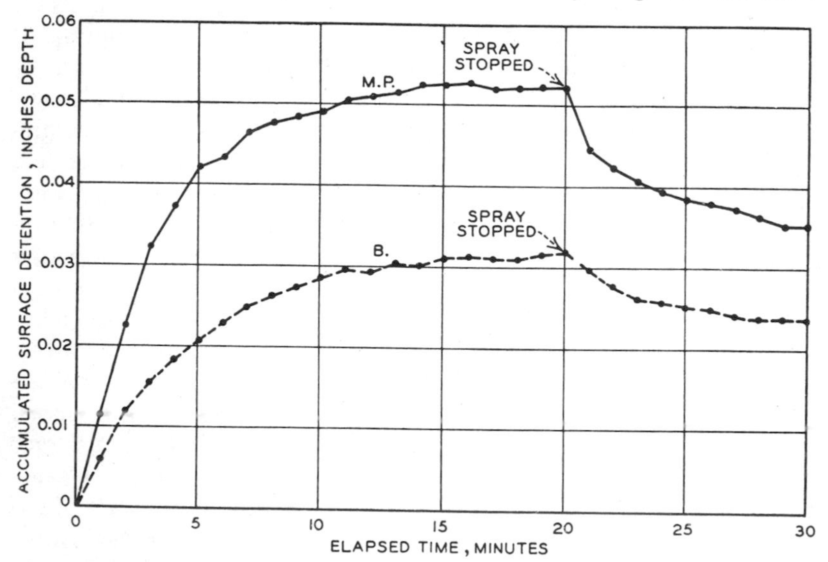

FIG. 11.—Cumulative depth of water retained on plants of Monterey pine and *Baccharis* with application of water spray. [*After Grah and Wilson* (139).]

plant and is 147 per cent of the dry weight of the needles. After the spray was shut off, the dripping and evaporation continued at a decreasing rate.

The usual measurement of interception involves a comparison of the catch of rain or snow under the crowns with that in large openings between crowns or at the edge of a wooded area or above the crowns. The comparison may also be made by the use of the same gage or gages for sufficient periods of time before and after cutting the trees. Variations in crown density between crowns of different kinds of trees and even under different parts of the crown of the same tree are almost infinite, and consequently the catch in the rain gages is correspondingly varied. For that reason it is even more necessary than in the case of measurement of precipitation in the open to obtain a large number of gage measurements. To minimize this number and still obtain reliable averages, Wilm (366) has used a sampling procedure such that the whole area was divided into a number of blocks, each of which was sampled by two or more randomly located samples in each period of time. The periods of time could be years, seasons, months, storms, or days, as desired. In each time period in each block, two or more samples were taken under the canopy and paired with the same number in the nearest open area of sufficient size that the gage catch was not affected by the trees.

Trough gages of sufficient length integrate the variations in precipitation under a forest canopy to the degree that they extend under the crowns and across openings in a way representative of the area.

In any case the catch of precipitation in rain gages under the forest will be subject to a correction for the water that runs down the stems of the trees and is never recorded in a rain gage. This water will be referred to as "stem flow." The amount of the stem flow is a correction to be deducted from figures of interception. Stem flow has been measured by placing metal collars in notches in the bark around the stems of trees and collecting and measuring the flow from these collars. Watertight box platforms to include the stems of several trees with rubber collars to intercept the stem flow have also been used for the measurement of net interception, corrected for stem flow.

Interception is expressed either as the difference between the catch in the open and that under a canopy or as the percentage which that difference is to the catch in the open. Thus if the precipitation in the open is recorded as 10 in. and under the crowns as 8 in., the interception is either 2 in. or $10 - 8/10 = 20$ per cent. Because the wind velocity is greater in a large open area where a Weather Bureau exposure can be obtained, the catch in the open area used as a basis for this comparison should be taken from recordings with a shielded gage, which should be comparable with the amounts of catch in the forest, where the errors from wind are very small. Otherwise, in the example cited if the true precipitation in the open was 11 in., then the correct interception would be 3 in. or $11 - 8/11$ giving 27 per cent instead of 20.

In the process of interception the small drops of fine misty rains are combined into larger ones that have to be a certain size before they will drip from the leaves. On the other hand, in heavy rains of large drops the foliage breaks the force and changes the velocity of fall and size of the drops. The net effect in heavy showers is probably to reduce materially the impact of the drops at the ground surface.

54. Analysis.—Interception was analyzed by Horton (167), who estimated that a leaf of oak or aspen may retain over 100 drops of water. If a tree with 500,000 leaves has an average of 20 drops per leaf, each drop having an average diameter of ⅛ in., the tree would contain 5.92 cu ft of interception storage water. If the crown diameter is 40 ft with a projectional area of 1,256 sq ft, the interception storage would amount to 0.06 in. of water.

The total interception is the sum of the interception storage and current evaporation and may be expressed by the equation,

$$I = S + KET$$

where I is the depth of water intercepted, S is the depth of interception storage, both I and S on the projectional area of the crowns, K is the ratio of the evaporating leaf surface to the projectional area E is the evaporation rate in depth per hour during the rain, and T is the duration of the storm in hours. From the foregoing equation it is evident that the depth of interception loss increases as each of the other factors increases. S and K are functions of the amount of foliage and size of crowns.

When, however, the interception is expressed as a percentage, the equation becomes

$$I\% = 100 \frac{S + KET}{iT}$$

Thus when i is the same the percentage loss decreases as T, the duration of the storm increases, and inasmuch as the numerator of the fraction is independent of the rain intensity i, the interception percentage decreases as the intensity of the rain increases, when the duration does not change.

On the basis of these relations and measurements under the crowns of individual more or less open-grown trees of several species in New York State, Horton derived a series of equations expressing for different species the interception in inches depth on the projected areas of woods as a function of the amount of precipitation per shower P_s. P_s was determined as the inches of precipitation over the watershed divided by the number of showers for the months when the trees were in leaf. When his equations are adjusted for stands of trees, P_s is substituted for iT, and the KET term is both multiplied and divided by P_s, they take the general form

$$I = S + \left(\frac{KET}{P_s}\right) P_s$$

The net rainfall reaching the ground under the crowns $(P_s - I)$ may be obtained directly by transposing the equation to the form

$$P_s - I = \left(\frac{1 - KET}{P_s}\right) P_s - S$$

For any given stand, the interception storage S and the proportion of the precipitation in any shower evaporated before it reaches the ground, KET/P_s, may both be considered constants. Thus the interception without correction for stem flow is a linear function of the precipitation per shower. The values of S vary from 0.01 to 0.05 and for KET/P_s, from 0.01 to 0.23 for a wide variety of species and stands.

For two stands of young ponderosa pine in Colorado, Johnson's data (191) give

$$I = 0.04 + 0.06P_s$$

and

$$I = 0.02 + 0.11P_s$$

In the partly evergreen brush type, composed of species of *Ceanothus*, oak, and buckeye, at North Fork, California, for the winter months Rowe (301) found a linear trend that can be expressed as

$$I = 0.02 + 0.23P_s$$

Definite evidence that both constants decrease with the density and volume of the stand is provided in the work of Niederhof and Wilm (276) in lodgepole pine in Colorado. They measured the precipitation under the crowns and in the openings for residual stands after different degrees of cutting. Their equations transposed to give interception are given in Table 16.

TABLE 16.—REGRESSION EQUATIONS FOR RESIDUAL STANDS OF LODGEPOLE PINE AFTER DIFFERENT DEGREES OF CUTTING

Residual stand per acre				Regression equation
Basal area, sq ft	M bd ft	Trees per acre	Stand-density index	$I = S + \left(\frac{KET}{P_s}\right) P_s$
159	12	382	310	$I = 0.029 + 0.20P_s$
96	6	223	188	$I = 0.015 + 0.13P_s$
66	2	206	134	$I = 0.007 + 0.11P_s$
65	4	181	134	$I = 0.013 + 0.09P_s$
40	0	147	86	$I = 0.015 + 0.01P_s$

Both the interception storage and the evaporation factor decrease with the decrease in density of the residual stand although not in direct proportion to that decrease. As more data become available it may become possible, for a given species, to predict the magnitude of these two constants from some measure of density of stand.

The amount of precipitation required to fill the interception storage capacity will evidently be greater than the storage capacity itself,

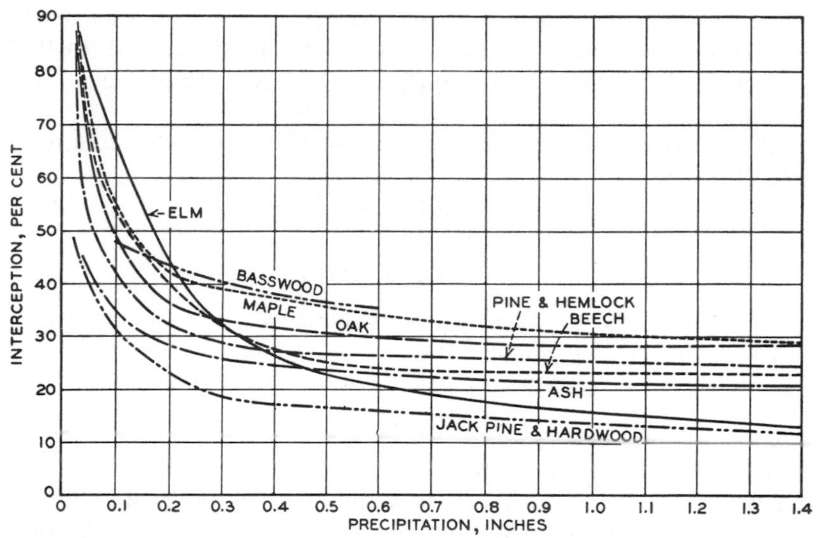

Fig. 12.—Percentage interception for different species in relation to precipitation per shower. Corrected for stem flow except jack pine and hardwood. [*From Horton* (167) *and Mitchell* (256).]

because of losses by evaporation and dripping during the wetting period.

When the interception is expressed as a percentage of the precipitation, the trends become curvilinear. The relation of interception, expressed as a percentage of the precipitation, to the amount of rainfall per shower is shown in Fig. 12. In addition to Horton's data (167) a curve is also included from Mitchell (256) for jack pine and partially cut northern hardwoods in Wisconsin. The close correspondence of the values for two distinctly different stands, so that they may be represented by a single line of trend, is doubtless to be explained as a coincidence in which the partial cutting of the larger denser hardwoods reduced the intercepting foliage to approximate equivalence with that of the smaller denser stand of light-foliaged jack pine.

In general, these curves without corrections for stem flow indicate interception of 40 to 100 per cent of the precipitation in showers of less than 0.1 in. and of 10 to 40 per cent in showers of over 0.4 in. with intermediate values for the intervening range.

55. Stem Flow.—The correction for stem flow is usually small relative to interception but not necessarily negligible, as its frequent omission would suggest. As would be expected, stem flow has been found to increase with the amount of precipitation per shower, and, for some species and stands at least, this relation is linear. Stem flow starts only after a certain amount of rain has fallen so that the Y intercept is negative. The equation for stem flow as a function of precipitation per shower would then have the form

$$S_f = bP_s - a$$

The amount at which stem flow starts may be as low as 0.01 in. for the smooth-barked species like beech or as high as 0.7 in. for rough-barked species. From the equation, this amount may be obtained by making S_f equal to 0, and solving for P_s so that

$$P_s = \frac{a}{b}$$

The regression of stem flow in cubic inches on rainfall per shower in inches depth as determined for the three species at 9,500 ft elevation in Colorado by Wilm and Niederhof (371), was

For lodgepole pine,

$$S_f = 3583P_s - 1{,}090$$

For alpine fir,

$$S_f = 1603P_s - 387$$

For Engelmann spruce,

$$S_f = 1035P_s - 301$$

The differences between species were significant, although for any of them the depth of stem flow for all the storms in a season was less than 0.01 in. over the projectional area of the crowns. Stem flow started only when the rain exceeded about 0.3 in.

For the brush species at North Fork, California, most of them with smooth bark and several steeply sloping stems converging at the base, Rowe's data (301) give the equation

$$S_f = 0.15P_s - 0.01$$

where S_f is in inches depth.

In a plantation of Canary pine, 25 years old, at Berkeley, California, the equation for the stand was

$$S_f = 0.03P_s - 0.02$$

The rough bark and numerous horizontal or slightly drooping branches doubtless cause the low stem flow of the pine compared with the brush. For individual trees in the same stand, the constants varied from 0.005 to 0.11 for the slope and from −0.002 to −0.036 for the Y intercept. Attempts to relate this variation to characteristics of the trees revealed no relation to diameter, height, crown length, or crown area. However, a relation was found to the excess or deficit in height relative to that of the surrounding trees. The relatively tall and relatively short trees showed high stem flow of 2.5 to 9 per cent compared with 1 per cent for those of average height.

A nearly linear relation between annual volume of stem flow in gallons S_f and basal area of individual trees in square feet A, which could be expressed by the equation

$$S_f = 9 + 707A$$

has been found by Wicht (364) for *Populus canescens* in South Africa in a year with 35 in. precipitation. He suggests that the stem flow from sample trees, by direct proportion, is to the stem flow of all the trees on an acre as the basal area of the sample trees is to the total basal area on the acre. This simple relation should be confirmed before it is used for other species and in other localities.

Similarly Hoppe's data (159) indicate for European beech that the volume of stem flow increases linearly with the stem basal areas and crown areas of the trees. On the other hand, the stem flow expressed as a percentage of the precipitation decreased with increasing crown area at a decreasing rate from 19 per cent at less than 108 sq ft to 15.5 per cent at 150 sq ft.

The magnitudes and variation of stem flow with species, for any given rainfall, are indicated by the same relative positions with respect to smoothness of bark and stem flow of Horton's trees. The order is beech, maple, ash, elm, pine, basswood, hemlock, oak, and shagbark hickory, as shown in Fig. 13. The stem flow for beech is the maximum, reaching 10 per cent in storms of 1.3 in. precipitation, whereas the other species range from 0.6 to 5 per cent. For *Cryptomeria* in Japan, Hirata (154) reports stem flow beginning with precipitation of 0.2 in. and reaching a maximum of 11 per cent in rains of 4.6 in. per 24 hours. Shortleaf pine stands, 25 years old, in North Carolina yielded 1 to 5 per cent of the precipitation as stem flow (266). Mature lodgepole pine in Colorado yielded less than 0.1 of one per cent of the precipitation as stem flow. A 32-year stand yielded 1.5 per cent while a near-by stand of aspen of the same age yielded only 1.1 per cent (114). This low stem flow for the smooth-barked aspen is an exception to the evidence that stem flow is large from smooth-barked species.

The stem flow for 86-year-old beech in Austria (159) began at less than 0.1 in. precipitation and reached as much as 21 per cent of the precipitation in rains of 1 in. The average for the beech stand for all the rains was 15.4 per cent. For 60-year spruce, stem flow began with

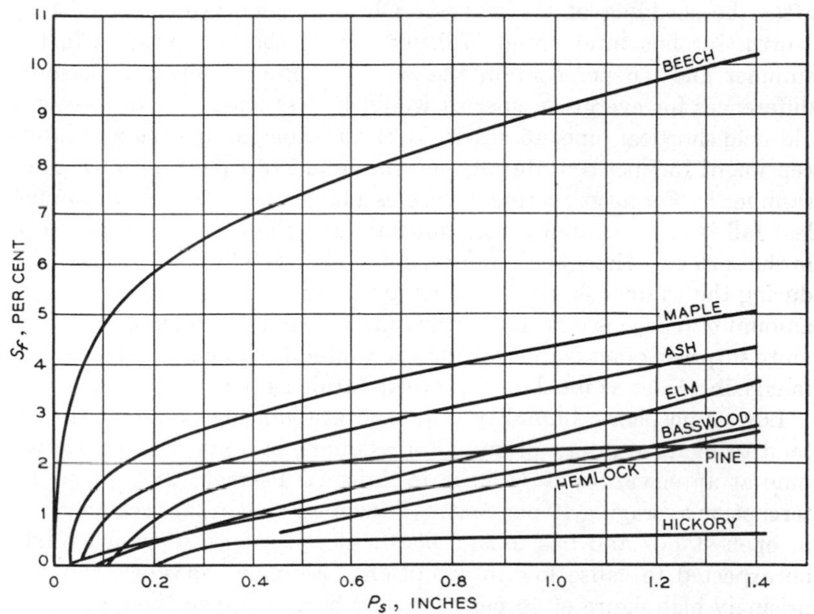

FIG. 13.—Percentage of stem flow for different species in relation to precipitation per shower. [*From Horton* (167).]

the rains of 0.1 to 0.2 in., reached a maximum of 5 per cent in heavier rains, and averaged 2.3 per cent. The rough-barked pine began to yield stem flow in rains of 0.3 in., showed a maximum of about 2 per cent with an average of 0.7 per cent. It is interesting to notice in this study that in rains of over 0.6 in. the percentage stem flow exceeded the percentage interception for the beech. Stem flow may also exceed interception at times in the brush types in California. Ordinarily, stem flow is only a fraction of the interception.

56. Interception and Its Variations with Season and Climate.—The actual magnitudes of interception expressed as a percentage of the precipitation vary with the season of the year, with regional differences in fogs and evaporation, with differences in the age and density of the stands, and with species, types, and successional stages of development.

The variation with season in the deciduous species between winter and summer obviously is associated with their loss of foliage and also with reduced evaporation, and the interception is therefore lower in winter. In a sugar maple–hemlock forest in northern Wisconsin, Mitchell (256) recorded interception of 24.6 per cent during the spring when the trees were in leaf as compared with 15.8 per cent in the fall after the shedding of the leaves. Old mixed hardwoods in western North Carolina intercepted 17.0 per cent of the precipitation in the summer and 6.6 per cent in the winter.[1] Presumably the seasonal differences for evergreen species would be less marked. Records for old-field shortleaf pine, 45 years old, in the same locality showed interception of 15.7 per cent during the winter and 16.1 per cent during the summer.[1] For most evergreen species a large proportion of the annual leaf fall is concentrated in the autumn, and the new foliage develops in the spring. There is, therefore, definitely less foliage on such species during the winter period than during the summer. This difference in amount of foliage is reflected in the difference in the interception. The more surprising fact is that during the winter deciduous forests without foliage intercept as much of the precipitation as they do.

Low atmospheric humidity and high evaporating power of the air tend to increase interception. For example, in a stand of ponderosa pine at an elevation of 7,250 ft in Arizona, Pearson (282) found interception averaging 40 per cent. Ponderosa pine in that region grows in open stands and has foliage of low density, both of which might be expected to cause low interception. Actually, however, the surprisingly high figure of 40 per cent may be ascribed to the high evaporating power of the air in that region.

57. Variations with Stand Density and Age.—Numerous examples might be cited of the variation in interception with crown density, both between species and between different stands of the same species. In the chaparral at 6,000 ft elevation in the San Bernardino Mountains, Munns found average interception of 31 per cent under the dense California scrub oak as compared with only 3 per cent under the notably thin-foliaged chamise. In Maine, interception under spruce and balsam fir was 37 per cent, whereas the admixture of light-foliaged paper birch with the spruce and fir reduced the interception to 26 per cent. Interception of snow at an elevation of 4,500 ft in southern Idaho was found by Connaughton (95) in an old stand of ponderosa pine with a partial understory to be 30 per cent, in a similar stand

[1] Unpublished data furnished through the courtesy of C. R. Hursh and H. J. Loughead, U.S. Forest Service, Appalachian Forest Experiment Station.

without the understory, 25 per cent, and in an open stand of ponderosa and lodgepole pine 20 to 30 ft high, only 5 per cent. With a crown density of 0.65, a mature stand of Douglas fir in Washington intercepted 34 per cent of the seasonal rainfall, while a 25-year-old stand with dense canopy, intercepted 43 per cent (323).

Application of the regression equations for lodgepole pine in Colorado for a rain of 1 in. gives the interception percentages for the residual stands of different densities, as in Table 17.

TABLE 17.—RELATION BETWEEN INTERCEPTION AND MEASURES OF STAND DENSITY

Interception for a rain of 1 in., %	23	15	12	10	3
Trees per acre	382	223	206	181	147
Basal area, sq ft/acre	159	96	66	65	40
Stand-density index	310	188	134	134	86

The relation between density and interception is evident although it is not linear. In these data, interception is more closely related to basal area and stand-density index than to the number of trees per acre.

Variations with age of the stand may be illustrated by the following figures for beech in Switzerland (381).

Age, years	20	50	60	90
Mean annual interception, %	2	27	23	17

The maximum percentage interception occurred at the age of 50 years, which was also the age of the culmination of the current annual increment. Although the trends of interception and increment do not exactly correspond, they afford an indication that the interception in a well-stocked stand is similar to the curve of current growth, which in turn is a function of the amount of foliage.

58. Variations with Species and Stages of Natural Succession.—The differences in interception between stands of different species and of different stages in natural succession would be expected to increase as a particular species or type represented a more advanced stage in the succession. If figures for seasonal interception are pieced together to represent a successional series, although taken from different localities, jack pine in Wisconsin, an intolerant pioneer species, intercepted 21 per cent (256); white and red pine in Ontario, 37 per cent (43); maple–beech climax in New York, 43 per cent; tolerant climax hemlock in Connecticut, the maximum interception of 48 per cent (258).

59. Variations within a Stand.—In addition to the variation in interception between different stands of trees, there are also wide variations within a single stand, depending upon the position with respect to crowns and openings and to different parts of the crowns of single trees. In general the interception is greatest close to the boles, becomes less under the central and outer parts of the crown, and is still less under the edge of the crown, where there is some concentration of dripping like that from a peaked roof. Finally, in the openings between the trees the interception varies with the size of the opening. In a small opening it may reach values comparable with those under the crowns and, in the larger openings with diameters approximating the height of the trees, may become zero or even negative on the basis of comparisons with a large open area. The following tabulation of selected data from Munns[1] gives an indication of these differences for a mature Jeffrey pine stand at an elevation of 6,000 ft in the San Bernardino Mountains of southern California.

Rain per shower, in.	Percentage interception			
	At base of tree	Under heavy crown	Under light crown	Under edge of crown
0.01	100	100	100	81
0.06–0.10	94	84	68	48
0.11–0.30	74	48	27	5
0.51–1.00	53	33	16	4

The descending sequence of percentage interception from stem to edge of crown and into an opening in each class of rainfall and the lower percentages in the heavier rains are illustrated with exceptions in Fig. 14 for a 25-year-old plantation of Canary pine in California.

In a dense stand of white and red pine 40 years old, in Ontario, for the period from May to October, Beall (43) reported interception of 57 per cent 1 ft from the bole, 27 per cent under average crowns, and 16 per cent in small openings.

The radial distance r from any point in an opening to the edge of the nearest crown in mature lodgepole pine has been incorporated as an independent variable with precipitation P in the open in a multiple regression equation by Niederhof and Wilm (276). When their equation is converted to express interception I in inches depth, it becomes

$$I = 0.234P - 0.102r + 0.423$$

[1] From the manuscript "Studies of forest influences in California," in the files of the U.S. Forest Service, through the courtesy of the author, E. N. Munns.

This may be interpreted by the statement that seasonal interception decreases about 0.1 in. for each increase of 1 ft in radius of opening up to 18 ft. At 18 ft, interception becomes zero, and at the edge of a crown where r is 0, interception becomes 1.82 in. or 30 per cent of the seasonal rainfall of 6 in. Again this is a specific case. The

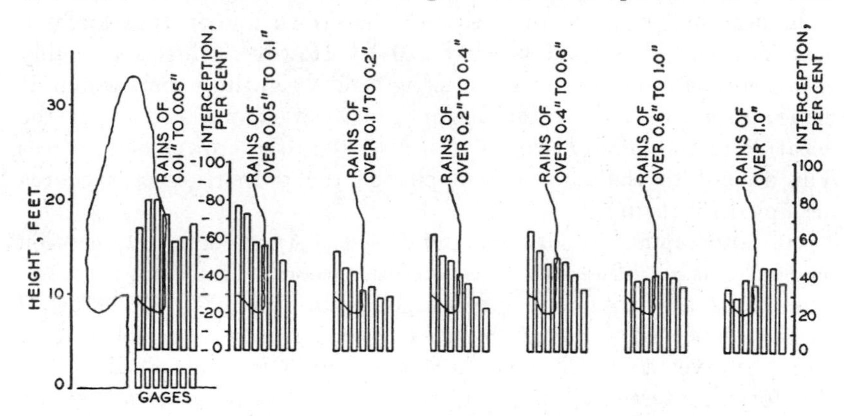

FIG. 14.—Percentage interception in different locations with respect to stem and crown of Canary pine for rains of different amounts.

regression coefficients would doubtless change for other types or localities or conditions such as direction of prevailing wind. Data are not yet available to evaluate the magnitude or direction of the resulting changes.

The range of the differences in percentage interception in different positions under the crowns in a forest stand is so wide that a large number of records of rainfall would be required even in a limited area to arrive at a reliable average figure for interception over any given area of forest.

Another method of determining average interception for a forest area would be to weight the figures for interception from gages sampling different positions with respect to the boles, crowns, and openings by the proportion of the total area occupied by the different parts of the crown and by the openings. For example, in a 25-year stand of Canary pine it was found that the zone within 1 ft radius of and including the trunks of the trees occupied 6.8 per cent of the area, the zone 1 ft wide under the edges of the crowns, 30.0 per cent; the zone between these two under the middle portions of the crowns, 23.7 per cent; and the openings between the trees, 39.5 per cent. If then, the interception in rains of 0.2 in., close to the trunks, is 65 per cent; under the crowns, 50 per cent; under the edges of the crowns,

47 per cent; and in the small openings, 31 per cent; these figures can be weighted by the percentages of area occupied by the different zones, and the average interception determined for the area as a whole. The interception determined from gages in fixed positions with respect to the trees varies also with the direction and velocity of the wind. It is highest under the leeward side of the crown and in that location increases with the wind velocity (204). However, this is probably a less serious source of error in using such weightings than would at first appear, because variations in the wind change somewhat the location of the different zones with respect to the crowns of the trees but do not to the same extent change the proportion of the area occupied by them.

60. Interception by Grasses and Herbs.—Appreciable interception of rainfall is not limited to trees and other woody vegetation. Grasses and herbs also intercept surprisingly large amounts. Undisturbed grass of species of *Avena*, *Stipa*, *Lolium*, and *Bromus*, allowed to grow up over a trough gage 10 ft long and 4 in. wide at Berkeley, California, intercepted 25.9 per cent of the 32.52 in. seasonal rainfall recorded in a similar gage around which the grasses were kept clipped. The deduction for stem flow was not determined. Bluegrass in Missouri during the month prior to harvest intercepted 17 per cent of the rainfall, according to Musgrave (268). He also noted for several crop species that the interception increased as the plants matured and as the density of the cover increased.

In a series of rains from June to August in Nebraska, the interception in inches depth by big bluestem was a linear function of the precipitation per shower, which can be represented by the equation

$$I = 0.02 + 0.66P.$$

The interception storage of 0.02 in. is like that for trees, but the evaporation term, two-thirds of the rainfall, is much larger and can hardly be ascribed to evaporation even in a dry climate. Clark (89) states that, from observations during the experiments, the amount of stem flow appeared to be small, but it seems likely that such a grass with many leaves and stems converging toward the bases like the shrubs would yield large amounts of stem flow and thus explain in part the high regression coefficient.

The summary of interception and stem-flow figures for a wide variety of stands in Table 18 further illustrates variations associated with differences in age, maturity, density, successional stage, and geographic location. The percentages of interception are not corrected

TABLE 18.—SEASONAL INTERCEPTION AND STEM FLOW AS PERCENTAGES OF THE PRECIPITATION

Type or species	Age or size	Place in succession	Locality	Intercep-tion, %	Ref.
Hemlock	Mature	Climax	Conn.	48	(258)
Douglas fir	25 yr	Climax	Wash.	43	(323)
Hemlock	Mature	Climax	N. H.	38	(258)
Spruce–fir	Mature	Climax	Maine	37	(265)
Hemlock	Mature	Climax	Adirondacks, N.Y.	34	(258)
Douglas fir	Mature	Climax	Wash.	34	(323)
Hemlock	Mature	Climax	Ithaca, N.Y.	31	(258)
Spruce–fir–paper birch	Mature	Climax	Maine	26	(265)
White pine–hemlock	Mature	Climax	Mass.	24	(265)
Western white pine–western hemlock	Over-mature	Climax	Idaho	21	(265)
Maple–beech	Mature	Climax	N.Y.	43	(258)
Mixed	Mature	Climax	N.Y.	40	(167)
Maple–hemlock	Mature, cutover	Climax	Wis.	25	(256)
Beech–birch–maple	Mature	Climax	Ontario	21	(43)
Ponderosa pine	Mature	Preclimax	Ariz.	40	(282)
Lodgepole pine	Mature	Preclimax	Colo.	32	(276)
Ponderosa pine	Mature	Preclimax	Idaho	27	(95)
Jeffrey pine	Mature	Preclimax	S. Calif.	26	(ms)
Lodgepole pine	32 yr	Preclimax	Colo.	23	(114)
Ponderosa pine	Mature	Preclimax	Idaho	22	(95)
Ponderosa pine	Young	Pioneer	Colo.	18	(191)
Calif. scrub oak	6 ft	....	S. Calif.	31	(ms)
Mixed brush	Mature	Preclimax	North Fork, Calif.	19	(301)
White pine–red pine	40 yr	Preclimax	Ontario	37	(43)
Jack pine	50 yr	Pioneer	Wis.	21	(256)
Shortleaf pine	45 yr	Pioneer	N. C.	16	(ms)
Quaking aspen	32 yr	Pioneer	Colo.	16	(114)
Chaparral, mixed	6 ft		S. Calif.	17	
Maple–hemlock	Mature (after leaf fall)	Climax (under-stocked)	Wis.	16	(256)
Hemlock	Mature	Climax	N.Y.	13	(258)
Oak–pine	Open, second growth	Pre-climax	N. J.	13	(376)
Ponderosa–lodgepole pine	25 ft	Preclimax	Idaho	8	(95)
Beech–maple	Mature	Climax	N.Y.	6	(258)
Chamise	6 ft	Pioneer	S. Calif.	3	(ms)

for stem flow. The sequence is that of descending percentages and also from climax to pioneer types in succession, with exceptions to bring the same or similar types near together.

61. Interception: Summary.—1. Interception of rainfall by vegetation represents a loss of precipitation that would otherwise be available to the soil.

2. The amount of interception is a function of the storage capacity of the surface of the vegetation, the evaporation rate during precipitation, and the amount of precipitation per shower.

3. Interception storage varies from 0.02 to 0.1 in. per shower.

4. The percentage of total precipitation loss decreases as the amount of precipitation per shower increases, from about 100 in showers that do not exceed the interception storage capacity to a nearly constant rate in heavy rains of between 10 and 40, depending on the character of the forest cover.

5. Monthly or seasonal interception losses vary according to the distribution of the precipitation.

6. Stem flow, a reduction factor for interception loss, is a small part of total precipitation, varying from 1 to 16 per cent. Both volume and percentage increase from 0 in light showers, as the amount of precipitation per shower increases.

7. Stem flow tends to increase with excess or deficit in relative height in comparison with neighboring trees and, in deliquescent species, with number and size of diverging branches.

8. Interception losses tend to be less in winter than in summer, the difference being small for evergreen and probably averaging about 50 per cent of total precipitation for deciduous species during the leafless period.

9. Interception losses may be large in regions of high evaporation and are usually low in regions where they are compensated by fog or cloud drip.

10. Interception losses vary with crown density: well-stocked stands intercept more precipitation than those understocked; stands at ages between the closing of the canopy and culmination of the current annual increment intercept more than those younger or older.

11. Interception losses vary with species and forest types because of differences in thickness and density of foliage and crowns. Hence, tolerant species intercept more than intolerant, climax more than preclimax, and mesophytic more than xerophytic.

12. Interception varies within a stand, being greatest near the stems and least under the edges of the crowns and in the openings.

CHAPTER XII

FOG DRIP

In foggy weather in coastal fog belts like that occupied by the redwood, or in the mountains among the clouds, the dripping of water from the foliage of trees is a common phenomenon. The term "fog drip," including "cloud drip," is simple and expressive, although the phenomenon has also been referred to as "occult condensation" and as "horizontal precipitation." The process results from the deposition of small drops of atmospheric moisture moving horizontally by contact on the surfaces of the foliage, where in time they combine to form larger drops that run off and fall to the ground. The contributions of water by fog drip have been given little study and are apparently widely variable in different regions and localities. Swiss and French observations have indicated discharges of streams larger than the amounts of precipitation falling on the drainage basins and the excesses were attributed to occult condensation.

62. Relation to Interception.—Fog drip, inasmuch as it represents measurable precipitation under the tree crowns when none is collected in the open, may be considered as negative interception. In other words the interception in any area will be subject to a reduction for any fog drip that may be caught beneath the trees. This phenomenon may be illustrated by figures recorded by Moore (258) in New York where, during the summer season, the computed interception for a hardwood–hemlock forest was 3 per cent but after eliminating days with fog amounted to 11 per cent. Similarly, in a pure stand of hemlock the total interception amounted to 9 per cent, whereas, omitting days with fog, it became 13 per cent. In stands of *Abies firma* in Japan at an altitude of 1,566 m, during two foggy months Hirata (154) recorded excesses of 10 and 11 per cent for the catch of rainfall under the forest as compared with that in the open, that is, negative interception.

63. Experimental Measurement.—Fog drip has been measured experimentally several times. In South Africa 3-in. rain gages, one with reeds projecting 1 ft above the gage and so arranged that moisture deposited upon them would run into the gage, caught 79.8 in. in the period from December to February while an ordinary gage caught 4.9

in. A repetition of the experiment by Phillips (285) at an altitude of 1,725 ft used 5-in. gages in which branchlets of broad-leaved conifers were arranged in wire frames projecting 12 in. above the gages. In the period from June to May the control gage caught 52.0 in. and the gage with the vegetative screen 94.6 in., or 182 per cent as much. The excess of the screened gage was recorded in every month except February, when the rainfall was predominantly normal showers and heavy downpours. In the other months most of the precipitation came as fine misty rains that are likely to give a subnormal catch in an unshielded gage at the same time that they contribute to fog drip. It was pointed out in comment on the foregoing experiment that the result was caused largely by the small size of the gage and the relative importance of cutting off the normal fall of the drops to leeward of the gage. Thus the percentage of excess should be greatest when the ratio of wind velocity to rate of fall of the raindrops is greatest. The same reasoning applies also to trees and the fog drip from them.

In Maryland, DeForest (109) repeated the experiment using 3-in. gages, one with a wire screen and one with a wire screen supplemented by artificial reeds of tin bent lengthwise at an angle of 135 deg. With natural rains the screened gages gave an excess over the normal of about 30 per cent. Several hundred tests with artificial rains gave excesses from 1 to 85 per cent, and the excess tended to increase with the rainfall.

An extension of the same idea has been used as the basis for a fog meter by Rubner in Germany (303). He arranged series of vertical aluminum rods 1 m long above the orifices of rain gages. One gage had 56 rods with a surface area equal to one-third of the horizontal area of the orifice. Another had 28 rods so that the ratio of surface to horizontal was one-sixth. On the basis of records from April to November for 6 years the average seasonal catch of fog drip in the gage with 56 rods was 3.5 mm and in that with the 28 rods, 2.0 mm. He considered that the relation between 3.5 and 2.0 was sufficiently close to that between one-third and one-sixth that fog drip could be considered to be directly proportional to the ratio of surface to horizontal area. Under a spruce forest on foggy days the daily catch of fog drip was from 0.1 to 0.3 mm or from two to twenty-four times that in the fog meter. In this forest the ratio of leaf area to ground area varied from 4 to 19 with an average of 8. Since 8 is to $\frac{1}{3}$ as 24 is to 1, the proportionality of fog drip to the ratio of vertical to horizontal area was confirmed. He concluded that, as a result of fog drip, the precipitation under forest cover in foggy weather might exceed that in the open by from 30 to 50 per cent.

Similarly, 4 years' records at 2,500 ft elevation in the Taunus Mountains by Linke (222), where there are some 200 foggy days each year, showed a gage catch near the edge of the forest averaging 157 per cent of that in the open, and toward the interior 123 per cent. The maxima were 300 and 260 per cent, respectively.

64. Magnitudes and Relations.—The differences in the soil-moisture content of samples of the surface foot of soil under the trees and in the open in the hills east of Berkeley, California, on July 31, 1927, were ascribed by Means (244) to fog drip from the trees on the western exposure toward the ocean fogs. Under plantations of Monterey pine and eucalyptus 15 to 20 ft high at 1,600 ft above sea level, the percentage soil-moisture content under the trees varied from 22.9 to 28.5, and 10 ft from the trees in the open, from 7.7 to 9.4. The difference would be equivalent to from 2 to 3 in. of water. Obviously evaporation and factors other than fog drip would affect such a comparison of soil moisture, but the common observation of wet places on the banks and puddles in the road under the trees after foggy nights would indicate that a part of the large difference was actually attributable to fog drip. More recently rain gages under a plantation of Canary pine at about 800 ft elevation on the same west slope have never yielded more than 0.01 in. of fog drip in rainless periods. This suggests that fog drip on the windward side of ridges becomes increasingly important toward the crests. Confirmation is found in the records taken on Mt. Wilson in southern California, and summarized in Table 19.

The gages were on the summit of the mountain where the trees were fully exposed to the fog and cloud-bearing winds from the south and southeast. The records under the trees represent the resultants of interception when rain or snow fell from clouds above the mountain and fog drip either with or without rain when the clouds or fog extended to and below the crest. Under the low-crowned *Ceanothus* (*C. leucodermis* Greene), fog drip and interception were nearly equal. The fog drip increased to 25 and 38 in. as the height of the trees increased to 40 and 80 ft. In other words the fog drip is a function of the length of crown exposed to the horizontally moving droplets of fog. Under the 80-ft pine the fog drip was twice as great under the leeward as under the windward side of the tree, presumably because the wind moved the drops to leeward as they fell. These results are probably extremes only representative of the top of the mountain.

At 2,500 ft elevation at Henninger Flats, also on the south slopes of Mt. Wilson, records[1] by the Los Angeles County forestry department in 1930 and 1931 showed excess of fog drip over interception in only two out of six gages under pines and cedars 30 to 40 ft high. The fog drip in these two did not exceed 12 per cent of the precipitation in contrast with the more than 100 per cent excess on the crest.

Another example of the amount of water that fog drip may add to the soil comes from the Cascade Head field station of the Pacific Northwest Forest Experiment Station.[2] Under a dense stand of Sitka

TABLE 19.—FOG DRIP UNDER VEGETATION AT 5,850 FT ALTITUDE ON MT. WILSON, CALIFORNIA

Location and cover	Gage catch,* in.	Fog drip, in.
October to May, 1916–1917		
Open	22.81	0
Under dense ceanothus, 8 ft high	22.67	0
Under dense canyon live oak group, 45 ft high	47.74	24.93
Under bigcone spruce, 40 ft high	48.05	25.24
Under ponderosa pine, 80 ft high	60.53	37.72
January to May, 1918		
Open	27.23	0
Under ponderosa pine, S side, windward	52.51	25.28
Under ponderosa pine, E side, windward	56.62	29.39
Under ponderosa pine, N side, leeward	85.38	58.15
Under ponderosa pine, W side, leeward	87.50	60.27

* Records furnished through the courtesy of the late W. P. Hoge and reproduced with the permission of the Mt. Wilson Observatory.

spruce and western hemlock 85 years old, near the Oregon Coast, in 18 weeks aggregating 142 days from May, 1940, to December, 1941, the excess of fog drip over interception averaged for three rain gages under the trees was 11.23 in. The rate of excess averaged 0.08 in. per day. The precipitation in the open for the same 142 days was 25.19 in., so that the 11.23 in. of fog drip represented a 44.6 per cent excess. In a single week when the gages in the open showed only 0.01 in., those

[1] Unpublished data furnished through the courtesy of the Department of Forestry of Los Angeles County.

[2] Unpublished data furnished through the courtesy of T. T. Munger, Pacific Northwest Forest Experiment Station, U.S. Forest Service.

under the trees caught 0.78 in. Thus fog drip may be a major source of moisture on ridges near the coast in the Sitka spruce and redwood regions.

65. Fog Drip: Summary.—1. In regions of frequent fogs or fine misty rains, as near oceans and in some mountains and mountain valleys, fog drip may at certain seasons increase the precipitation reaching the ground by amounts up to two or three times the precipitation in the open.

2. The amount of fog drip increases with the elevation of the crowns above the ground, with the area of vertical exposure of foliage surface, and with ratio of foliage to ground area.

3. The amount of fog drip from isolated trees is greater on the leeward than on the windward side.

4. In forests it is greatest near the coastal or windward edge and decreases toward the interior.

CHAPTER XIII

EVAPORATION AND CONDENSATION

Evaporation of moisture from free water surfaces, snow, or soil into the atmosphere is an important source of loss of water for vegetation or for human use. Evaporation used in this sense has a distinct form, interception, which although largely a process of evaporation, has been considered previously. It is also distinct from transpiration, which will be taken up later. The term "evaporation" has been used in the literature of engineering and hydrology to include two or all three of these processes, and in using figures for evaporation from different sources it is essential to know what is included in the meaning of the term. The reverse process of condensation goes on with evaporation, and evaporation is ordinarily used to mean the net amount by which evaporation exceeds condensation in any given period of time.

As a loss of water it is usually essential to know or to be able to estimate the magnitude and rate of that loss either locally in a given area or drainage basin, or for comparative purposes between different localities or regions.

66. Measurements, Factors, and Formulas.— Evaporation is affected directly or indirectly by atmospheric pressure, wind velocity, atmospheric humidity, temperature, and solar radiation. The last three are interrelated, and where only one or two of the three factors are used in an expression for evaporation, it is usually assumed that the one or two provide a sufficient expression for the combined influence of all three. Atmospheric pressure becomes important principally where comparisons of areas at different altitudes are involved. In addition to the foregoing factors, evaporation is strongly affected by the wetness of the evaporating surface. It is obviously unlikely that any instrument will give the same combination of these factors and their intensities as will be found in natural sites, and hence no instrument is likely to give an accurate measurement of the water loss from soil or plant surfaces. The climatic factors involved in evaporation are continually changing and cannot be standardized for natural sites, but the wetness of the surface can be so standardized to a considerable extent, if the evaporation is measured from a free water surface of uniform area and exposure. Such a measure is applicable to the loss from lakes and reservoirs and may provide a standard of comparison for evaporation from other surfaces. Standardized methods for the measurement of evaporation from a free water surface in a pan 10 in. deep and 48 in. in diameter have been described by the U.S. Weather Bureau (192).

The totals of evaporation for 24-hour or longer periods integrate the several climatic factors and also their hourly, daily, or periodic fluctuations. The use of a free water surface is desirable for this reason, and also because the results are directly comparable with records of precipitation. The latter advantage does not obtain in the case of Livingston or other atmometers, which aim to simulate the surfaces and exposures of plants. The records from such instruments usually cannot be translated reliably into inches depth of water evaporated, and no two such instruments give the same results for any given place or time. However, weekly evaporation records from a Bates evaporimeter converted to inches depth of water E_B, when plotted over the corresponding depths from an adjacent standard Weather Bureau pan E_w for the season of 1939–1940 at an elevation of 5,000 ft in the San Dimas Experimental Forest, indicated a well-defined linear trend that could be represented by the equation

$$E_B = 0.338E_w - 0.045$$

This was in the range from 0.5 to 3.0 in. weekly evaporation from the pan. Both evaporimeter and pan were in the open.

Records of evaporation from free water surfaces in the United States range from 1 to 14 in. per month. On the basis of such data, combined with records of the climatic factors, numerous empirical formulas for the prediction of evaporation from the physical elements have been developed and may be found in the following references (249, 251). The one suggested by Horton (173) has advantages both on theoretical grounds and for practical use. If E represents evaporation in inches per 24 hours; B_0 the barometric pressure of 29.9 in. of mercury at sea level; B, the barometric pressure at the actual elevation; e_w, the vapor pressure of saturated air at the temperature of the water surface in inches of mercury; e_a, the actual vapor pressure or the saturated pressure at the temperature of the dew point; ϵ, the base of natural logarithms equal to 2.718; v, the wind velocity in miles per hour; k, a constant with a value of approximately 0.2; and C, a constant for standard Weather Bureau pans equal to 0.4; then

$$E = C\frac{B_0}{B}(e_w - e_a) + C(1 - \epsilon^{-kv})\, e_w$$

In this equation the term for vapor pressure difference $(e_w - e_a)$ represents rate of diffusion of water-vapor molecules between the water surface and the adjacent atmosphere, and this process varies inversely as the atmospheric pressure. The term in which wind velocity appears as a negative exponent expresses the rate of removal of water vapor by the wind and is not influenced by the atmospheric pressure. This wind term in the equation takes values varying from 0 to 1, as the wind velocity increases from 0 to infinity. If the temperature of the water surface is greater than the temperature of the adjacent air, the vapor-pressure difference and E will be positive, and evaporation is indicated. On the other hand, if the temperature of the surface is less than the temperature of the adjacent air, which may often be the case with snow or with cold spring water, the vapor pressure difference becomes negative. If its numerical value exceeds that of the wind term, E also will be negative, and condensation will result in the addition of water to the cold surfaces of lakes, streams, soil, ice, or snow. As the wind velocity increases the evaporation will increase but condensation will decrease, according to the formula. This agrees with the usual observations that water evaporates most rapidly in high winds, and conversely that dew and frost are deposited mostly during still nights. Examples of evaporation computed by Horton's formula for different conditions of weather and vegetative cover are given in Table 20.

TABLE 20.—EVAPORATION OR CONDENSATION AT SEA LEVEL COMPUTED BY HORTON'S FORMULA FOR DIFFERENT COMBINATIONS OF PHYSICAL CONDITIONS

	Hot dry day		Rain	Snow	Snow cover	
	Open	Forest			Open	Forest
Temperature of the water surface, °F	75	70	70	32	32	32
Temperature of the air, °F	90	85	60	60	41	42
Relative humidity, %	20	30	100	100	70	69
Wind velocity, mph	20	3	2	2	2	0
Evaporation, in./24 hr	0.574	0.281	0.183	−0.110	0.025	−0.002

The assumed figures for temperature, relative humidity, and wind velocity are such as may frequently be found under the specified conditions. The computed effect of forest cover in reducing by one-half evaporation on a hot, dry, windy day is striking. Another notable contrast is between evaporation of almost 0.2 in. in saturated air when the temperature of the water surface is 70°F compared with condensation of more than 0.1 in. on a snow surface. Forest cover may under certain conditions make the difference between evaporation and condensation.

On the basis of experiments with free water surfaces of limited area under controlled conditions, Hickox[1] developed the formula

$$E = C(e_w - e_a)\frac{v^{3/4}T^{3/8}}{D^{1/4}B^{1/4}}$$

where T is absolute temperature and D is diameter of the surface. According to this formulation, evaporation varies with the vapor-pressure difference and with fractional powers of wind velocity and absolute temperature and varies inversely with the one-fourth power of diameter of surface and of barometric pressure. Evaporation does not continue to increase with wind velocity indefinitely nor to decrease as the diameter of the surface increases toward the dimensions of lakes and reservoirs.

Another formula for evaporation based on physical theory has been developed by Leighly (218), with the thought that it should be applicable to evaporation from moist surfaces such as soil or leaves, as well as from water surfaces. Subsequently it has been tested and slightly supplemented by Martin (240) to give in general form

$$E = \frac{CK(e_w - e_a)v^r}{l^n w^m}$$

When the constants were determined by experiments with moist blotting-paper leaves of different dimensions, it became,

$$E = \frac{0.73K(e_w - e_a)v^{1/2}}{l^{0.3}w^{0.2}}$$

where E is evaporation in grams per square centimeter per hour; K is a coefficient including diffusion in the same units; e_w is the vapor pressure at the evaporating surface; e_a is the vapor pressure at the outer limit of the boundary layer, both in millimeters of mercury; v is the wind velocity in centimeters per second; l is the length of the evaporating surface parallel to the direction of the wind; and w is the width of the surface perpendicular to the wind, both in centimeters. If the evaporating surface is circular or square the evaporation is inversely proportional to the square root of the diameter or side, respectively. The experiments covered a range of wind velocity

[1] Hickox, G. H., "Evaporation from a limited free water surface," Univ. Calif. Ph.D. thesis in mechanical engineering, 1940.

only up to 5.5 mph, and it is unlikely that the exponent, ½, is equally applicable for considerably higher velocities. Another difficulty lies in the determination of the thickness of the boundary layer, at which distance from the surface, temperature and humidity must be measured. In Martin's experiments the thickness varied with the humidity and with the position of the evaporimeter, in the range of 0.8 to 1.5 cm. The measurement of the temperature at the evaporating surface might require the use of a thermocouple.

An error that is not negligible will usually be involved in assuming that the saturation deficit of the atmosphere is equal to the vapor-pressure difference between the wet surface and the air. The former will usually be either colder or warmer, and its temperature must be measured in order to obtain the correct vapor-pressure difference term for use in this or in Horton's or other formulas.

If the temperature of the water or wet surface is equal to the temperature of the wet-bulb thermometer, as may often happen during rain, then evaporation will be proportional to the wet-bulb depression. This may yield a closer estimate of evaporation, for example in the case of an atmometer, than would the use of saturation deficit based on the dry-bulb temperature.

Another method of approach to the prediction of evaporation has been made by Cummings on the basis of the energy balance (106). His formula equates the rate of energy removal by evaporation to the rate of energy supply by solar radiation, minus the loss by back radiation and corrected for change in heat storage in the water and leakage through the walls of the container. The formula is

$$E = \frac{R - R_b - S - C}{L(1+r)}$$

where E is the depth of water evaporated per unit of time, R is the solar radiation in calories per minute per square centimeter of horizontal surface, R_b is the back radiation to the sky, S is the heat stored in unit time in a column of unit area and depth, and C is the heat leakage through the walls of the container, L is the latent heat of vaporization of water in calories per cubic centimeter, and r is called "Bowen's ratio" and is derived from temperature and vapor-pressure differences as follows:

$$r = 0.46 \left(\frac{t_w - t_a}{e_w - e_a} \right) \frac{760}{B}$$

where t_w is the temperature of the water surface in degrees centigrade, t_a is the temperature of the air, e_w is the vapor pressure of water in the air saturated at the temperature of the water surface, e_a is the vapor pressure at the temperature of the air, and B is the barometric pressure, the last three terms in millimeters of mercury. If temperature in degrees Fahrenheit and pressure in inches of mercury are used in Bowen's ratio, the two numerical values become 0.01 and 29.92, respectively.

The solar radiation varies from zero to 70 or 80 cal per hr per sq cm or at the maximum, about 750 cal per day. If all this radiant energy was expended for evaporation without loss, the maximum rate of evaporation would be 0.05 in. per hour, or 0.5 in. per day, or about 15 in. per month. The latent heat of vaporization of water at usual temperatures is about 580 cal per cu cm. If the evaporation is expressed in inches, 1,500 cal per sq cm are used in the evaporation of 1 in. of water. The range that the other factors in the equation may take under ordinary conditions has been indicated by Cummings (107), who considers that useful approximations can be made even when exact determinations of all factors are not feasible. The evaporation of 0.5 in. per day is obtained by dividing the latent heat of vaporization, 1,500 cal per in. per sq cm, into the solar radiation of 750 cal per day per sq cm. This formula suggests pointedly that the differences in evaporation in forests of different species or density will be related to the amount of solar radiation penetrating the canopy.

The term "evaporation opportunity" is sometimes used in a specific technical sense to represent the ratio of the maximum loss by combined evaporation and transpiration to the evaporation from a large free water surface. In this sense the application of the term will be considered after the subject of transpiration.

The recent method of determination of "evaporation" from land and water surfaces by Thornthwaite and Holzman (342), which actually is most useful for the determination of combined evaporation and transpiration can also be used for evaporation from a water or snow surface or bare soil if no vegetation is present. The following formula has been derived to express the relations:

$$E = \frac{K^2\rho(H_{q1} - H_{q2})\,(v_2 - v_1)}{\left(\ln \dfrac{h_2}{h_1}\right)^2}$$

where E is the evaporation (plus transpiration) or condensation in inches per hour, K is von Kármán's dimensionless coefficient, having a value of 0.38; H_q is specific humidity, grams per kilogram; ρ is density of the air, pounds per cubic foot; v is velocity of the wind, miles per hour; ln is natural logarithm; h is height in feet; 2 as a subscript indicates the higher elevation; 1 as a subscript indicates the lower elevation. The equation is equally applicable when cgs units are used for E, ρ, and v.

According to the equation of state,

$$BV = \frac{RT}{m}$$

where B is atmospheric pressure; V is specific volume, which equals $1/\rho$; R is the universal gas constant; m is the gram-molecular weight of dry air; and T is absolute temperature, which equals t (Fahrenheit) + 459.4. Hence,

$$\rho = \frac{Bm}{RT}$$

When vapor pressures e are substituted for specific humidities, barometric pressure becomes unnecessary as a factor. If these substitutions, including the values of the constants, are made the equation for evapo-transpiration becomes

$$E = \frac{120.3(e_1 - e_2)\,(v_2 - v_1)}{\left(\ln \frac{h_2}{h_1}\right)^2 (t + 459.4)}$$

where e_1 and e_2 are in inches of mercury.

The use of these formulas at times when the air is unstable will give results for evaporation that are too high. Condensation at night when the air is stable will be too low. They should be more nearly correct for periods of 1 day or longer. Further study of the vertical distribution of wind velocity will probably show how corrections can be made. The equipment required to obtain the necessary data includes two hygrothermographs installed at different levels, the upper one 25 or 30 ft above the ground and the lower one 2 to 5 ft, depending on the roughness of the surface, two anemometers at corresponding levels, and a barograph. One difficulty to date has been in obtaining a sufficiently accurate measure of humidity. The hygrothermographs have not proved satisfactory, and the development of better instrumentation is being studied.

When the installation is set up over an area of bare soil or a water or snow surface, the results would be exclusively for evaporation or condensation. When, however, as has been the case with the installations to date, there is vegetation growing beneath the instruments, the calculated losses include both evaporation and transpiration. The method requires that the instruments be surrounded by a considerable area having the same conditions of surface whether with or without vegetation, because the humidity will evidently be affected by the water vapor carried an appreciable distance by the wind. Thus instruments on the shore of a lake would give representative figures for evaporation from the water surface only when the wind was blowing from the lake toward the shore.

67. Dew and Frost.—The deposition of dew or, at temperatures below freezing, frost are similar phenomena caused in these instances by

TABLE 21.—MONTHLY CONDENSATION IN INCHES ON GRASS-COVERED WEIGHING MONOLITH LYSIMETERS AT COSHOCTON, OHIO (144)

Month	No. Y103* 1943	No. Y103* 1944	No. Y102† 1944	Average
January	0.72	0.73	1.10	0.85
February	0.42	0.88	0.66	0.65
March	0.46	0.62	0.39	0.49
April	0.25	0.49	0.32	0.35
May	0.09	0.17	0.11	0.12
June	0.12	0.17	0.12	0.14
July	0.10	0.07	0.20	0.12
August	0.12	0.07	0.19	0.13
September	0.38	0.20	0.43	0.34
October	0.43	0.45	0.76	0.55
November	0.40	0.39	0.81	0.53
December	0.65	‡	‡	0.65
Total	4.14	4.24	5.09	4.92

* Keene silt loam.
† Muskingum silt loam.
‡ Snow drifting vitiated the record.

condensation of the moisture from the atmosphere in contact with foliage or surfaces below the temperature of the dew point. This frequently happens as a result of cooling by radiation during still, clear nights. Dew and frost in northern latitudes are mentioned by Zon (381) as amounting to 0.4 to 0.8 in. annually. Dew from the grass in meadows in England is said to contribute 1 in. of water annually, and Brooks (54) has suggested that dew from trees would not exceed that from grass.

One of the few series of measurements so far available comes from the lysimeters of the Soil Conservation Service at Coshocton, Ohio. The monthly amounts of condensation for two different years and for lysimeters with two different soil types in the same year are given in Table 21. These lysimeters were constructed around undisturbed blocks of soil and rock, 14 ft long, 6.22 ft wide and 8 ft deep and placed on scales that recorded the weight every 10 minutes and were sensitive to a weight equivalent to 0.01 in. depth of water. There is no apparent reason why the measurements of condensation should not be representative of unenclosed soils with similar vegetation in the locality. They show an annual cycle with maximum of over 1 in. in January and minima of 0.07 to 0.12 in. in the summer. The annual totals are from 4 to 6 in., depths that are considerably greater than have formerly been ascribed to dew and frost constituting a noteworthy supplement to precipitation.

Figures for condensation on snow to be taken up in the next chapter are of course examples of the deposition of frost. If a rate of 0.002 in. gain in water equivalent of the snow per day is multiplied by the 120 days of a 4-month snow season, the seasonal total is 0.24 in. but this is only the amount by which condensation exceeded evaporation for that season. The amount of frost, if it could be segregated, would be much larger on many days, but the number of days on which excess condensation occurred would be smaller.

68. Evaporation: Summary.—1. Annual evaporation from a free water surface in the United States varies regionally from over 90 in. in the Southwest to less than 20 in. in the Northeast or Northwest, ± 10 per cent for any one year. Monthly evaporation may vary from as much as 14 in. in summer to less than 1 in. per month in winter. These amounts approximate the theoretical maxima of evaporation from saturated soils.

2. Evaporation integrates the periodic variations of pressure (altitude), wind, solar radiation, temperature, saturation deficit of the atmosphere contiguous to the evaporating surface, and wetness of the evaporating surface.

3. Evaporation increases with vapor-pressure difference and altitude and up to a certain point as the wind velocity increases, the other factors being constant in each instance.

CHAPTER XIV

SNOW

The snow occurring in much of the forested part of the United States is an important source of water. The snow is a form of temporary storage of water that contributes to floods and to prolonged spring and summer flow.

The proportion of the precipitation falling as snow provides a means of evaluating the importance of snowfall in different regions. A map giving lines of equal percentages of precipitation as snow, prepared by Brooks and others (56), shows that the area of more than 20 per cent snow includes northern New England and New York, the northern Lake states and most of the Rocky Mountain and Sierra Nevada–Cascade ranges. Much of the western mountain area is within the line of 30 per cent with a single island of more than 40 per cent in the Wasatch Range in Utah, where, according to estimates by Clyde (91), 60 to 90 per cent of the annual precipitation falls in the form of snow.

69. Measurements.—The measurement of snowfall is subject to wider variation than that of rain because of the large horizontal component in the fall of snow even in light winds. For snow the ordinary rain gage, even when used without the funnel top, gives results from 20 to 50 or even 70 per cent low. Snow bins 5 to 10 ft square protected by wire screens or shielded gages give better results. Both Nipher and Alter shields have been recommended for snow. Snow boards 18 to 24 in. square covered with white flannel or painted white on a rough surface, and laid on the ground or on the surface of the old snow, can be used with an inverted rain-gage can or other core cutter with which the snow can be measured for depth, collected, and weighed. When snow and rain fall together, a part or all of the rainfall is lost from board measurements. Losses also occur in mild weather if the snow on the boards is not measured promptly after each storm.

Although figures for snow usually are expressed in units of depth, a more useful measure is the water equivalent of the snow. This may be defined as the total water content of the snow expressed in inches depth. The relation between the water equivalent and the depth of the snow provides a measure of the density of the snow layer. Density of snow ρ_s may be defined as the volume of water contained in a unit volume of snow or the ratio of the water equivalent d of a snow core to its depth D_s. These three expressions are related by the equation

$$\rho_s = \frac{d}{D_s}$$

From this relation any one of the three expressions may be obtained when the other two are given.

A related but distinct term is the "water content" of the snow, which is the amount of liquid water in the snow layer and may be expressed as the ratio of water to snow as a percentage by weight or volume. The complement of the water content is the percentage of ice in the snow and this is called the "quality" of the snow. Quality is determined by the use of a calorimeter, in which case it is the ratio of the heat of melting of snow in calories per gram to 80 cal per g, the latent heat of fusion of pure ice. If the temperature of the snow is below freezing the quality is 100 per cent or a calorimeter determination may give more than 100 per cent. Quality is useful both as a measure of the wetness of the snow and as a factor in the amount of heat required to melt the snow.

The percentages of snow in the precipitation falling at different altitudes on the west slope of the southern Sierra Nevada and in the mountains of southern California derived from Weather Bureau and Snow Survey records are shown in Table 22. Less than 5 per cent of the precipitation is snow at an elevation of 2,000 ft, 45 per cent at 6,000 ft, and 80 per cent at 8,000 ft. When the water equivalents of the snow are weighted by the percentages of area occupied by the different zones, the areal percentages of water equivalent of the snow are obtained as shown in the last column of Table 22. Although the figures are partly estimated it is evident that the heavily forested zones between 4,500 and 8,500 ft are the important ones as sources of water from snow.

In order to convert the commonly published figures of depth of snowfall to water equivalents, it is necessary to know the density. For this purpose it is often assumed that the density of freshly fallen snow is uniformly 10 per cent. Actually, the density varies in different storms and also systematically with altitude. Again according to the data from California a linear trend is indicated from an average density of 12 per cent at 2,000 ft to about 6 per cent at 10,000 ft.

Field methods in the measurement of snow on the ground, developed primarily in the snow surveys, consist in taking cores of snow by means of sampling tubes. The depth of snow is read on the graduated scale on the outside of the tube before it is lifted out of the snow and after lifting, tube and snow core are weighed on a balance graduated to read directly in inches of water equivalent.

TABLE 22.—SNOWFALL IN THE ZONES OF VEGETATION ON THE WEST SLOPE OF THE SOUTHERN SIERRA NEVADA

Vegetation	Altitudes, ft	Areas, %	Water equivalent of snowfall		
			Precipitation, %	Inches	Weighted by area, %
Grassland	50– 800	28	0	0	0
Woodland and brush	800– 3,200	24	0	0	0
Ponderosa pine	3,200– 4,200	8	19	8	7
Mixed conifer	4,200– 6,500	20	37	17	36
Fir and lodgepole pine	6,500– 8,500	15	68	24	39
Subalpine	8,500–11,500	5	95	33	18
Totals	....	100	......	....	100

The magnitude of discrepancies in the measurement of snow with different types of gages and by the differences between samplings with tubes after successive storms is suggested by Garstka's figures in Table 23 (132).

TABLE 23.—COMPARISONS OF WATER EQUIVALENT OF SNOWFALL IN GAGES AND FROM SNOW SURVEY DATA IN WOODED AND OPEN AREAS IN MICHIGAN IN JANUARY, 1943

Cover	Recording gage, in.	Standard gage, in.	Gage with Nipher shield, in.	Snow survey, in.
Cultivated	0.63	0.91	1.30	1.57
Wooded	1.11	1.16	1.57	1.76
Percentage relatives (Nipher shield = 100)				
Cultivated	49	70	100	121
Wooded	71	74	100	112

The recording and standard gages give measures of snowfall that are seriously low. A gage with a Nipher shield may also give results that are too low, although other factors affecting snow-survey measurements preclude a rigorous comparison. The smaller differences between methods of measurement in the wooded than in the cultivated area probably reflect the reduced velocities of the snow-bearing winds, although the comparisons are based only on storms when the winds did not cause drifting. The two areas were 10 miles apart so that the greater snowfall in the woods cannot be ascribed with certainty to the influence of the leafless forest.

The depth of snow alone is not a reliable measure of its water content or yield, because depth of snow may change from three different causes, namely, (1) the addition of new snow by precipitation, (2) a decrease associated with an increase in density, and (3) a decrease by evaporation. Precipitation has been considered previously.

70. Density of Snow.—The density of freshly fallen snow has been shown to vary from 4 to 17 per cent. As the snow lies on the ground after falling, its density increases until old snow may have a density of from 20 to 50 per cent and, after a heavy rain, even as high as 90 per cent. These changes in density with exposure are associated with changes in the water content and structure of the snow layer, which in turn are caused by the diurnal and other changes in radiation and temperature. Between day and night there will often be sufficient differences in temperature to subject the snow to alternate thawing and freezing. In this process during the day, water from the upper layers of the snow under the influence of the sun's radiation flows down over the crystals in the lower layers. During the night these films of water freeze again. Thus, the fine particles of fresh snow become enlarged and solidify into larger icy particles that are denser and pack more closely than did the new crystals. The old snow layer in this way becomes much denser than it was in its original condition.

The increases in density of the snow during the winter as a result of this process are indicated in all series of successive measurements during the snow season. In Douglas fir forest and in open areas in the Yakima watershed in Washington, Griffin (141) found densities averaging 31 per cent on Apr. 10 which increased to 40 per cent through May and to 50 per cent on June 25. The rising trends were similar for the forested and open sites, but the changes under the forest lagged behind those in the open. In other words the density of the snow in the forest remained about 14 per cent lower than that at the same date in the open until after the snow in the open had disappeared.

Several thousand measurements of snow density in the mountains of western Nevada and adjacent California show similar increasing densities with the advance of the season. Between 9,000 and 10,000 ft on the Wasatch Plateau in Utah, Clyde (91) recorded an increase in density from 39 per cent on Apr. 21 to 49 per cent by Apr. 27 during a period of warm weather while the depth of the snow decreased 4 in.

Drainage of water from the snow began only when the density reached about 50 per cent. On May 2 the density had decreased to 40 per cent and remained almost constant at that figure until May 20 even during heavy rains. It may be concluded that water does not begin to drain from snow until the whole layer has reached a density of 40 to 50 per cent, and thereafter the drainage tends to be continuous unless interrupted by fresh snow or prolonged cold weather.

71. Influence of Forest on Density of Snow.—The relation of snow density to the density of the forest is indicated by a large number of measurements in the Sierra Nevada of California (13). The average density in the open was 42.4 per cent; in thin forest, 41.1 per cent; and in dense forest, 38.0 per cent. The relations for different periods and for different elevations are summarized by the following relatives in which the density in the open is represented by 100 per cent.

Period	Elevation, ft	Forest	
		Thin, %	Dense, %
February–March	5,000–7,000	99.6	96.1
April–May	5,000–7,000	99.3	95.5
February–March	7,000–9,000	95.8	77.2
April–May	7,000–9,000	94.0	90.7

These figures would indicate that the density of the snow is less as the density of the cover is greater and that the cover has more effect in this respect above 7,000 ft than below.

On the west slope of the Sierra Nevada between 5,000 and 6,500 ft, the seasonal increase in density of the snow from 10 or 15 per cent after it falls, to 40 or 50 per cent in March or April, tends to be quite uniform at a rate of 0.3 to 0.5 per cent per day. Irregularities in the trends occur with falls of fresh snow or rain. The rate of increase is slightly greater in the open than under the forest but the range from 0.3 to 0.5 per cent includes 10 different conditions of forest type, age, size, and density in two seasons.

The maximum rates of increase in density for periods of 2 to 5 days are much higher than the seasonal rates. The medians of the five highest rates in the two seasons were close to 2 per cent per day in the forested areas, with higher rates of 3 and 2.5 per cent in clearings and with a lower rate of 1.7 per cent under dense white fir. Thus under continued favorable conditions a fall of fresh snow of density of 15 per cent might increase to 45 per cent in as little as 15 days. In locations exposed to high winds a rapid increase in density of the snow is ascribed to the effect of the wind by Church (87). As an extreme example he mentions an increase of 8 per cent in density in a single night in Greenland.

72. Evaporation from Snow.—Snow as water in the solid state or a mixture of solid and liquid is subject for the most part to the same principles controlling the evaporation and condensation of moisture as from a free water surface. The heat equivalent or latent heat of fusion for dry snow when its quality is 100 per cent is 80 cal per g, or 80 cal are used to melt 1 g of snow, which is the equivalent of 1 cm depth of water on an area of 1 sq cm. Eighty calories for 1 cm depth become 200 cal required to melt 1 in. water equivalent of snow. If the quality of the snow is less than 100 per cent, proportionately less heat will be used. Similarly the latent heat of vaporization for water is 600 cal per g, or 1500 cal for 1 in. depth. The sum of the two, 1700 cal, will be required to evaporate 1 in. water equivalent of snow.

Conversely, when 1 in. of water is condensed from the vapor to the liquid state, 1500 cal are released, and since 200 cal are required to melt 1 in. water equivalent of snow, the condensation of 1 in. water equivalent will cause the melting of snow to a depth of 1500/200, or 7½ in. water equivalent. Actually, the phenomenon of condensation of water vapor on the snow occurs only in small amounts, but, whatever the amount of condensation, if the snow temperature is 32°F, 7½ times that depth water equivalent of snow will be melted.

When the snow surface has a temperature of 32°F, which is the case during most of the melting period, it becomes possible to express in a simple equation the relation between relative humidity or vapor pressure and temperature when evaporation (or condensation) is zero. This trend becomes linear when the relative humidity is plotted on a logarithmic scale and may be expressed by the equation

$$\log \mathrm{H}_r = 2.52 - 0.017t$$

where H_r is percentage relative humidity, and t is temperature in degrees Fahrenheit. Any combination of relative humidity and temperature that results in an intersection above the line plotted from this equation indicates condensation, and any intersection below the line indicates evaporation, provided the air temperature is above 32°F. From this relation and from the fact that snow temperatures often are below 32°, especially at night, it becomes apparent that there will be frequent occasions when condensation will take place on a snow surface, thus adding to the total water content of the snow and at other times tending to balance losses by evaporation.

73. Influence of Forest on Evaporation from Snow.—The few actual records of evaporation from snow, and particularly in relation to forest cover, vary widely. At 7,000 ft in Utah, measurements of evaporation in glass jars, from Nov. 7 to May 4, by Baker (29) indicated a mean daily evaporation of 0.017 in. For the 180 days of the snow season, the total amounted to about 3 in., or 14 per cent of the snowfall. As the air temperature increased, the evaporation increased at an accelerated rate. Subsequently at 8,700 ft in the same locality in May, Croft (105) recorded average daily evaporation of 0.04 in. On May 8, cores in pans exposed to free wind movement showed evaporation of 0.05 in. while those with no wind movement showed only 0.02 in. The difference between sun and shade was only 0.002 in.

The rate of evaporation from snow from December, 1917, to February, 1918, in Ohio was found (175) to be 0.023 in. per day.

Determinations of evaporation from snow in pans at the crown level of the trees in the Lake Tahoe region of Nevada from November to March by Church (88) showed total water evaporated in an open meadow to be 8.5 in.; in a semi-open pine forest, 4.8 in.; and in a small fir glade, 2.4 in. In inches per month the corresponding figures would be 1.7, 1.0, and 0.5, and, in inches per day, 0.057, 0.032, and 0.016.

Near the headwaters of the Colorado River in Colorado the evaporation for four seasons averaged 2.23 in. (353). Most of this evaporated in the months of April and May. Starting with evaporation of 2 in. on a commercially clear-cut area of lodgepole pine, it was estimated that evaporation would decrease linearly with increasing stand density to 0.8 in. in the uncut forest.

Measurements for several years from pans set in the snow at some 20 stations, representing a range of conditions from open meadow to dense forest, at elevations between 5,000 and 6,000 ft on the Stanislaus National Forest in California, showed a median evaporation for all stations and years of 0.007 in. per 24 hours. For 1936 the median was 0.013, and in 1938, −0.002. The negative value in the latter year means that condensation exceeded evaporation. For individual forest types the medians ranged from a maximum of 0.12 in. per day in 1936 under a mature open stand of ponderosa pine to −0.002 in. per day in 1938 in dense stands of white fir, red fir, and ponderosa pine. Records taken morning and evening at a limited number of stations showed that evaporation exceeded condensation in 83 per cent of the day periods and that condensation was in excess 72 per cent of the night periods. Apparently, the evaporation from snow on the Pacific slope of the western mountains is decidedly less than that in the interior. The difference is presumably associated with the greater content of water vapor in the air carried by the prevailing onshore winds from the Pacific Ocean.

74. Accumulation.—The accumulation of snow is obviously related to the distribution and density of the crowns of the trees and the openings between them. Snow is subject to interception just as is rain, and it has been shown that the magnitudes of interception of snow are not very different from those of rain. Wet snow will tend to collect in the crowns of evergreen trees more than would rain. It is thus exposed to more rapid evaporation. Part of this snow clinging to the crowns falls or drips after the precipitation ceases, but the magnitude of this drip when it has been measured in gages rarely amounts to more than a few hundredths of an inch.

Snow, being porous and flakelike, has a larger horizontal component in its fall than rain and is more subject to movement by the wind, as it in turn is modified by the friction of the forest. Thus it might be expected that the variations in accumulation of snow in relation to the crowns and openings between crowns in the forest would show similar and perhaps even greater variation than was found in the case of rain.

In a series of experiments with model slat fences in a wind tunnel, Finney (122) determined relations of the eddies and the drifting of flake mica and balsa sawdust to leeward of fences of different heights. For fences of 50 per cent density (slat width equal to space between slats), the distance l_0 to the leeward ends of the eddies and of the drifts nearly coincided and increased with the height h of the fence according to the equation

$$l_0 = 15h$$

The eddy area did not vary with wind velocity in the range of from 10 to 30 mph.

The distance l_m to leeward of the fence at which the maximum depth of drift was found increased with the height h of the fence and with the wind velocity v and with decreasing density of the "snow." For the balsa sawdust of 0.03 density, in winds of 10 mph,

$$l_m = 3.5h$$

In winds of 25 mph

$$l_m = 10h$$

For the latter wind velocity the coefficient became about 4 for the flake mica of density 0.20.

For a 10-ft fence with vertical slats the distance to the maximum depth of drift as a function of wind velocity was

$$l_m = 0.8v + 12$$

When the slats were horizontal, however, the equation became,

$$l_m = 0.5v + 35$$

Good examples of excess accumulation of snow are provided by records from windbreaks in the northern Great Plains. In the lee of a shelterbelt in South Dakota at the end of February there was 100 in. of snow, whereas to windward and beyond the influence of the belt to leeward, the depth of the snow was less than 20 in. The maximum drift was at three times the height of the trees and the effect was noticeable as much as eight times the height of the trees to leeward (335). Maximum drifts accumulate to leeward of dense low-crowned belts of 1 to 3 rows of young trees or of shrubs like willow or Siberian pea tree (Fig. 18, page 163). Older belts without foliage near the ground allow so much wind to pass beneath the crowns that little or no drift is formed. Belts of 10 or more rows with shrub rows to windward tend to trap the snow within the belt although a small amount of excess accumulation may extend a considerable distance to leeward. Multiple belts may combine the foregoing effects. For example, a 30-year-old windbreak 30 ft high in North Dakota composed of three rows of willow in the windward belt and three rows of willow, one of boxelder, and one of green ash in the leeward belt, with a treeless lane 110 ft wide between them, trapped 60 in. of snow in the lane and formed a drift 35 in. deep within the height of the trees to leeward compared with a depth of less than 10 in. beyond the influence of the break (38).

The effect of interception of snow by the trees in shelterbelts is usually displaced to leeward where it is obscured by drifting. Otherwise, the phenomena of accumulation of snow are similar to those in extensive forests.

In general, the accumulation of snow under the crowns where interception is at a maximum is less than it would be if the trees were not there, whereas the accumulation in openings between the crowns, where wind velocities are reduced, or eddies form, is greater than where the wind has full sweep. If the openings between the crowns are small the excess accumulation will also be small or negative, and apparently the greatest excess accumulation is found in openings where the distance across the opening is at least equal to the height of the surrounding trees.

Examples of the variation in water equivalents of the snow accumulation on the ground are indicated in Fig. 15. If the 11-in. average water equivalent of the snow in an 80-acre clearing is taken as the basis of comparison, we find under the crowns and close to the trunks of dense white firs less than 2 in. water equivalent but in the openings between the trees as much as 17 in. water equivalent.

In mature lodgepole pine 70 ft high near Fraser, Colorado, Wilm and Collet (369) found that, from under the crowns out into the

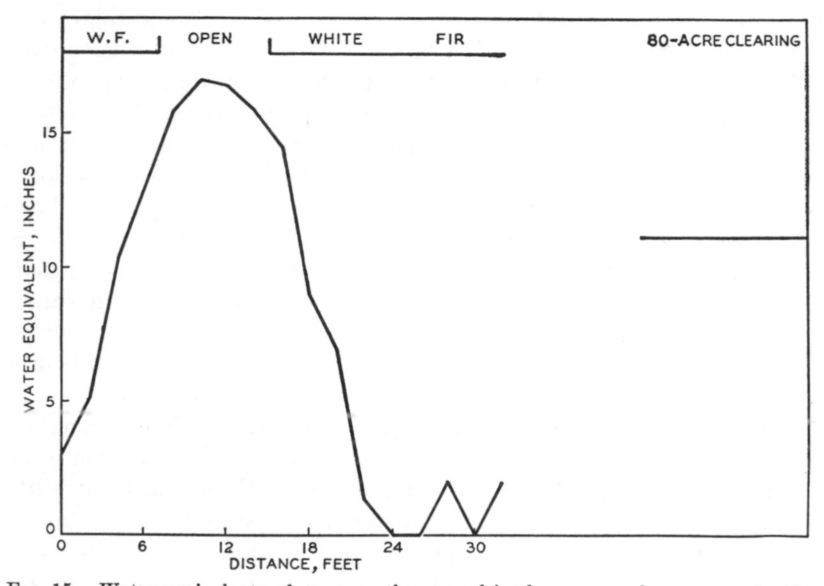

FIG. 15.—Water equivalents of snow on the ground in the open, under crowns of white fir, and in an opening between crowns.

openings between the trees, the water equivalent of the snow increased linearly at the rate of 1 in. in 15 ft. The relation was maintained from a distance of 30 ft beneath to a distance of 30 ft beyond the edges of the crowns into the openings. Thus the maximum accumulation was not yet reached in an opening of diameter equal to 6/7 the height of the trees.

The evidence from the multiple-belt windbreaks suggests that an opening two or three times the height of the trees may be most effective in trapping a maximum accumulation of snow.

75. Melting of Snow.—The melting of snow is often used to express any decrease in depth. Its meaning has been restricted by Clyde (91) to the water that actually drains from the snow cover. Confusion will be avoided by defining the term as the change from the solid to the liquid state. Obviously, melting is a resultant of the balance between gains and losses of heat. The gains are derived from turbulent air exchange, condensation, incoming radiation, warm rain, and the soil. Incidentally, snow reflects about 75 per cent of the radiation. The losses of heat result chiefly from outgoing radiation and evaporation. Means of evaluating each of these sources, including allowances for cloudiness and the albedo of the snow, have been suggested by Wilson (373). His figures for extreme conditions give a total of 3.8 in. water equivalent melted in ½ day.

Often snow melts first at the surface, and the water thus produced in the surface layers percolates downward, filling the capillary storage and increasing the density up to 40 or 50 per cent before drainage from the snow begins. After drainage starts it may continue independently of ordinary temperature changes in the air above the snow surface, indicating that the melting at this period results from the hydrostatic pressure of the water in the saturated layer at the bottom of the snow pack and in capillary films over the snow crystals.

The rate of melting after maximum density was reached, was found by Clyde (91) to be 1.4 in. water equivalent per day for the period from Apr. 27 to May 9. During the 16 days of the melting period there was a cumulative total of 318 day-degrees F above freezing and 17.2 in. water equivalent of snow melted, so that the melting was at the rate of 0.054 in. of water per day-degree F. In laboratory tests the corresponding rate varied from 0.046 to 0.083. Horton (166) gives 0.05 for the rate of melting in the sun in New York State. Heavy rains, ordinarily associated with temperatures above freezing, may start or accelerate the rate of melting.

The phase of melting of the snow in which water passes down into the soil is definitely influenced by the forest cover. The influence may be expressed in two forms: one, the actual rate of drainage, and the other, the retarded date of disappearance of the snow.

Measurements of the reduction in water equivalent of the snow layer during the comparatively short period of rapid melting in April and May at 5,000 ft on the western slope of the Sierra Nevada indicated rates of melting from 0.7 to 2.0 in. water equivalent per 24 hours. These rates are comparable with the rate of rainfall in moderate to heavy storms and probably do not exceed the infiltration capacity of these forested soils. This suggestion tends to be confirmed by the observation that surface runoff does not appear on the bare slopes immediately below a rapidly melting snow layer, except where the water is concentrated in an impervious drainage channel.

The amount on the ground at any time is the resultant of the snowfall that accumulates beneath the crowns, and the losses by evaporation and melting. Inasmuch as the snow on the ground is a form of temporary storage of water, the storage equation applicable to other forms of storage is also useful here. According to that equation the amount of water at the beginning of any period, plus additions and minus losses, must be equal to the amount at the end of the period. If the water equivalents of the snow at the beginning and end of a period are denoted by S_1 and S_2, the precipitation by P, interception by I, evaporation by E, and the melting by Q, the storage equation is

$$S_1 + P - I - E - Q = S_2$$

Ordinarily the amounts of snow on the ground can be easily measured, and the precipitation and interception determined by gages in the open and under the forest. If evaporation is measured the equation may be solved for the amount of melting Q, and takes the form

$$Q = P - (I + E) + (S_1 - S_2)$$

Similarly, if the melting or runoff is measured, the equation can be solved for evaporation or for any one of the terms when the others are measured.

The storage equation, when solved for runoff, represents a balance between the additions in the form of snowfall or rain and the losses in the form of interception and evaporation. However, it has been shown previously that interception and evaporation vary in opposite directions with changes in the density of the forest cover. The interception is greater as the density of crowns or stands increases, and evaporation from the snow decreases as the density of the canopy increases. Under different conditions of forest cover, therefore, an interplay between the effects of interception and evaporation on the accumulation of snow may be anticipated. Where the interception is high and the evaporation low, the accumulation would tend to be low if interception exceeded evaporation; and where interception was low the accumulation would be high unless evaporation exceeded interception. There are various examples from different parts of the country where these relationships are illustrated in different degrees by actual records.

If the data from northern Arizona (184) are analyzed according to the different densities of crown cover of the ponderosa pine, the average departures from the open park as a basis of comparison are shown in Table 24. In the openings and where there was little overhead cover the accumulation was from 7 to 15 per cent in excess of that in the park, whereas under the partial to complete crown cover, the accumulation was less, and in the case of the complete cover of dense groups almost 32 per cent less than that in the park. In other

TABLE 24.—INFLUENCE OF CROWN COVER OF PONDEROSA PINE ON THE ACCUMULATION OF SNOW AT 7,250 FT NEAR FLAGSTAFF, ARIZONA, IN THE WINTERS OF 1910–1911 AND 1912–1913 (184)

Character of Crown Cover	Average Departure from Open Park for 2 Years, %
None	+ 15.6
Little	+ 7.4
Partial	− 5.8
Complete except to the NE	− 13.1
Complete	− 32.5

words, where interception was large the accumulation of snow was low, whereas, in the openings between the trees or under a slight cover an excess of snow was deposited.

TABLE 25.—WATER EQUIVALENTS, DENSITIES, AND MELTING OF SNOW UNDER DIFFERENT TYPES OF COVER AT 4,500 FT ELEVATION IN CENTRAL IDAHO, 1931–1933 (95)

Cover	Departures from open,* in. water equivalent	Relatives, %	Densities of snow, %	Number days later than open to disappearance
Cleared bare ground	0	100	39.7	0
Sagebrush, 2.5 ft high	+0.1	101	39.4	0.3
Ponderosa and lodgepole pine, 20–30 ft high	−0.9	94	38.2	5.5
Ponderosa pine, partially cut	−1.5	89	....	8.3
Ponderosa pine, virgin	−2.7	81		
Ponderosa pine, mature:				
Without understory	−3.5	75	33.3	3.6
With understory	−4.2	70	32.0	8.3

* Average depth at time of maximum accumulation for the 3 years was 14.2 in.

Similarly, influences of different degrees of cover are illustrated by Connaughton's work in central Idaho (95), from which the figures in Table 25 have been derived. In this case the departures from the open in inches water equivalent of snow became greater, as the density of the stand increased, to a maximum of −4.2 in. under mature ponderosa pine, running 60,000 fbm to the acre with an understory. This represents 70 per cent of the average water equivalent of the snow measured on cleared, bare ground. The same work also illustrates the progressive effect of increasing crown cover on the density of the snow from almost 40 per cent in the open to 32 per cent under the mature pine with an understory.

These conflicting effects of crown cover on accumulation, interception, and evaporation lead to variations and sometimes apparent inconsistencies in the rate of disappearance of the snow. In the foregoing case, notwithstanding the greater accumulation and density in the open, the snow disappeared first in the cleared area as a result of more rapid evaporation and melting. Both in the partially cut stand and in the mature stand with an understory, the date of disappearance was 8 days later than in the open. In this case the greater accumulation in the partially cut stand was just balanced in the effect on the time of disappearance by the lower evaporation and melting under the pine with an understory.

TABLE 26.—ACCUMULATION, DENSITY, AND RATE OF DECREASE OF SNOW UNDER DIFFERENT CONDITIONS OF COVER AT ELEVATION OF 6,225 FT, LAKE TAHOE, CALIFORNIA (87)

Cover	Maximum accumulation, Mar. 11–14, in. of water	Density of snow, %	Daily rate of decrease next week in. of water
Mature fir with glades	16.5	38.6	0.29
Open pine and cedar	13.5	38.8	0.31
Pine and fir	12.1	38.6	0.27
Open meadow	11.7	39.3	0.33
Fir	11.0	36.2	0.20
Dense fir	10.7	35.2	0.23

Similar relations are illustrated by a large series of measurements made by Church (87) on the east side of Lake Tahoe. The figures for the maximum accumulation in inches of water, the percentage density of snow, and the rate of loss during the following week are shown for the six conditions of cover including an open meadow in Table 26. Again the maximum accumulation is in the open stands where the trees or groups have openings or glades between them, and the minimum in the dense fir stands where the interception is at a maximum and exceeds the excess accumulation in the openings. The water equivalent in the open meadow had an intermediate value. The density was highest in the open meadow, slightly less in the forest types of low density, and least in the dense fir type. The rate of disappearance as indicated by the daily loss in inches during 1 week, reached a maximum of 0.33 in. in the open, varying from 0.27 to 0.31 in the open forest, and from 0.20 to 0.23 in the dense forest. Church concluded from his work that the mature forest with many glades, forming what he compares to a honeycomb arrangement of trees and openings, is the ideal for the conservation of snow. Furthermore, he suggests that the glades or openings should be of such extent in

TABLE 27.—RELATION OF CROWN DENSITY TO WATER EQUIVALENT AND RETARDED DISAPPEARANCE OF SNOW IN MIXED DOUGLAS FIR AT 3,000 FT ELEVATION, NEAR YAKIMA, WASHINGTON

Crown density	Water equivalent of snow, in.	Retarded disappearance, weeks
(Open) 0	0	0
0.49	4.4	2.0
0.60	21.5	3.3
0.77	27.8	4.0

relation to the height of the trees that the sun would not reach the snow surface. Finally, he considered that mountain hemlock and fir are the most efficient trees in conserving snow. These are species whose dense foliage reaches close to the ground, and protects the snow from insolation and evaporation.

Examples of the effect of forest cover in three areas of the Cascade range in Washington are provided by Griffin (141). His figures indicate the retention of considerable snow under Douglas fir, mountain hemlock, and lodgepole pine for a period sometimes as long as 42 days after the snow in the open had disappeared. The average retardation was 17 days. Furthermore, although there appeared to be no relation between the amount of snow conserved and either altitude, aspect, or degree of slope, there was a relation to crown density. The figures averaged for 18 stations are given in Table 27. The amount of snow remaining on the ground after the open stations became bare increased progressively as the density of the canopy increased up to 0.77, where the snow had a water equivalent of 27.8 in. In the same way the time after the disappearance of the snow in the open was increased to a maximum of 4 weeks in the densest stands. Griffin pointed out that the snow in the forest tends to be deepest and last longest in the small openings of the denser stands.

Another example comes from the White Mountains in New Hampshire.[1] The water equivalent of the snow and the runoff in inches depth were measured for three separate weeks in April on two basins similar in topography and geology except that one was 80 per cent virgin forest and the other 100 per cent cut over and burned. The

TABLE 28.—RATES OF DECREASE IN WATER EQUIVALENT OF SNOW (d_1-d_2) AND DISCHARGE OF STREAMS q IN APRIL ON A BASIN WITH OLD SPRUCE AND FIR COMPARED WITH A BASIN AFTER CUTTING AND BURNING IN THE WHITE MOUNTAINS, NEW HAMPSHIRE

	Spruce–fir, in. per day	Burn, in. per day	Week in April
d_1-d_2	0.20	0.29	2– 9
q	0.16	0.25	
d_1-d_2	0.72	1.05	14–17
q	0.45	0.72	
d_1-d_2	0.29	0.50	22–29
q	0.43	1.00	

figures are given in Table 28. In all three weeks the snow disappeared more rapidly in the burned drainage basin, as indicated by the changes in water equivalent. Similar relations appeared from the measurements of runoff of the two brooks, but in the third week, from Apr. 22 to 29, the difference between them was accentuated to the extent of a 0.57 in. excess in daily runoff from the burned area.

The depth of water d melted by rain when the rainfall P in inches and its temperature t (approximately that of the wet bulb) are known is

$$d = \frac{P(t-32)}{144}$$

where 144 is the latent heat of fusion of ice in Btu per pound. The depth of rain which will melt any given depth of snow D_s can be computed by substituting $D_s\rho_s$ for d, transposing, and solving for P.

76. Influences of Forest on Snow: Summary.—1. Evaporation from snow in the open may vary from less than 0.01 in. to more than 0.07 in. of water per day or from 0.2 in. to 2 in. per month.

[1] Data supplied through the courtesy of N. C. Hoyt of the U.S. Geological Survey.

2. The accumulation of snow in the forest as compared with open areas is less directly under the crowns because of interception and other factors, and greater in the openings because of reduced wind velocity.

3. The combined effect of these conflicting factors for an area of forest as a whole, is likely to be (*a*) a net gain in accumulation in old forests of moderate to low density and in young stands before crown closure and (*b*) a net loss in evenly spaced pole forests at ages of maximum crown density and interception.

4. Intermittent melting at the surface on sunny days results in increased density but not in runoff into the soil until the density reaches 40 to 50 per cent.

5. The density of snow tends to decrease from a maximum in the open as the density of forest cover increases.

6. The rate of melting and date of disappearance of snow are retarded, the latter sometimes as much as six weeks, by forest cover. These influences increase with increase in the density of the forest.

CHAPTER XV

INFLUENCES OF FOREST ON SOIL EVAPORATION

The subject of evaporation in effect forms a transition from the consideration of influences of vegetation on climatic to those on soil factors.

Actually the soil is the medium through and in which the most important forest influences become apparent, notably infiltration, runoff, and erosion. The soil factors that may be involved and may differ in the different horizons of the soil profile from the forest floor to the parent material include texture, colloids, structure, depth, moisture, air, temperature, inorganic constituents, reaction, organic matter, and soil life.

The means by which the forest may affect the soil include the roots, the forest floor, the shading effects, the interception of precipitation by the crowns, and the sheltering effect on wind movement. If a diagram is prepared to show all the connections between the foregoing vegetation and soil factors, the complexity of the interrelations at once becomes apparent. Four factors have particular interest, namely, evaporation, soil temperature, organic matter, and moisture.

77. Evaporation from Soil.—In forest influences the evaporation from the soil is ordinarily of more interest than that from a free water surface. The relations become more complex, because the evaporation from the soil varies with a number of soil factors, in addition to the meteorological ones previously considered. These factors include (*a*) soil texture and retentive capacity and (*b*) the moisture content and distribution at different depths to the water table. Under the latter, three conditions may be distinguished: first, when the soil is saturated to the surface, or when the water table is at the surface; second, when the water table is below the surface but the surface is within the reach of capillary rise from the water table; and, third, when the water table is below the reach of capillary action.

Some evidence is available to show that the evaporation from a soil saturated to the surface corresponds quite closely to that from a free water surface. The ratios between the two have been found to vary from 0.8 to slightly over 1. The latter case can easily be conceived in areas where a recent fire has left a surface layer of blackened organic material that would be highly absorptive of heat and thus accelerate the evaporation process.

An analysis of evaporation from the soil by Woodruff (377) indicates stages of decreasing rates as the soil becomes drier. In his experimental sands the first stage when the surface was submerged gave a ratio of evaporation from the sand to evaporation from a free water surface of 1.0. The second stage included the period while the surface was moist and water was being withdrawn from the larger pores near the surface. The ratio was 0.79. When the surface was moist and the water was being drawn from larger pores beneath the surface the ratio was 0.71. When the surface was moist and the smaller pores were being emptied it was 0.22. After the surface became dry the ratio was less than 0.03, and changed at a decreasing rate. For each of the four earlier stages the ratio and consequently the rate of evaporation was essentially linear.

After the surface became dry the evaporation could be plotted as a linear function of the depth of the dry layer. In symbols,

$$E = CD$$

where E is evaporation, D is the depth of the dry layer, and C is a constant. Introducing time as a variable, the relation of depth of dry layer to time t was derived as

$$D^2 = kt$$

The values of the square of the depth of the dry layer plotted over time gave a linear trend. In other words, the depth of the dry layer varied directly with the square root of the time.

This analysis indicates that the movement of water through soil in the liquid phase ceases when the surface becomes dry. Thereafter the loss of water is the result of evaporation by diffusion beneath the surface at a rate inversely proportional to the square of the depth of the dry layer. This means that it will require three times as long for the second inch to become dry as for the first. The limiting moisture content between movements in the liquid and vapor phases is the field capacity or capillary saturation.

When the soil contains only hygroscopic moisture, the surface layer may absorb moisture from the air at night and give it off again during the day. This diurnal exchange of water vapor has been found in India (238) to extend to a depth of 10 mm in sands and to 40 mm in clay soils.

Evaporation from the soil is a change of soil moisture with time. Soil moisture percentage M has been estimated by means of a multiple regression equation in logarithmic units with mean air temperature t in degrees Fahrenheit for the 21 days prior to the moisture determination and the number of days N since a rain sufficient to wet the surface 3 in. of soil to field capacity as independent variables. For the upper 3 in. of a bare sandy clay loam at 2,750 ft elevation near North Fork, California, Anderson (12) obtained the equation

$$M = 912.8t^{-1} N^{-0.19}$$

for days without soil freezing or snow cover. Differentiating gives

$$\frac{dM}{dN} = -176.7t^{-1} N^{-1.19}$$

$-dM/dN$ is the rate of loss of soil moisture with time, which is rate of evaporation. If we convert to logarithmic form and let $E_{\%}$ represent rate of evaporation in percentage of soil moisture per day,

$$\log E_{\%} = 2.25 - \log t - 1.19 \log N$$

The constants in this equation should of course be confirmed or modified for other localities. The last equation is applicable for values of N between 2 and 10 days. The 21-day period for temperature and time since rain might be shorter or longer with better results for other data. If soil temperatures were available they would probably be more closely correlated with changes in soil moisture or evaporation than would air temperatures. Subject to such modifications the method of determining rate of evaporation by differentiating a multiple regression equation with percentage of soil moisture as the dependent and time since rain as an independent variable has much promise.

For many purposes it is essential to have evaporation expressed in inches depth of water or convertible to those terms. This conversion can readily be made when the percentage moisture content M or change in moisture content of the soil during a given period $(M_1 - M_2)$ and the volume weight ρ_a and depth of the soil layer D are known. The relation can be expressed by the formula

$$E = \frac{(M_1 - M_2)\rho_a D}{100}$$

Extreme examples indicating the range of monthly evaporation that might be found for different soils may be taken for illustration, assuming that evaporation will reduce the moisture from the field capacity to the wilting point. In a sand with a volume weight of 1.5, field capacity of 7 per cent, and wilting point of 4 per cent, if 12 in. depth of soil is susceptible to evaporation, the depth evaporated would be 0.48 in. At the other extreme a clay-loam soil with a maximum volume weight of 1.4, field capacity of 37 per cent, and wilting point of 20 per cent would be subject to evaporation of 2.8 in. of water. If the two soils were wetted to field capacity and dried to the wilting point three times at 10-day intervals during the month, then the monthly evaporation might vary from about 1.4 in. for the sand to 8.5 in. for the clay loam. It is considerations of this kind that have led to the statment that, if all the rains in southern California occurred in storms of 0.5 in. 1 week apart, evaporation would account for the total rainfall.

In this way, when volume weight of the soil is known, the $E_{\%}$ of the foregoing multiple regression equation may be converted to evaporation in inches depth of water per day. This change modifies only the value of the numerical constant in the equation. Total evaporation for an interval of time may then be obtained by integrating the equation between two values of N.

Freezing of the soil draws water to the surface forcibly from depths as great as 3 ft, and upon thawing, this water, often in excess of field capacity, is readily subject to evaporation. When this effect of freezing on evaporation in inches is evaluated by adding to the regression analysis another independent variable, F, the sum of the daily depths of freezing in inches depth for the period N, the equation becomes, for values of N between 3 and 14 days,

$$\log E = 1.30 - \log t - 1.19 \log N + 0.48 \log F$$

Calculated and observed evaporation agreed closely, and under conditions of frequent freezing and thawing both were more than three times those without freezing. Thus the evaporation in inches per day from bare soil with freezing was 0.10 in.; without freezing, 0.03 in. and under brush cover where no freezing occurred, 0.025 in. (12).

In general, where the water table is within the reach of capillary rise, but below the surface, the evaporation is less than that from a free water surface and has been shown to decrease as the depth to the water table is increased. The evaporation is greatest from compact silty soils because they are favorable for high capillary lift of water.

The commonest case is that of well-drained soils where the water table is below the reach of capillary action and where there is evidence that losses from evaporation below a depth of 8 in. are extremely slow (357). If it is desirable to estimate largely, the depth of soil affected might be increased to 12 in. A large part of the loss by evaporation under these conditions will take place in the first 5 days after a rain.

Evaporation from a free water surface cannot give a precise measure of the loss from soil and vegetation, and any measure from an artificial

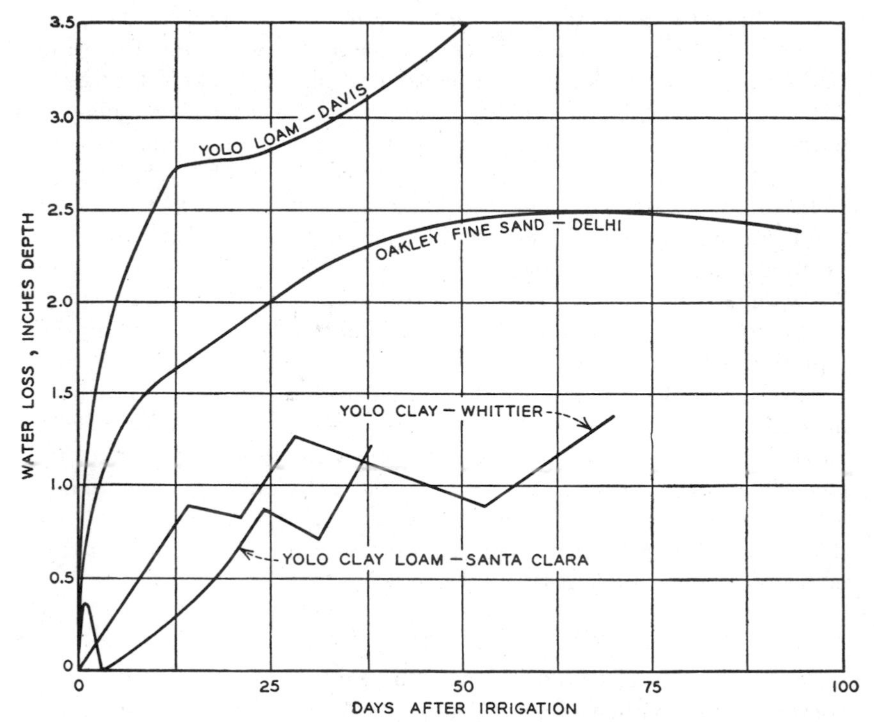

Fig. 16.—Cumulative evaporation from bare soils after irrigation in California. [*After Veihmeyer* (357).]

surface like that of an atmometer reflects only the atmospheric conditions and does not take into account the varying wetness of the natural surface. For this reason such measures cannot be used without corrections to determine the residue of precipitation available for stream flow after deducting the losses from the soil by evaporation.

The evaporation losses from bare soil after applications of water by irrigation in California are shown in Fig. 16, where the water loss in inches depth is plotted over the number of days after irrigation (357). The wide differences between different soil types are striking, although the relationship to texture is not consistent in that the

Oakley fine sand lies between the Yolo loam and the Yolo clay and clay loam. However, all the trends are consistent in portraying high evaporation for the first few days after irrigation, after which the rates decrease and become very slow after 30 to 60 days. The ranges represented by these soil types for different periods may be summarized as follows: in 3 days from 0.2 to 1.6 in., in 9 days from 0.3 to 2.5 in., and in 30 days from 0.8 to 3.0 in. Successive weighings of soil in tanks at Davis, California, were plotted as a smooth curve that can be expressed approximately by the equation

$$E = aT^{1/4}$$

where E is evaporation in inches depth, T is the number of days, and a is a constant that, for these data, has a value of 0.5. In experiments with soil in smaller containers the exponent was ⅓ instead of ¼. In other studies average seasonal rates of about 0.025 in. per day have been found. A total of 3.94 in. evaporated from bare soil between October and April in 155 dry days in Ontario, California (50). Depths of evaporation from soil over whole drainage areas have been estimated by Meyer (251) to vary from 4 in. annually for the Pitt River in California to 10 in. for the Tombigbee River in Alabama.

The rate of fall of the water level in initially saturated soil as a result of evaporation proceeded rapidly for the first 10 days, when the levels below the surface were as follows: for a coarse sand, 14 in.; for a fine sand, 28 in.; and for a heavy loam, 32 in. After 10 days the rate of fall was exceedingly slow for a 6 months' period. This is the basis for the statement that the depth to which evaporation is effective increases as the soil texture becomes finer. This statement is not in conflict with the finding that water rises more quickly in coarse-textured soils but ultimately reaches a greater height in fine-textured soils through the process of capillarity.

78. Effects of Forest and Vegetation on Evaporation from Soil.—It has been shown that, on forested as compared with unvegetated areas, the solar radiation, the maximum temperatures, the vapor-pressure differences, and the wind velocities are reduced. The reductions are greater as the density of the vegetation increases. Evaporation decreases with a decrease in each of these factors. Therefore, it may be expected that there will be a decrease in evaporation as the density of the cover increases.

In the simple case of windbreaks, Bates (36) has shown that the reduction in evaporation is proportional to the density of the windbreak, and also that the reduction in evaporation and the distance at which such reduction is found increases with the wind velocity (Fig. 17). Thus the area of greatest reduction will be found close in the lee of a dense grove, while it may be as much as five times the height to leeward of a narrow shelterbelt with few lower branches.

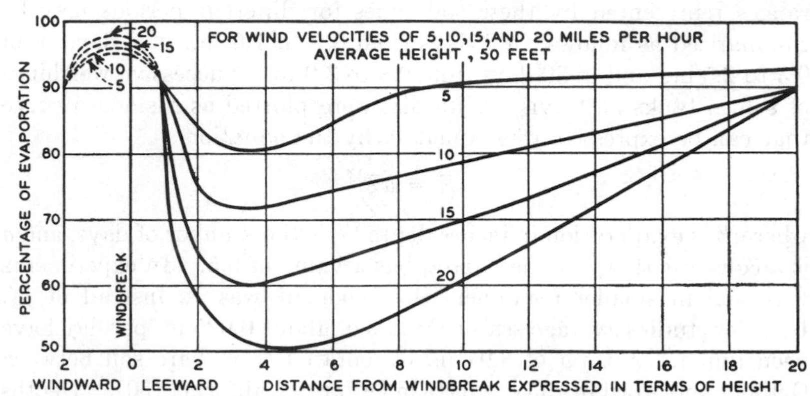

FIG. 17.—Relative evaporation in relation to wind velocity and distance from a one-row cottonwood windbreak. [*From Bates, United States Forest Service* (36).]

Interesting examples of relative evaporation under different kinds of cover are reported from some of the older work in Germany (121) where measurements were made from May to September.

Evaporating Surface	Relative Evaporation, %
Free water	100
Saturated bare soil in the open	93
Free water in the woods	36–39
Saturated bare soil in the woods	35
Saturated soil under litter in the woods	13

Records from lysimeters 21 in. deep at North Fork, California, for 1934–1936, with average precipitation of 48 in., gave average annual evaporation from bare soil of 17.1 in. compared with 12.3 in. from the same soil covered with 2 in. of ponderosa pine litter (348).

Differences between species are illustrated by German records for 4 years of monthly mean evaporation from sandy soil at capillary saturation under an open cover (116).

Kind of Litter	Relative Evaporation
None	100
Beech	55
Spruce	40
Pine	33

When a litter layer of equal depth of 5 cm was used for the three species, the beech and spruce showed relative evaporation of 11 per cent and the pine 15 per cent, as compared with 100 without litter.

The relative evaporation at Lincoln, Nebraska, from a silt-loam soil for 4 selected days in August, 1938, was as follows (304):

Condition at Surface	Relative Evaporation, %
From bare soil (0.19 in. equals 100)	100
Bare soil shaded	64
Bare soil shaded and sheltered from wind	47
Straw mulch, 1½ in.	27
Soil mulch, 1½ in.	9

As in Germany the depth of the straw mulch had little effect after it was increased above 3 in., corresponding to 8 tons per acre. Coarse sand or gravel has been shown to be as effective as soil, straw, or litter in reducing evaporation.

Comparisons between different conditions of cover are not the same when measured by different methods. This is evident from observation of the different surfaces and exposures presented by water

TABLE 29.—RELATIVE EVAPORATION IN OPEN, FOREST, AND CHAPARRAL BY DIFFERENT METHODS FROM APRIL TO NOVEMBER AT 6,000 FT ELEVATION IN SAN BERNARDINO MOUNTAINS, CALIFORNIA

Cover	Water surface, %	Atmometers, %	Soil boxes	
			%	Depth, in.
Open	100	100	100	3.5
Jeffrey pine	88	83	66	2.3
Chaparral	75	62	37	1.3

surfaces, atmometers, and soil surfaces. A comparison by three methods was made by Munns over a period of 4 years, with results shown in Table 29.[1]

The table includes for the soil boxes the total depth of evaporation for the period from April to November. The boxes of galvanized iron contained 1 cu ft of soil placed in them in layers corresponding to its natural stratification. They were inserted within closely fitting similar boxes which were buried in the soil. The results were considered somewhat unsatisfactory by Munns, because the water did

[1] From the manuscript "Studies of forest influences in California," in the files of the U.S. Forest Service, through the courtesy of the author, E. N. Munns.

not drain off freely after heavy precipitation. However, the figure of 3.5 in. for evaporation in the open corresponds well with other figures from bare soil in the same general region. The outstanding point in the table is that the forest and chaparral cover reduced the evaporation from soil relatively much more when measured by the soil boxes than by the water surfaces or the atmometers.

Comparisons at 5,100 ft elevation in the San Gabriel Mountains for one season using Bates evaporimeters showed 63 per cent evaporation under chaparral relative to that in the open as 100 per cent. The relation when weekly amounts were plotted gave a linear regression represented by the equation.

$$E_c = 0.68E_0 - 0.04$$

where E_c is inches per week evaporated under the chaparral, and E_0 inches per week in the open.

Comparisons of evaporation from the soil under a cover of grass and of oak woods was made in an unpublished report by Li. Samples were taken to a depth of 14 in. periodically during the spring period near Berkeley, California. For the 25 days from Apr. 25 to May 20 the evaporation amounted to 0.57 in. in the woodland and 0.53 in. in the grassland, corresponding to rates of 0.023 and 0.022 in. per day. The moisture content of 19.7 per cent in the woodland as compared with 14.3 per cent in the grassland on Apr. 25 probably accounts for the fact that the loss from the grassland did not exceed that from the woodland.

The relation to natural succession and incidentally to types, tolerance, density of canopies, and stands is illustrated in the following figures of relative evaporation derived from atmometer measurements in eastern Washington and adjacent Idaho (362).

Type	Relative Evaporation, %
Bunch grass-rim rock	100
Prairie	61
Ponderosa pine	43
Western larch–Douglas fir	32
Western redcedar	29

The later the stage in succession, the lower is the evaporation.

Many comparisons have been made, usually by the use of atmometers, between the evaporation under different conditions and types of cover relative to the evaporation in the open as 100. A variety of such figures from different sources are arranged in generally descending order in Table 30. The lowest reduction corresponding to relative evaporation of 94 is recorded for chamise at 6,000 ft in the San Bernardino Mountains; the greatest, with a relative of 10 comes from a dense swamp forest of gum, ash, and yellow poplar on Long Island.

TABLE 30.—EVAPORATION IN DIFFERENT FORESTS RELATIVE TO THAT IN THE OPEN AS 100 FROM ATMOMETER AND EVAPORIMETER DATA

Forest type	Density and age	Location	Relative evaporation	Ref.
Chamise		6,000 ft, S. Calif.	94	
Jeffrey pine	Open	6,000 ft, S. Calif.	91	
Jeffrey pine	Group	6,000 ft, S. Calif.	86	
Jeffrey pine	Heavy shade	6,000 ft, S. Calif.	79	
Jeffrey pine	Reproduction	6,000 ft, S. Calif.	72	
Manzanita–ceanothus		6,000 ft, S. Calif.	71	
Ponderosa pine		7,250 ft, Ariz.	70	(282)
California scrub oak		6,000 ft, S. Calif.	68	
Jeffrey pine	30 ft high	6,000 ft, S. Calif.	68	
Mixed chaparral		6,000 ft, S. Calif.	67	
Maple–birch	Partially cut	N. Y.	68	(332)
Aspen		Mich.	57	(143*a*)
Oak–chestnut	Open	N. Y.	50	(346)
Aspen	0.8 density	Minn.	48	(143*a*)
Ceanothus velutinus		Idaho	45	(359*a*)
Jack pine	0.5 density, 25 ft high	Minn.	67	(143*a*)
Pitch pine	Open	N. Y.	60	(346)
White and red pine	Young growth	Mich.	54	(143*a*)
Eastern hemlock		N. Y.	52	(258)
Western white pine	Half cut	Idaho	47	(187)
Ponderosa pine		E. Wash.	43	(362)
Red pine	0.8 density	Minn.	40	(143*a*)
Western larch–Douglas fir		E. Wash.	32	(362)
Hemlock–hardwood		N. Y.	39	(258)
Maple–beech		N. Y.	38	(258)
Oak–chestnut	Dense	N. Y.	33	(346)
Western redcedar		E. Wash.	29	(362)
Western white pine	Uncut	Idaho	25	(187)
Chestnut–ash–basswood	Dense	N. Y.	14	(346)
Gum–ash–yellow poplar	Swamp	N. Y.	10	(346)

In general, species and types of xerophytic character with light foliage, open stocking, and from regions of high evaporation show the lower reductions in evaporation. Climax types reduce the evaporation more than preclimax. The mesophytic, well-stocked, or dense forest types give the greatest effect.

79. Soil Evaporation: Summary.—1. Evaporation as measured by losses from soil surfaces is usually less than from free water or wetted instrumental surfaces, and the differences become progressively greater as the surface soil is drier.

2. The depth to which evaporation is effective in lowering the free water level in initially saturated soils increases as the soil texture becomes finer from about 14 in. for coarse sand to over 32 in. for heavy loam.

3. Water rises more quickly by capillarity in coarse- than in fine-textured soils, but the evaporation losses from any soil are extremely slow after the surface 8 in. becomes dry.

4. Evaporation from bare soil decreases as the amount of moisture in the surface foot decreases and as the depth to the water table increases.

5. Insofar as vegetation reduces wind velocity, it tends to reduce evaporation also and in proportion to its density.

6. The percentage reduction of evaporation by windbreaks increases with the velocity of the wind in the open at least up to 20 mph and may be as much as 40 per cent or even 70 per cent in extreme cases.

7. The evaporation from soil covered by forest floor is 10 per cent to 80 per cent of that from bare soil. The reduction varies with the kind of floor and increases as the floor becomes thicker at least up to a thickness of 2 in.

8. The evaporation from water or wetted surfaces is decreased by vegetative cover progressively more by mesophytic and climax than by xerophytic and preclimax types; more by dense- than by light-crowned species; more by tolerant than by intolerant species; more by well-stocked than by understocked stands; more by stands near the physical rotation age than those older or younger; and more by deciduous than by evergreen species during the warm season.

CHAPTER XVI

SOIL TEMPERATURE

The temperature of the soil is determined fundamentally by the balance between the gains and losses of heat. The gains result directly or indirectly from the heating effects of the sun, and it has already been shown how radically vegetative cover may reduce the solar radiation reaching the surface of the ground. However, there are several factors other than vegetation that influence soil temperature and frequently obscure the effect of the vegetation. These other factors include the amount of dust and water vapor in the atmosphere, the latitude, slope, aspect, altitude, distribution of land and water, and nature of the soil.

TABLE 31.—SPECIFIC HEATS OF MINERAL AND ORGANIC MATERIALS

Soil	Specific heat			
	By weight	By volume		
	Dry	Dry	50% saturated	Saturated
Sand	0.20	0.30	0.51	0.72
Organic matter	0.42	0.15	0.53	0.90

80. Relations to Properties of Soil.—Taking up first the effects due to the nature of the soil there are differences in specific heat between different soils and also in the specific gravities. The latter tend to counterbalance the differences in specific heat when they are expressed on a volume basis. Because water has the highest specific heat, the content of moisture in the soil will also modify its specific heat. Examples of specific heats for sand and organic matter, expressed both by weight and by volume and at different moisture contents, are shown in Table 31.

The general principles governing the variation of soil temperature with air temperature, radiation, moisture, and slope are available in textbooks of soil science (42). Many of them are illustrated in the following comparative material from forested and open areas.

81. Influence of Forest on Soil Temperature.—The forest cover influences soil temperature by interposing a canopy between soil and sky, which affects (*a*) solar radiation, (*b*) surface radiation, (*c*) air temperature, (*d*) duration of snow cover, and (*e*) evaporation. The forest, by breaking the force of the wind, affects (*a*) wind movement and (*b*) evaporation. The forest, by intercepting precipitation, affects (*a*) the depth and duration of snow cover and (*b*) the content of

TABLE 32.—FLOOR AND SOIL TEMPERATURES IN IDAHO WESTERN WHITE PINE TYPE OF DIFFERENT DENSITIES FOR JULY AND AUGUST, 1930–1933

Depth	Temperature	Clear cut, °F	Departures from clear cut	
			Half cut	Uncut
Forest floor.....	Absolute maximum	158	−36	−64
Forest floor......	Mean maximum	127	−35	−50
Soil at 1 ft.......	Mean	60	− 6	− 8

soil moisture. The forest, as a source of the forest floor, affects (*a*) the depth and character of the A_0 horizon, (*b*) the content of soil moisture, (*c*) the content of organic matter, (*d*) the color of the soil surface, and (*e*) surface radiation. It is usually impossible, with existing data, to separate clearly the foregoing factors in their influence on soil temperature. However, some of them can be distinguished in the following data.

The effect of different degrees of cutting on the temperatures in the forest floor and at a depth of 1 ft in the mineral soil are indicated in Table 32 (187). The extremely high temperatures with the maximum

TABLE 33.—INFLUENCE OF UNCUT AND PARTIALLY CUT DOUGLAS FIR FOREST AT 9,000 FT ELEVATION IN COLORADO ON SOIL TEMPERATURES AT 1- AND 4-FT DEPTHS (34)
(Maximum departures, in mean monthly temperatures, forest from open, in °F)

Depth, ft	January		February		June–July	
	Uncut	¾ cut	Uncut	¾ cut	Uncut	¾ cut
1	+4.8	+2.9	......	......	−6.9	−4.2
4	......	......	+2.4	+0.6	−5.3	−2.3

of 158° in the forest floor is reduced by 50° in the uncut forest, whereas at 1-ft depth the reduction is only 8°F.

Comparisons between winter and summer, different degrees of cutting, and depths of 1 and 4 ft may be drawn from Table 33.

The following indications may be derived from these figures. Forest cover reduces summer temperatures and increases winter temperatures at both 1- and 4-ft depths. The lag in response increases with depth. The influence of the forest in both seasons increases with the density of the cover and decreases with depth in the soil.

During the growing season in hemlock–hardwood forests in New York, Moore (258) found the maximum soil temperatures at 6 in. were reduced 11° by the cover and the minimum, 3°. At 18 in. the reductions were 6° for both maximum and minimum.

In comparing the temperature of the litter with that of the air, Stickel (332) found that for an increase of 1°F in air temperature the litter temperature in the open increased 1.2° and in the forest 0.8°.

TABLE 34.—EFFECT OF CANOPY AND FLOOR OF WHITE PINE IN NEW HAMPSHIRE ON MEAN DAILY MAXIMUM SURFACE TEMPERATURES IN °F

Cover	Forest floor		Departure, litter-covered from bare surface
	Removed	Left	
Opening in full sun........	107	83	−24
Under full stand..........	75	65	−10
Departure, shade from sun.	−32	−18	−42

At 1 in. depth in the soil the corresponding increases were 0.65 and 0.60°. The soil temperature in the forest was less responsive to change in air temperature than in the open. However, the relation between air temperature and soil temperature must be looked upon as largely indirect because of the heating effect of solar radiation and the consequent convectional currents rising from the surface. Air of lower temperature than that of the soil, usually during the night, evidently could contribute to reduced soil temperatures, or moist air of higher temperature than that of the soil could increase the soil temperature by condensation. Additional evidence is contained in data by Shreve (318) from north and south aspects at corresponding elevations near Tucson, Arizona. He found that both maximum and minimum temperatures on the south aspects were higher than on the north, the maximum by from 13 to 20°, and the minimum by from 4 to 11°. This was during the spring and summer at a depth of 3 in. He concluded that insolation is more important than air temperature in determining soil temperature.

Openings in a forest stand and also the removal of the forest floor may make marked differences in the surface temperatures, as shown in Table 34 from data by Toumey and Neethling (345). The temperature in the shade was as much as 32°F lower than that where the sun shone on the bare ground. The removal of the forest floor increased the temperature in the sun by as much as 24°. A maximum difference of 42° is indicated by the removal of both shade and floor. The figures indicate that the crown canopy is more effective than the forest floor in reducing maximum surface temperature.

During late summer and fall, similar comparisons of maximum and minimum temperatures at depths of 3 and 12 in. by Shreve (319) in a redwood forest near Carmel, California, indicated lowering of both maxima and minima by the forest cover. The differences were greater at the 3-in. than at the 12-in. depth. Removal of the 5-in. leaf mulch decreased the maxima and the minima at the 12-in. depth but increased the minima at the 3-in. depth. The departures are shown in Table 35.

TABLE 35.—DEPARTURES OF MAXIMUM AND MINIMUM TEMPERATURES, °F, FOREST FROM OPENING IN SUN, IN REDWOOD FOREST NEAR CARMEL, CALIFORNIA (319)

Date	Depth in soil, in.	Maximum		Minimum	
		5-in. leaf mulch	Bare surface	5-in. leaf mulch	Bare surface
Aug. 18–26	3	−12	−27	−10	−8
Sept. 29–Oct. 6	12	− 4	− 8	− 3	−6

The accumulation of oak litter held in place by herbs as compared with bare soil in central Arizona as reported by Hendricks (148), for October and November resulted in an increase of the average daily minimum by 5°F and a reduction of the average daily maximum by 7°F. Consequently, the daily range was reduced by 12°.

In a study of soil temperatures in a mixed plantation of white and red pine 28 years old in Connecticut, MacKinney (234) found that the litter reduced the daily variations in maxima by 50 per cent in autumn and by 85 per cent in spring. The minimum temperature fluctuations were reduced about 75 per cent in both fall and spring. Consequently the diurnal range of surface temperature was also reduced.

Month	Open	¾ *h*	½ *h*	¼ *h*
July	87.7	84.8	79.0	70.8
August	75.7	73.8	67.7	62.8

Measurements at 1-in. depth during the summer in Minnesota (352) indicate the distances in multiples of the heights of the trees to the north of the shading point to which the shadow of the crowns may affect the maximum temperatures, as illustrated by the foregoing tabulation. If the temperatures are plotted over multiples of the height of the trees, the trend for July is nearly linear so that for a change of one-tenth unit of height there is a change of 2.8°F. For August the trend is linear so that for a change of one-tenth unit of height there is a change of 2.2°F. The trends indicate that the effect of the shadow during those months extended to a point about seven-eighths of the height of the trees.

82. Freezing and Thawing of Soil.—From the viewpoint of forest influences the most important aspect of soil temperature is the freezing and thawing of the soil and the consequent effect on permeability, infiltration, and erosion as a result of water movement after thawing. Freezing is the process of formation of ice crystals from liquid water. The amount of the ice formed depends upon the amount of water available. The formation of ice involves a considerable suction force that is usually sufficient to draw water from any sources with which there is a capillary connection. If the soil contains an excess of free water or if ground water is within reach of capillary action, ice strata may form near the surface as in the phenomenon of frost heaving. The expansion associated with frost heaving may be much greater than could result from the expansion of the original moisture content of the soil. The increase in water content of the surface soil as a result of the formation of ice layers may be as much as 20 to 30 per cent by weight.

The forest cover and floor have marked effects upon the dates of freezing and thawing and upon the depths to which freezing penetrates. In the ponderosa pine type at 7,200 ft elevation in Arizona, Jaenicke and Foerster (184) found that, on Feb. 18, when the frost disappeared from the ground on the south side of the trees, there remained 13 in. frozen soil under the crowns on the north side, 21½ in. in the openings, and 29½ in. in the adjacent park. During the period from Feb. 18 to Mar. 27 the depth of the frozen ground decreased by 14½ in. in the openings, by 13 in. under the north side of the crowns, and by 9 in. in the park.

Similarly in Connecticut the depth of penetration of frost was less under forest cover and the dates of freezing were later as shown in Table 36. The pine with its forest floor reduced the depth and retarded the date of freezing more than the deciduous types, but it also retarded the date of complete thawing by 13 days compared with the plowed field.

The depth of freezing under forest may be greater than in the open where coniferous forests are growing on peat or poorly drained soils. Thus in New Hampshire, Belotelkin (45) in three winters found average maximum penetration of frost in an open field at 15.0 in., under hardwoods at 5.0 in., in a spruce flat at 15.7 in., and in a spruce swamp at 19.0 in. The contrast in influence between hardwoods and spruce was ascribed to the much thinner snow cover attributed to the greater interception by the dense evergreen canopy and the consequent reduction in the insulating effect of the snow. Lower minimum temperatures associated with cold-air drainage into the spruce flat and swamp

TABLE 36.—EFFECTS OF FORESTS ON FREEZING AND THAWING OF SOIL IN CONNECTICUT (195)

Cover	Maximum depth of frost, in.	Freezing retarded, days	Date of last frost
Plowed field	10.1	0	Apr. 11
Mixed hardwood	5.1	21	Apr. 6
Red maple	4.1	31	Apr. 11
White pine	2.9	44	Apr. 24

may also have contributed to the difference. Not only did the soil freeze deeper in the spruce types but also complete thawing was recorded in the spruce flat 26 days later and in the spruce swamp 55 days later than in the open. The extreme figures in one of the three years were 47 and 79 days later, respectively.

In a 28-year red pine plantation in Connecticut, MacKinney (234) found that the forest floor delayed the freezing of the soil from Dec. 3 to Jan. 4 and decreased the maximum depth of freezing from 8 to 5.8 in. He also pointed out that the forest floor affected the frozen soil by keeping it porous, loose, and permeable when the bare soil became solid and impermeable.

Thawing of frozen ground may take place both from above after the snow is gone and from below. Near La Crosse, Wisconsin, Scholz (310) found that 10 in. of frozen soil in a bluegrass pasture thawed both from above and below while the 4 in. of frozen soil in an adjacent ungrazed woodlot thawed only from below. Under forests of white pine and of red maple and also in a plowed field in Connecticut, the depths of the last frost were recorded about midway between the surface and the winter's maximum depth of frost penetration (195).

Similar results were recorded in the open and in the spruce types in New Hampshire (45).

Striking differences in penetration of frost in relation to a shelterbelt are illustrated in Fig. 18 (335). However, it is evidently the snow drift trapped by the shrub rows that has more influence on the depth

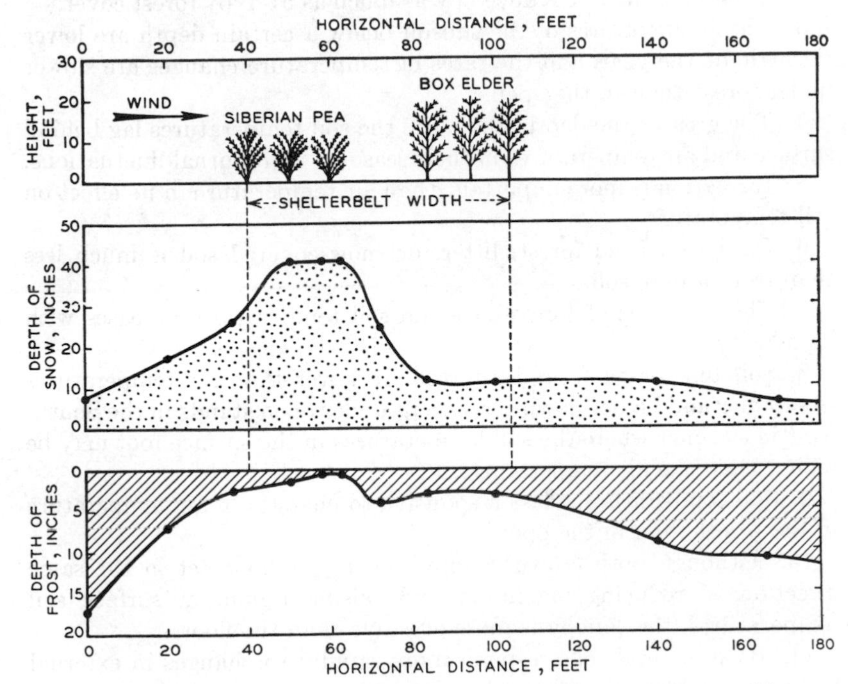

FIG. 18.—Accumulation of snow and penetration of frost as affected by a multiple-row shelterbelt. [*From Stoeckeler and Dortignac* (335).]

of freezing than any direct effect of the shade or litter. Where the snow was 41 in. deep, only 1 or 2 in. of soil were frozen. Away from the shelterbelt where the snow was only 6 to 8 in. deep the freezing extended 12 to 17 in. deep.

83. Influences of Forest on Soil Temperature: Summary.—1. Soil temperatures under forest cover in most forest regions are cooler in spring and summer and warmer in fall and winter than in the open. Forest cover usually reduces the maxima and to a less degree increases the minima, except in the Southwest at high elevations where the minima may be increased more than the maxima are reduced. Thus in general the ranges of diurnal and seasonal fluctuations in soil temperature are reduced by forest cover.

2. These influences are largest near the surface and become less with increasing depth of soil. In places the effect of the forest is noticeable to depths of from 12 to 30 ft. The mean daily maxima in the forest floor in summer may be reduced by as much as 51°F by forest cover.

3. The temperatures of the subsoil below a certain depth are lower throughout the year, and the rates of temperature changes are slower under forest than in the open.

4. The greater the depth, the more the soil temperatures lag behind surface and air temperatures in their seasonal and diurnal fluctuations.

5. Insolation is more important than air temperature in its effect on soil temperature.

6. Radiation from forest, litter, or snow-covered soil is much less than from a bare soil.

7. The influence of forest cover on soil temperature increases with its density.

8. Soil temperatures under forest are lower than air temperatures in summer and higher in winter, except under open stands in the Southwest in summer where the soil temperatures in the surface foot may be higher than the air temperatures.

9. Soil temperature is less responsive to changes in air temperature in the forest than in the open.

10. Although both crown canopy and forest floor act in the same direction in reducing maximum and raising minimum surface soil temperatures, the canopy is more effective than the floor.

11. Bare mineral soils respond more rapidly to changes in external factors than litter-covered soils.

12. A layer of loose snow and litter is more effective as insulation than either alone or than compacted snow.

13. Forest floor retards the date on which the underlying mineral soil begins to freeze.

14. Forest cover generally reduces the depth to which soil freezes, more on the south than on the north sides of trees and more under the crowns than in the openings. It may advance the date on which frost disappears by as much as 5 weeks. Locally, reduced snow cover because of interception by dense evergreen stands may reverse the foregoing relations.

CHAPTER XVII

LITTER AND FOREST FLOOR

The forest floor includes the surface accumulation of organic matter above the mineral soil. Thus it corresponds to the A_0 horizon of the soil profile. It is derived from the fall of dried leaves, twigs, bark, flowers, fruits, and from animal remains, all in various stages of disintegration and decomposition.

Three layers of the forest floor are often distinguished, according to their position and degree of decomposition. The litter, or *L* layer, comprises the uppermost undecomposed material. The duff, or *F* layer, is the intermediate, partially decomposed layer in which the structure of the constituent materials is still evident. The leaf-mold, or humus or *H* layer, is the lower layer next to the mineral soil, where the materials no longer show their original structure. These layers may be quite distinct or they may be difficult to separate, and the leaf-mold layer may merge with the mineral soil by a gradual transition.

84. Annual Accumulation of Litter.—The annual accumulation of organic materials corresponds roughly to the litter layer, but it is not possible to collect the litter as a separate layer and be sure that all or only the annual accumulation has been included. It becomes necessary, therefore, to place on the ground beneath the trees burlap strips or trays or traps in which the material can be collected at yearly or shorter intervals.

The amount of the annual accumulation as it varies with the vegetation and site has considerable interest as one of the factors in the development of the forest floor and its effectiveness as a protective cover. The rate of accumulation varies in different seasons within the year. Mixed hardwoods in New Hampshire deposited 70 per cent of the annual total in October, and gray birch 45 per cent in August. The percentages represent the maxima in any one month (271). Longleaf pine in northern Florida showed two maxima, one of 19 per cent of the annual in October and the other of 40 per cent in May, June, and July.

Within the same stand and unrelated to the growth with age, there may be considerable variations in the amount of the annual leaf fall in different years. This is illustrated by Table 37 from the chaparral of the San Dimas Experimental Forest in Los Angeles County.

The annual accumulation for the ceanothus–chamise type in 1936 was twice that in 1935. On the other hand the difference for the same 2 years for the evergreen oak at 2,000 ft higher elevation was only 40 per cent. Similar comparisons in Germany (11) for Scotch pine in one instance indicated that, in 7 years of record, the maximum was three times the minimum.

TABLE 37.—ANNUAL ACCUMULATION OF LITTER IN THE CHAPARRAL, SAN DIMAS EXPERIMENTAL FOREST, LOS ANGELES COUNTY

Year	Ceanothus*–chamise, 16 years old	Canyon live oak, 55 years old	Manzanita†
	Metric tons per acre		
1935	0.7	0.5	0.9
1936	1.4	0.7	1.3
1937	0.9	0.6	1.2
1938	1.1	0.7	1.1

* *Ceanothus crassifolius* Torr.

† *Arctostaphylos glandulosa* Eastw.

The variation between species and forest types is not large or well defined. In the southern Appalachians, a stand of all-aged hardwoods deposited 1.81 metric tons per acre annually (179). A mixed stand of pine and oak of 0.9 density gave 1.40 metric tons the first year after burning, and 1.18 after raking (322). Second-growth slash pine yielded 1.6 metric tons per acre. Second-growth longleaf pine gave from 1.1 to 1.4 metric tons (153), whereas old growth in Florida that had not been burned for 42 years yielded only 0.45 metric tons per acre. A number of collections from stands of red pine, jack pine, and white pine of ages from 30 to 250 years, and with densities from 0.7 to 0.9 in northern Minnesota varied only between 1.03 and 0.90 metric tons annually (11). The conclusion of early European workers that in even-aged, well-stocked stands on an average site there is not much difference in the annual accumulation between spruce, beech, and pine, between coniferous and deciduous species, or between light- and heavy-crowned species, seems to be confirmed.

85. Variation with Site.—The annual accumulation does vary markedly with site quality. This is illustrated by the figures in Table

38 for Norway spruce (116) and European beech (262). The values for site III are taken as 100 in computing the relatives.

For beech the accumulation on site I is four times that on site V. The relatives of annual growth are slightly less sensitive to differences in site quality than are those for annual accumulation.

In stands of sugar maple, beech, and associated species in New York, Chandler (76) found 1.33 metric tons of fresh litter on Ontario loam of site index 65, compared with 1.16 tons on Lordstown stony silt loam of site index 49. Similarly, white oak in Illinois produced annually 1.5 metric tons per acre on Bluford silt loam and 2.0 tons on Clinton

TABLE 38.—ANNUAL ACCUMULATION OF LITTER ON DIFFERENT SITE QUALITIES FOR SPRUCE AND BEECH IN GERMANY

Site quality	I	II	III	IV	V
Norway spruce, 18–40 years, metric tons per acre	1.9	1.6	1.4	1.3	
Beech, metric tons per acre	1.2	0.9	0.6	0.5	0.3
Relatives for beech	208	156	100	78	42
Relatives of current annual increment at 80 years for beech	161	127	100	76	53

silt loam (237). However, when the annual accumulation was expressed per unit of basal area of the stands, both soil types had the same dry weight of 48 lb of litter per sq ft of basal area. To what extent such a relation between the annual litter and basal area is generally applicable must await additional confirmation. It seems unlikely that it would be evident in stands of different ages.

Ordinarily, the site quality for any given species or forest type becomes lower with successively higher locations on a slope. It is doubtless this relation that underlies the statement of European workers that the annual accumulation decreases with increase in altitude.

86. Variation with Age.—The variation in the annual accumulation of litter with the age of the stand would be expected to be related to the trend of current annual growth with age. It has been shown in several instances that the maximum amount of foliage on the trees in a stand, expressed either in weight or in area, tends to be greatest at about the age of the culmination of the current annual growth. The same relation between the annual accumulation of litter and age of the stand is indicated in the following figures for Norway spruce and beech in Germany (116).

ANNUAL ACCUMULATION, METRIC TONS PER ACRE

Age	45	75	105
Beech	1.68	1.65	1.63
Spruce	1.60	1.61	1.32

Age	30	50	70	90	100+
	Metric Tons per Acre				
Spruce	1.26	1.50	1.54	1.46	1.38

The maximum accumulation for spruce is doubtless between 50 and 70 years. For beech it is probably close to 45 years. In both cases the maxima occur at ages not far from the culmination of the current annual increment.

During the years when the crowns of a young stand are closing, the annual accumulation increases more rapidly than earlier or later. The following data for a Norway spruce plantation in Germany on site I, spaced 4.3 by 2.6 ft, illustrate the point (116).

Age	18	19	20	21	22	23
Metric tons per acre	0.3	1.2	2.2	3.0	2.8	3.8

The crown closure was probably complete at the age of 20 or 21 years. Incidentally, 3.8 metric tons per acre is a maximum figure for annual accumulation.

In Japan, well-stocked stands of *Cryptomeria Japonica* on site I produced the following metric tons dry weight annually in different age classes (279).

Age class, years	0–15	16–30	31–44	45–100	100+
Metric tons per acre	0.32	1.22	2.40	1.95	2.21

Presumably a larger number of samples would have resulted in reversing the order of the values for the two older classes.

Collections of *Chamaecyparis obtusa* from the same source indicate the variation with density of stand within the same site (II) and age class, 15–20 years.

Trees per acre	160	800–1,100	1,200–1,400	1,400–1,600	1,900+
Metric tons per acre	0.93	0.88	1.12	1.02	0.40

Understocking or overstocking tends to reduce the annual accumulation in comparison with a rather wide range of well-stocked stands.

The annual accumulation of litter from the herbaceous or grassy vegetation after a forest stand is cut may exceed that from the trees before cutting. Thus the annual litter from a dense stand of beech in Germany was 1.21 metric tons per acre, and, after clear cutting, the invading grass (*Calamagrostis* sp.) produced 1.77 metric tons (375). The comparison is for litter only and might be reversed if the annual increment of wood were included.

87. Total Forest Floor.—Although the annual accumulation of litter is obviously a function of the forest stand, the total forest floor, in-

Forest type	Forest floor, metric tons/acre	Ref.
Birch, sugar maple, and spruce, N. H.	119.2	(229)
Old-growth spruce, fir, and birch, Minn.	87.5	(6)
100-year white pine, hemlock, and sugar maple, Conn.	53.7	(229)
Old-growth paper birch and sugar maple, Minn.	33.5	(10)
Old growth bigtree, Calif.	26.1	
Shortleaf pine of 0.9 density, southeastern U.S.	2.9	(322)
Old-growth longleaf pine, 42 years unburned, Fla.	1.7	(153)

cluding litter, duff, and leaf mold, is a resultant of the leaf fall and current accumulation plus the action of factors of the environment in weathering and as they affect the composition, abundance, and activity of the fauna and flora and the biological and chemical processes of decomposition (361). Thus the total amount of forest floor should be much less than the sum of the annual accumulations up to any given age. It will also be subject to much more variation with environmental factors than was the case with the annual accumulation.

The amount of forest floor varies with the latitude and altitude and the associated differences in climate, soil, and vegetation. Some examples of the extremes in reasonably well-stocked stands will illustrate the wide range of this variation.

At the low end the series might be completed by continuing into the tropics, where the observation has frequently been made that there is little or no forest floor. The decrease in the amount of forest floor from north to south as conditions of temperature become more favorable for decomposition is evident. However, there may be exceptions to this general relationship in localities of warm temperature where prolonged dry seasons, otherwise favorable, retard the processes of decomposition. Thus the types in southern California produce quite large amounts of forest floor. For example, at 5,000 ft in Los Angeles County a stand of canyon live oak unburned for more than 50 years had 12.2 metric tons per acre of forest floor, and manzanita (*Arctostaphylos glauca*) had 21.2 metric tons. Even 14 years after a heavy fire the average for chaparral at elevations of about 2,900 ft was 5.3 metric tons.

The large amounts of forest floor in the form of raw humus under plantations of pine in southern France and under spruce in the German lowlands, where the trees were planted outside their natural range, provide other examples of large accumulations resulting from slow decomposition.

Where the annual accumulation is about the same the foregoing differences must be attributed to variation in the rate of decomposition, slow in cool or dry climates where the accumulation exceeds decomposition during at least part of the year, and rapid in the warm, moist climates where decomposition is favored.

88. Variation with Site.—The amount of forest floor also varies with site quality. One example may be drawn from the southern pine region where the dry weight of floor in closed stands of longleaf pine, 25 years old and 14 or 15 years unburned, varied with the soil type as follows (153):

Soil Type	Dry Floor, Metric Tons/Acre
Blanton fine sand	9.0
Orangeburg fine sandy loam	14.2
Norfolk fine sandy loam	15.6

The Blanton fine sand represents a poorer site quality than the other two closely related soil types.

In the southern Appalachian hardwoods, topographic situations correspond to site qualities and the following amounts of forest floor reflect the differences in site (179):

Site	Metric Tons/Acre
Upper slope, south aspect	1.4
Upper slope, north aspect	4.6
Lower slope, north aspect	6.0
Lower slope, south aspect	7.4
Cove	4.1

The series is arranged in ascending order of site quality and the amount of floor increases from the poor to the better site with the exception of the cove where there is a decrease. Presumably the smaller amount of floor in the cove sites is again a matter of more rapid decomposition, associated with a difference in type, notably an increase in the proportion of yellow poplar.

Another example may be taken from data for aspen in northern Minnesota as shown in Table 39. In these stands of similar age and basal area, the amount of forest floor increases with increase in site index and in current annual growth. However, the increase is not directly proportional to either one. In general, it may be concluded that for the same type and age the amount of floor tends to be greater on the better sites.

TABLE 39.—AMOUNT OF FOREST FLOOR IN RELATION TO SITE QUALITY OF ASPEN IN NORTHERN MINNESOTA (7)

Site index	Floor, metric tons/acre	Current annual growth, cu ft/acre	Basal area, sq ft/acre	Age, years
74	19.5	103	112	38
58	16.4	68	104	40
47	10.9	41	106	40

89. Variation with Age.—The age of the stand will also affect the amount of forest floor, at least up to the age where something approaching equilibrium between accumulation and decomposition is reached. This may be shown first by the change in depth with increasing age. In stands of loblolly pine 10, 20, and 70 years old originating on old fields of Georgeville stony clay loam of the Duke Forest, North Carolina, starting with a depth of 1.25 in. at 10 years, the rate of increase to 70 years, based on only three points, was linear with a slope of 0.026 in. annual increase (94). Without samples from stands between 20 to 70 years it is not certain that the trend would actually be linear or that a maximum higher than the 2.80 in. at 70 years may not have been reached at an earlier age more nearly corresponding to the culmination of growth.

Depth measurements of the humus layer, excluding the litter, in the Connecticut River drainage (261) showed a similar linear trend for old-field white pine from 1.6 in. at 20 to 2.4 in. at 70 years, a rate increase of 0.016 in. annually. In the several types of hardwood, hemlock, spruce, and fir the depth of the humus layer decreased at a decreasing rate for the first 30 years after cutting and then increased again up to 60 to 100 years. For the hemlock–spruce–fir mixtures the annual increase in depth was 0.097 in., and for the sugar maple types, 0.11 in. Both were linear for the ages over 40 years. In the temporary hardwood types there was only a slight increase up to 55 years.

On the same soil type and undisturbed by fire, slash pine 9 years old had 7.8 metric tons, and slash pine 30 years old, 14.6 metric tons per acre of forest floor. Longleaf pine unburned for 11 years had 9.1 metric tons, and that unburned for 14 years had 15.6 metric tons per acre. Both stands were 25 years old and growing on the same soil type. However, the increase in weight of forest floor is more

TABLE 40.—OVEN-DRY WEIGHTS PER ACRE OF FOREST FLOOR OF SASSAFRAS, BLACK LOCUST, AND SHORTLEAF PINE AT DIFFERENT AGES IN INDIANA, OHIO, AND ILLINOIS

Age class, years	Sassafras	Black locust	Shortleaf pine
	Metric tons/acre		
2–5	2.3	2.4	
6–10	2.5	3.9	1.2
11–16	2.9	3.8	8.1*
17–20	3.4	5.6	
21–35	3.1	4.6	9.4

* Scotch pine.

than could reasonably be attributed to 3 years' additional accumulation (153).

In natural stands of sassafras and black locust, and in plantations of shortleaf pine spaced 6 by 6 ft, the changes in dry weight of forest floor with age are indicated in Table 40 (21). The dense rapidly growing stands of locust and sassafras deposited a considerable layer of floor within 5 years. Apparently the maximum accumulation was reached at about 20 years. The slower growing more widely spaced pines did not form a good forest floor until after the tenth year, but from then on their floor was more than double that of the deciduous species.

Three years after a fire that destroyed a 25-year-old stand of black locust growing on blow sand in central Illinois, the new crop of sprouts had produced 1 metric ton per acre of forest floor, determined by loss on ignition (142).

Plantations of several coniferous and broadleaved species at Berkeley, California, at 30 years of age were still increasing in the amount of forest floor at an accelerating rate as indicated by the fact that the periodic annual increase was in excess of the mean annual

(198). In this way the periodic annual increase, determined by dividing the difference in weight between two periods by the number of years intervening, automatically takes into account the effect of decomposition. Approximate equilibrium between accumulation and decomposition for longleaf pine was attained in 8 to 12 years (153). Thus the amount of forest floor under old growth is not necessarily greater than under second growth. For example, longleaf pine 250 years old had 10.2 metric tons per acre while samples from stands 25 years old averaged 12 tons per acre.

90. Variation with Natural Succession.—There are marked differences in the amount of forest floor between different forest types and between different stages in a successional series, even when the types are growing on the same soil type. For example, on the volcanic plateau of the Lassen National Forest, Bodman (51) found 21 metric tons per acre under old-growth white fir and only 11 tons under an adjacent stand of preclimax ponderosa pine. However, the change with succession is not always an increase. In the Piedmont the succession from loblolly pine 70 years old to uneven-aged oak was associated with a decrease in depth of the A_0 horizon from 2.8 to 1.9 in. (94).

In northern Minnesota, samples from stands representing the four stages in the forest succession from jack pine to sugar maple–basswood, all growing on the same soil type, gave the following amounts of forest floor, in metric tons per acre (8): jack pine, 15.1; red pine, 17.0; white pine, 31.4; and sugar maple–basswood, 27.1. The progressive increase through the three stages of succession of the pines is counteracted in the case of the maple–basswood by more rapid decomposition. A slower rate of accumulation might also be involved.

91. Variation within a Stand.—The amount of forest floor will vary with the location beneath the crowns or in the openings between the trees. The comparison is evident in the following figures from the Stanislaus National Forest in California.

Type	Metric tons/acre	
	Under the crowns	In the openings
Ponderosa pine	37.0	8.7
Mixed conifer	35.9	12.0

In young stands before the crowns have closed, the floor covers only the area or less than the area of the vertical projections of the crowns. As the crowns close, the area covered by the forest floor and its amount per unit of land area increase quickly where rapidly growing species occupy a good site and more slowly for less rapidly growing species on poor sites.

92. Relation between Total Forest Floor and Annual Accumulation. For some purposes the relation between the total amount of forest floor and the annual accumulation may have considerable interest. This relation may be expressed simply as a ratio of total to annual. At one extreme in the case of a forest in the tropics, where the decomposition eliminated all the floor each year, the ratio would be one to one or even smaller. At the other extreme, with a maximum value for total forest floor of 120 metric tons per acre as found in the birch–sugar maple–spruce in northern New Hampshire, and if it is assumed that the annual accumulation does not exceed 2 tons per acre, the ratio would be 60. Actually these ratios commonly seem to vary between 3 and 15. Because the amount of annual accumulation does not vary widely, the ratio of total to annual becomes essentially a measure of the rate of decomposition. If equilibrium between accumulation and decomposition has been attained, the ratio then represents also the number of years for the establishment of that equilibrium.

If the total amounts of forest floor are determined periodically in the same stand where collections of the annual accumulation are made, the two measures may be compared in a way analogous to that between periodic annual and mean annual growth of forest stands to indicate the culmination of the mean annual increase, which should also be the age of maximum amount of forest floor. In the analogy, decomposition of the floor corresponds to mortality of trees. For example, a plantation of Canary pine at Berkeley, California, when 30 years old had 12.6 metric tons per acre of forest floor. Six years earlier it had 4.0 metric tons per acre. The annual accumulation collected in trays each of the 6 years averaged 2.7 metric tons per acre. This was the gross periodic annual increase. The net periodic annual increase was (12.6–4.0)/6 or 1.4 metric tons, leaving 1.3 metric tons per acre as the rate of decomposition. The mean annual increase in floor rose from 0.17 tons at 24 years to 0.42 metric tons per acre at 30 years but at that age was still well below the 1.4 tons periodic annual increase. Thus the culmination of the mean annual increase would not be attained for several years.

93. Influences of Litter and Forest Floor on Soil Productivity.—The evident influence of the forest floor is by the addition of organic matter containing nitrogen and other inorganic materials partly bound in organic complexes. The organic constituents usually recognized in analyses include a wide variety of substances, principally lignins, celluloses, hemicelluloses, sugars, starches, pectins, tannins, resins, fats, waxes, crude proteins, and amino acids. These substances are complex and undergo progressive changes after leaf fall through the agency of a large population of bacteria, actinomycetes, fungi, worms, insects, rodents, and other animals.

The composition of the litter and forest-floor material varies between different tree species and types, and consequently the rate of decomposition also varies. The litter of white ash, paper birch, yellow poplar, black locust, and hickory decomposes rapidly, whereas that of balsam fir, scarlet and black oak, beech, and red, pitch and white pine decomposes slowly (53, 250).

In considering the chemical constituents of leaf materials a distinction must be made between the green leaves and the dry leaves of autumn, because certain substances are translocated from the leaves to the wood parenchyma for storage before leaf fall. Nitrogen, phosphorus, and potassium for this reason are found in much smaller percentages in the freshly fallen than in the green leaves, whereas the percentage of calcium is maintained or even increases as the leaves mature.

In the contribution of the litter to soil productivity, nitrogen, potassium, phosphorus, and calcium are most likely to be deficient in the soil, in which case additions by the accumulation of litter are important. These constituents are extracted from the mineral soil by the roots and returned to the surface in the litter with the accumulation of leaves and other plant parts. Thus it may be anticipated that their amounts in litter will vary with species and with their availability in the mineral soil of the root zone.

94. Variation with Species and Soil.—The magnitude of the percentage differences between species and soil types for fresh litter may be illustrated by the figures in Tables 41 and 42.

The percentages are higher for all constituents and for both species on the Gloucester soil, suggesting that it is a better site. Red maple is more efficient than the pine in accumulating all four minerals, and the contrasts between species in the contents of phosphorus and potassium are strikingly large.

Comparisons of the percentages of nitrogen and lime for three species growing on two different soil types and in two different forest types on one of those soil types may be derived from recent work by Coile (92). Table 42 gives the figures. The Iredell loam is higher in bases than the Georgeville soil. Perhaps for that reason the percentages of CaO and of nitrogen are somewhat higher. However, in the case of red oak the percentages of lime are lower than for the Georgeville loam. Hickory grows naturally on soils well supplied with calcium, and its deep taproot system is apparently efficient in reaching and accumulating ample supplies. The hickory is notably higher than the oaks under all three conditions. The differences in the percentages of nitrogen whether the trees are growing in the stand of loblolly pine or in that of mixed oaks is small. Hickory has

TABLE 41.—PERCENTAGE MINERAL CONTENT OF FRESHLY FALLEN LEAVES OF RED MAPLE AND WHITE PINE ON DIFFERENT SOIL TYPES IN SOUTHERN NEW HAMPSHIRE (230)

Soil type	N		P_2O_5		K_2O		CaO	
	RM	WP	RM	WP	RM	WP	RM	WP
Brookfield loam	0.59	0.42	0.16	0.06	0.56	0.10	1.40	0.67
Gloucester sandy loam	0.60	0.53	0.44	0.12	0.64	0.24	1.73	0.91

a higher percentage in the pine stand and post oak slightly higher in the oak type. Similarly, the differences are conflicting in the case of the calcium. Again, the hickory has decidedly more in the pine type, whereas the red oak has slightly more in the oak type. Obviously, the differences in the contents of fresh litter resulting from differences in soil type and forest type are not simple.

Species have been classified according to data of the principal nutrient constituents of fresh litter in Table 43. The species are ranked in order from high to low within the classes.

From this table it is at once evident that the species having high contents of calcium, nitrogen, and phosphorus are those which are associated naturally with fertile soils, usually well supplied with calcium. On the other hand, the species in the low classification include most of the pines, which are commonly found on soils tending to be acid and low in lime content. To some extent the species high in calcium content are also high in nitrogen and phosphorus, but in this there are notable exceptions. For example, balsam fir, which is low in lime, is at the top of the list in nitrogen content. Green ash is low in nitrogen but high in phosphorus. In these relations, cause and effect cannot be separated with certainty, and particularly in the case of calcium. The species with high calcium content are those that are found naturally on mineral soils which are well supplied with calcium.

One application of information of the kind contained in Table 43 is obvious. If species are being selected for planting with the objective of improving the fertility of a poor or eroded soil, they should be those which have high contents of nutrient substances. At the same time such selections have to be limited to species that will survive

TABLE 42.—NITROGEN AND CALCIUM CONTENTS OF FRESH LITTER OF NORTHERN RED OAK, POST OAK, AND HICKORY IN DIFFERENT FOREST AND SOIL TYPES OF THE PIEDMONT PLATEAU IN NORTH CAROLINA (92)

Soil type	Georgeville stony clay loam		Iredell loam
Forest type	70-yr. loblolly pine	White–black–red oak	White oak–post oak
	Percentage of nitrogen		
Hickory	0.63	0.50	0.75
Northern red oak	0.58	0.56	0.64
Post oak	0.61	0.62	0.68
	Percentage of CaO		
Hickory	3.89	2.44	4.21
Northern red oak	1.88	1.90	1.82
Post oak	1.17	1.01	0.95

and develop on the soils to be planted. Black locust is an outstanding example of a species that stands near the top of the list as a soil improver, but it has been found to be a near failure on many of the poor eroded soils of the southeastern states where the choice of species that will survive and grow is almost limited to the shortleaf, loblolly, and slash pines.

95. Variation between Layers of the Forest Floor.—The chief criterion for the separation of the layers of the forest floor into litter, duff, and leaf mold is the degree of decomposition. The processes of decomposition are primarily oxidation of the organic matter and consequent progressive mineralization of the materials. Thus it would be expected that the percentages of mineral constituents would be successively higher in passing from the litter down to the leaf mold. This is well illustrated by the percentages of nitrogen in the floor from several different forest types in Oregon in Table 44 on page 180.

TABLE 43.—PERCENTAGE NUTRIENT CONTENT OF FRESH LITTER OF DIFFERENT SPECIES (8, 9, 53, 76, 77, 78, 92, 94, 103, 131, 153, 228, 230, 231, 288, 305)

CaO	N	P_2O_5	K_2O
High, 3 to 6.4%	High, 1.1 to 2.1%	High, 0.3 to 0.7%	High, 0.6 to 2.0%
Boxelder	Alder	Cucumbertree	Flowering dogwood
Basswood	Black locust	Flowering dogwood	Chamise
Green ash	Black walnut	Chamise	Hickory
Hickory	Balsam fir	Black cherry	Cucumbertree
Black locust	Bur oak	Sweet birch	Sweet birch
Flowering dogwood	Blue oak	American elm	Sugar maple
Blue oak	Yellow birch	American beech	American beech
Hophornbeam	Chestnut	White ash	Northern red oak
Black cherry	Eastern redcedar	Blue oak	Blue oak
Black walnut	Chamise	White oak	White oak
Yellow poplar	Black spruce	Red maple	Black cherry
Northern white cedar	White spruce	Sugar maple	White ash
			Basswood
			Red maple
Medium, 1.5 to 3%	Medium, 0.8 to 1.1%	Medium, 0.15 to 0.3%	Medium, 0.3 to 0.6%
Quaking aspen	Hophornbeam	Basswood	Quaking aspen
Eastern redcedar	Basswood	Hickory	American elm
Cucumbertree	Chestnut oak	Yellow poplar	Hophornbeam
White ash	Tamarack	Northern red oak	Red spruce
White spruce	Sassafras	Red spruce	Eastern hemlock
American elm	Eastern hemlock	Quaking aspen	Northern white cedar
Sugar maple	Shortleaf pine	Hophornbeam	
Bur oak	American elm	Balsam fir	
Sycamore	Boxelder	Eastern hemlock	
Sweet birch	Northern white cedar	Red pine	
Sourwood	Red spruce		
Yellow birch	Sycamore		
Blackgum	Virginia pine		
Pin oak			
Sweetgum			
Gray birch			
Paper birch			
Northern red oak			
White oak			
Red maple			
Black oak			
Sassafras			

TABLE 43.—PERCENTAGE NUTRIENT CONTENT OF FRESH LITTER OF DIFFERENT SPECIES (8, 9, 53, 76, 77, 78, 92, 94, 103, 131, 153, 228, 230, 231, 288, 305) —*Continued*

CaO	N	P_2O_5	K_2O
Low, 0.3 to 1.5%	Low, 0.4 to 0.8%	Low, 0.05 to 0.15%	Low, 0.05 to 0.3%
Balsam fir	White oak	White pine	Red pine
Scarlet oak	White ash	Longleaf pine	White pine
Chestnut oak	Black oak	Slash pine	Jack pine
Chamise	Hickory	Jack pine	Balsam fir
American beech	Cucumbertree	Northern white cedar	Longleaf pine
Eastern hemlock	Quaking aspen		Slash pine
Blackjack oak	Sweet birch		
Black spruce	American beech		
Chestnut	Sourwood		
Southern red oak	Flowering dogwood		
Shortleaf pine	Northern red oak		
Red spruce	Yellow poplar		
Post oak	Red pine		
Alder	Sugar maple		
White pine	Post oak		
Jack pine	Sweetgum		
Red pine	Loblolly pine		
Tamarack	Scarlet oak		
Pitch pine	Jack pine		
Virginia pine	White pine		
Longleaf pine	Southern red oak		
Slash pine	Green ash		
Loblolly pine	Blackgum		
....	Black cherry		
....	Red maple		
....	Blackjack oak		
....	Pitch pine		
....	Slash pine		
....	Longleaf pine		

The percentages of nitrogen increase from the litter to the duff to the humus with one exception. It appears from comparison of species that the presence of fir in the forest evidently results in higher percentages of nitrogen in the forest floor than are found under stands of redwood or of ponderosa pine in eastern Oregon. However, a mixture of shore pine with the fir along the coast gives even higher percentages than were found under pure stands of fir. The differences illustrated in Table 44, both between the different layers of the forest floor and between different forest types and localities are large enough to be significant in indicating the influence of the different species on the productivity of the soil.

Analyses of the constituents in the freshly fallen leaves and in the layers of the forest floor under the four stages in the forest succession in northern Minnesota provide additional evidence (8). In that region the forest succession from jack pine to red pine to white pine to sugar maple–basswood is well established, and in the particular area studied all four stages, undisturbed by fire for many years, were represented by well-developed stands growing close together, and all

TABLE 44.—PERCENTAGE OF NITROGEN IN THE FOREST FLOOR OF DIFFERENT FOREST TYPES IN OREGON (290)

Floor layer	Shore pine and fir	Eastern Oregon pine and fir	Coast fir	Redwood	Eastern Oregon pine
Litter.......	1.80	1.37	1.12	0.96	0.91
Duff........	2.83	1.71	1.39	1.06	1.16
Humus.....	2.66	1.97	2.07	1.46	1.04

four on the same soil type. The evidence that these types over a sufficient period of time would constitute a successional series on the same area was conclusive. The figures are given in Table 45.

Several trends may be observed in these figures. First, the progressive mineralization of the forest-floor material from the litter through the duff into the leaf mold is indicated by the progressive decreases in the percentages of volatile matter in these three layers. The volatile matter in the litter in general is not very different from that in the freshly fallen leaves, in both cases usually over 90 per cent, whereas in the leaf mold the percentages drop to between 44 and 58.

All the constituents in all the forest types with minor exceptions show a progressive increase in the percentage content from the surface litter to the leaf mold next to the mineral soil. The increase is more noticeable in the case of nitrogen than of the other substances and becomes still more striking when the nitrogen is expressed as a percentage of the organic matter rather than of the whole sample. Thus for jack pine 0.9 per cent of nitrogen in the organic matter of the litter becomes 2.1 per cent in the leaf mold, and in the maple–basswood type the change is from 1.3 to almost 3.6 per cent. In general, the litter tends to be a little higher in all constituents than the freshly fallen leaves.

The trends with the succession from jack pine to maple–basswood are even more interesting as indicating an intensification of the influence of the forest as the succession proceeds toward the climax. The percentage of volatile matter becomes less with the stages of the

TABLE 45.—PERCENTAGES OF CONSTITUENTS OF THE FOREST FLOOR IN THE DIFFERENT STAGES OF THE FOREST SUCCESSION ON THE SAME SOIL TYPE IN NORTHERN MINNESOTA (8)

	Jack pine	Red pine	White pine	Maple-basswood	Jack pine	Red pine	White pine	Maple-basswood
	Volatile matter				CaO in organic matter			
Freshly fallen leaves	95.85	95.73	95.67	91.56	0.90	1.41	1.59	4.31
Litter	95.36	96.60	93.15	88.19	1.15	1.08	1.51	5.12
Duff	84.18	88.15	81.67	82.37	1.39	1.42	2.28	6.65
Leaf mold	44.77	57.68	56.64	51.07	2.27	1.48	2.87	7.51
A_1 horizon, 0–3 in.	......	2.21	2.18	4.50	......	0.11	0.16	0.32
	P_2O_5 in sample				K_2O in sample			
Freshly fallen leaves	0.09	0.15	0.18	0.50	0.19	0.29	0.20	0.50
Litter	0.16	0.14	0.20	0.32	0.14	0.20	0.23	0.32
Duff	0.22	0.22	0.23	0.36	0.18	0.19	0.22	0.36
Leaf mold	0.15	0.22	0.23	0.36	0.14	0.19	0.24	0.46
	N in sample				N in organic matter			
Freshly fallen leaves	0.58	0.67	0.64	0.84	0.61	0.70	0.67	0.92
Litter	0.87	0.69	0.94	1.17	0.89	0.73	1.01	1.33
Duff	1.35	1.15	1.41	1.93	1.64	1.30	1.73	2.36
Leaf mold	0.98	1.07	1.41	1.82	2.10	1.84	2.49	3.56
Whole floor	0.91	1.07	1.15	1.71				
A_1 horizon, 0–3 in.	0.04	0.08	0.08	0.21				
A_2 horizon, 4–6 in.	0.02	0.03	0.03	0.04				
A_2 horizon, 7–12 in.	0.01	0.02	0.02	0.01				

succession, which obviously corresponds with the increases in the mineral constituents. Nitrogen, potassium, phosphorus, and calcium all increase with the stages in the succession, the most notable increases being in nitrogen and calcium, and in the change from white pine to maple–basswood. The old-growth white pine, where the samples were taken, already contained understory reproduction of maple, basswood, and other deciduous species, which doubtless accounts in part for the higher percentages in the white pine as compared with the red pine type. The most striking difference is in the duff content of lime, of which that of the jack pine contained only 1.4 per cent and that of the maple–basswood almost 6.7. The ecological hypothesis that each stage in a successional series tends to enrich the soil, thus promoting the invasion of the next stage, seems to be confirmed by the progressive increases in lime and nutrients. Similarly, the evidence of progressive oxidation and mineralization of the organic matter with depth is also clear.

A compilation of the percentages of nitrogen, calcium, phosphorus, and potassium for different types and localities is contained in Table 46. The types are arranged in descending order of the percentage nitrogen content. In general, the deciduous species tend to be high in the four constituents, and the pines low. There is considerable variation in the composition of the forest floor under the same forest type as exemplified by the figures for aspen and jack pine. These differences are doubtless associated with differences in soil and in the availability of the nutrient ions. On the shallow limestone soils of Dorr County, Wisconsin, under stands of northern white cedar and balsam fir, the *F* and *H* layers contained 2.17 per cent N and 5.88 per cent CaO (130). It is an interesting fact that the southern pines that grow so rapidly are at the bottom of the list as to content of nutrient substances. Another interesting point is the comparison of the composition of the white fir with that of ponderosa and Jeffrey pine, where the samples were collected from closely adjacent old-growth stands on the same soil. White fir floor contains much more nitrogen and lime and somewhat more phosphorus and potassium than the pine. The figures suggest that white fir is more efficient in extracting these constituents from the soil and returning them to the surface. In spite of its rating as an inferior species from the point of view of wood utilization, white fir may be an important member of the mixed-conifer type in maintaining and improving the productivity of the soil. Incidentally, as white fir represents a later stage in natural succession than the pine, this is another example of increasing beneficial influences on the soil with the successional development.

By applying the percentages of nitrogen or other constituents to the weights per acre of litter or forest floor, the amounts of nitrogen or other substances in the organic material on an acre may be obtained. These amounts can then be converted to dollars per acre of fertilizer value by the use of current costs of fertilizers. A value of

$2.30 an acre in 1924 for the annual fall of pine litter in northern Minnesota was derived in this way (11). Much higher values would of course be obtained for other types or for the total amounts of

TABLE 46.—PERCENTAGES OF NITROGEN, CALCIUM, PHOSPHORUS, AND POTASSIUM IN DRY FOREST FLOOR FROM STANDS OF DIFFERENT TYPES, AGES, AND LOCALITIES (6, 7, 10, 11, 21, 51, 153)

Type	Age	Locality	N	CaO	P_2O_5	K_2O
Black locust.	2–30	Ind., Ill., Ohio	2.81	3.22		
Quaking aspen.	. . . .	Minn.	1.84			
Birch–maple–basswood	Old	Minn.	1.71			
Sugar maple–basswood	65	N.W. Minn.	1.7	4.6	0.35	0.38
White fir.	300	Lassen National Forest, Calif.	1.7	3.4	0.38	0.18
Spruce–fir–birch.	. . . .	N. Minn.	1.64			
Quaking aspen.	55	N.W. Minn.	1.63	4.02		
White pine.	6–30	Ind., Ill., Ohio	1.59	2.40		
Sassafras.	2–30	Ind., Ill., Ohio	1.59	2.20		
Spruce–fir–birch.	. . . .	N. Minn.	1.52			
Birch–maple–basswood	Old	N.E. Minn.	1.43			
Ponderosa–Jeffrey pine	300	Lassen National Forest, Calif., site III	1.3	1.7	0.28	0.16
White pine.	250	N.W. Minn.	1.25	1.6	0.22	0.23
Jack pine–red pine.	50	E. Minn.	1.18	0.79	0.20	0.19
Longleaf pine.	. . . .	South	1.10	. . . .	0.29	Trace
Jack pine.	30	E. Minn.	1.03	0.83	0.20	0.17
Jack pine.	60	N.W. Minn.	1.0	1.1	0.18	0.15
Shortleaf pine.	6–30	Ind., Ill., Ohio	0.99	0.97		
Jack pine.	55	N.W. Minn.	0.98	0.99	0.16	0.15
Red pine.	200	N.W. Minn.	0.97	1.04	0.19	0.19
Red pine.	100	E. Minn.	0.88	0.75	0.14	0.12
Quaking aspen.	20	N. Wis.	0.87	1.45		
Red pine–white pine. . .	250	N.W. Minn.	0.86	0.94	0.14	0.14
Jack pine.	60	N.W. Minn.	0.82	0.79		
Shortleaf pine.	. . . .	South	0.71	0.27		
Slash pine.	. . . .	Fla., Miss.	0.54	0.49	0.09	0.05
Longleaf pine.	. . . .	Fla., Miss.	0.51	0.64	0.09	0.06
Loblolly pine.	. . . .	South	0.45	. . . .	0.18	

forest floor. Such figures have some interest, but as they are based on costs of fertilizers for agricultural use, they hardly represent anything that could be realized in increased growth of forest stands.

96. Effect on Mineral Soil.—If all the forest floor is consumed in forest fires the nitrogen is volatilized and is the chief loss. The other mineral constituents do not volatilize, but small losses may occur when partially burned particles are carried away by the wind. Phosphorus and potassium that have been bound up before a fire in unavailable forms of organic complexes are rendered temporarily more available and also more soluble by the fire. Thus they may contribute temporarily to increased growth of vegetation following the fire, but they are also subject to easy washing and leaching, so that the benefit is likely to last only for one or two seasons. In two regions where the subject has been studied the evidence as to the effect of fire seems to be conflicting. In Minnesota the burning of the forest floor had only detrimental effects (10). In the southern pine region, however, there seemed to be some small benefits after burning (152). The difference can probably be explained by the differences in the climates and conditions for decomposition in the two regions. In the north, decomposition is slow, nitrogen tends to accumulate in a considerable layer of forest floor, and the fire consumes much. In the south, with conditions favoring rapid decomposition, nitrogen and organic matter rapidly pass from the forest floor into the mineral soil, where they are below the reach of the effect of the fire, and consequently the amounts burned on the surface are small, and the beneficial effects come partly from the substances that have passed into the mineral soil by decomposition before the fire, and partly from the nonvolatile salts released by the fire.

The effect of fire on the nitrate nitrogen in the upper inch of soil on areas of chaparral in California has been clarified by Sampson (305), as shown by the following figures in parts per million for unburned and burned plots the first year after the fire.

Vegetation	Soil type	Nitrate nitrogen, parts per million	
		Unburned	Burned
Mixed chaparral	Mariposa silt loam	23	35
Chamise	Aiken clay loam	6	12
Chamise	Hugo clay loam	3	6

Although the amounts are small they almost doubled in the first year after the fire. However, the difference was much less in the second year and had disappeared by the third year after burning. Thus fire may result in a temporary increase in the fertility of the surface layer of the soil.

In Table 45 figures are given for nitrogen and lime in the mineral soil below the forest floor, so that the extent of the influence of the forest on the soil may be observed. The increases in the percentages of lime and nitrogen as were noted for the forest-floor layers are evident in the upper 3 in. of the mineral soil. The effect on the percentage content of nitrogen is small but still evident in the 4- to 6-in. layer but disappears below 6 in. The maple–basswood forest with the preceding stages in the succession represented a period of at least 600 years of occupation by the forest, and in this period the influence on the mineral soil detectable by analyses extended to a depth of 6 in. Obviously, the process of soil improvement by natural vegetation is not a rapid one.

Another example of the influence of the forest on the soil is available from the area of ponderosa pine in southern Idaho, where numerous samples were taken of the upper 4 in. of soil and analyzed for percentage of organic matter (266). The results were classified according to the degree of deterioration of the forage vegetation. Where the vegetation was good the average percentage content of organic matter was 10.5; where there was some deterioration, 4.8; and where deterioration had proceeded to the gully stage, only 2.4.

97. Effect of Roots.—The influence that vegetation has on the soil by the penetration of its roots is probably more important than is usually realized. Certain things are known about roots but usually not in quantitative terms. The roots give off carbon dioxide and thus tend to increase the acidity and the solvent action of the soil solution. By their penetration, expansion, and subsequent decay they create channels and change the soil structure. It has been mentioned that a channel 0.1 in. in diameter will carry 1 million times as much water as the disconnected soil pores of equal volume. In the process of decomposition, organic constituents are added to the soil. In places in the southeast where shortleaf pine has been cut and the roots have had time to decay, holes several inches in diameter and 4 or more ft deep remain as channels only partially filled with decaying wood.

There are a few indications of the amount of root material per unit of land area. The weights per acre of roots in plantations of jack pine with four different spacings and the comparison with weights of needles and tops are shown in Table 47. In these 17-year stands the weight of roots is not greatly different from the weight of needles, tending to be greater for the close spacing and less for the wider spacing in which the crowns had not yet closed. The roots, however, are equal to only from one-quarter to one-seventh of the total weight of the tops. Presumably, the ratio of top to root would increase with age.

Roots less than 0.3 in. in diameter had the following oven-dry weights in metric tons per acre in the upper 5.3 in. of Georgeville soil in the Duke Forest, North Carolina (93).

	Metric Tons per Acre
Lobolly pine, 10 years old	1.3
Lobolly pine, 70 years old	3.7
White oak–black oak–red oak, uneven-aged	4.9

Thus the amount of roots increased with age and with the forest succession.

Shortleaf pine 110 years old on White Store sandy loam had 1.8 metric tons, and red gum–yellow poplar, uneven-aged, on Congaree

TABLE 47.—DRY WEIGHTS OF ROOTS AND TOPS OF JACK PINE PLANTATIONS IN VERMONT, 17 YEARS OLD AND 13 FT HIGH, AT DIFFERENT SPACINGS (2)

Spacing, ft	Roots, metric tons/acre	Needles, metric tons/acre	Total tops, metric tons/acre	Ratio, top to root
2 by 2*	2.9	2.3	13.9	4.8
4 by 4*	1.6	2.3	11.1	6.9
6 by 6	1.7	1.8	7.4	4.4
8 by 8	1.2	1.7	6.9	5.7

* Crowns had closed.

silt loam, 1.6 metric tons per acre dry weight of roots less than 0.3 in. diameter in the upper 4.5 in. of soil. If these figures are compared with those preceding, the smaller values may be associated with the differences in forest types or soil types or both.

The perennial grasses develop roots and underground parts that have been shown by Weaver and Harmon (363) to amount to as much as 4.9 metric tons per acre dry weight in the upper 12 in. of soil for big bluestem and range from that to 2.5 metric tons per acre for needlegrass. The weight of roots in the case of big and little bluestem is more than three times the annual yield of hay. The roots of *Lespedeza sericea* have been found to weigh from 1 to more than 2 tons of dry matter to the acre in comparison with a yield of hay of 1½ to more than 2 tons of dry matter per acre in two cuts (287).

There may be a wide variation in the proportion of root material in the total weight of the plants under different site conditions. Cypress grown under pond conditions in the greenhouse developed only 31 per cent of root material, whereas the specimens growing under swamp and land conditions developed 56 and 53 per cent of their weight in roots (69).

98. The Time Element.—One of the important aspects of the influence of forest on soil is the time element. In how many years after a forest is planted on bare eroding land or abandoned fields will the effect of its protective cover become noticeable? Periodic determinations over a sufficient period of years on the same area are not yet available, but Billings (49) has reported on several soil properties in a

TABLE 48.—PROGRESSIVE INFLUENCE WITH AGE OF SHORTLEAF PINE ON PROPERTIES OF SURFACE 2 IN. OF GRANVILLE SANDY LOAM

Age, years	Depth, A_0, in.	Total clay, A_1 %	Organic matter, A_1 %	Moisture equivalent A_1, %	Volume weight	Water-holding capacity by volume, %
0	0	10.2	0.85	6.8	1.35	33.7
9	0.7	9.4	0.82	6.5	1.33	36.2
13	0.75	9.5	0.71	6.1	1.35	36.0
21	1.5	10.6	1.33	7.2	1.27	42.4
31	0.95	11.9	1.65	7.7	1.21	40.5
56	1.35	11.1	1.79	7.3	1.12	41.5
83	1.7	12.5	2.80	11.5	1.14	43.5
110	1.5	13.3	2.94	10.4	1.03	45.3

series of shortleaf pine stands to 110 years old, all growing on the same soil type and including samples from an old field representing the condition before the stands were established. The figures for the upper 2 in. of mineral soil and for the depth of the forest floor are given in Table 48. The depth of forest floor, the clay content, the organic matter, the moisture equivalent, and the water-holding capacity all increase with the age of the stand, while the volume weight decreases. However, there are irregularities in the trends, doubtless associated with minor differences in the sites, which tend to obscure the definiteness of the relation to age. It is quite evident that the 9- and 13-year stands have had little effect. It is equally evident that the stands 21 years of age and older have exerted noticeable influence. With respect to the depth of the forest floor and the water-holding capacity the 31- and 56-year-old stands have lower values than the 21-year, which suggests that the intensity of the influences may not be greatly augmented after the culmination of the current annual growth. However, the values for the stands of 83 and 110 years again show progressive changes with age. The differences in the table are all confined to the A_0 or A_1 horizons, and no such differences were found in the A_2 or deeper horizons.

99. Influence on Soil Structure.—The degree of aggregation or crumb structure is promoted by additions of organic matter and colloidal materials, in proportion to the amount added, for usual applications. Materials that decompose rapidly produce larger percentages and sizes of aggregates than inert substances. For example, the following linear regressions were obtained in adding 1, 2, 4, 6, and 8 tons per acre of sucrose s, alfalfa a, and wheat straw w to surface soil of Gilpin silty clay loam in West Virginia (58), when Y equals percentage of aggregates

Over 0.25 mm	Over 1.0 mm
$Y = 39 + 4.9s$	$Y = 17 + 4.4s$
$Y = 37 + 2.9a$	$Y = 17 + 0.9a$
$Y = 37 + 1.5w$	$Y = 16 + 1.0w$

The aggregation has been found to be greater in forested than in cultivated soils. In a comparison in Missouri (280), the forested soils had 32 per cent of aggregates greater than 0.1 mm, whereas the cultivated soils showed less than 13 per cent.

By the removal of the forest litter of red pine in lysimeters, similar results were obtained by Lunt (227). The percentage of aggregates in the upper inch of soil was reduced by 40 per cent in 3½ years. In the next 2 in. below, the decreases ranged between 11 and 24 per cent. Such changes are almost certainly influential both in the water relations and in the general productivity.

100. Nitrogen Fixation.—Another influence on the productivity of the soil is the definite improvement in nitrogen content that results from the fixation of atmospheric nitrogen in the nodules on the roots of members of the legume family. Species of the pea family that have been reported to have the root nodules include black locust, lead plant, wild indigo, dyer's greenweed, *Acacia armata*, red clover, white clover, yellow sweet clover, white sweet clover, bur clover, black medick, partridge pea, *Lespedeza hirta* and *L. striata*, *Desmodium canescens* and *D. canadense*, hairy vetch, spring vetch, *Lathyrus latifolius* and *L. sylvestris*, wild lupine, hog peanut, trailing wild bean, foxtail dalea, *Sesbania exaltata* Rydb., and *Lotus corniculatus*. Honey locust, redbud, Kentucky coffeetree, and some species of *Cassia* other than partridge pea do not have nodules. Outside the pea family, certain species of alder, ceanothus, *Elaeagnus*, and bayberry have nitrogen-fixing nodules (361). Root tubercles were found by Quick on seven species of ceanothus including the most common in the Sierra Nevada, and the growth of other species was increased by association with them. Several of the foregoing species are being used for soil stabilization or improvement.

Definite information as to the extent of this influence is available for the black locust in the Central states, indicating the increased nitrogen content of the soil and the effect on the growth of adjacent trees of other species. Near State College, Pennsylvania, Ferguson (120) reported from samples from the upper 6 in. of mineral soil the following percentage contents of nitrogen:

	N, Per Cent
Under locust	0.102
Under large catalpa	0.098
Under small catalpa	0.089

The height of the catalpa, a 13-year-old plantation, decreased at the rate of 0.67 ft per ft of distance as the distance from the locust increased from 20 to 60 ft. Similar effects were found by McIntyre and Jeffries (233) for Hagerstown clay loam.

	N, Per Cent
Under locust	0.146
12 ft from locust under catalpa	0.132
48 ft from locust under catalpa	0.126

In another example where plantations were growing on the shaly phase of Berks loam the following percentages of nitrogen were obtained:

	N, Per Cent
Under locust	0.23
6–18 ft from locust under catalpa	0.18
27 ft from locust under catalpa	0.17
59 ft from locust under catalpa	0.15

The effect of black locust upon the heights of an adjacent 20-year plantation of boxelder is reported by Young (380) to amount to a difference of 17 ft, the rows nearest the locust being 31 ft high and those beyond its influence 14 ft. The influence extended to a distance of 65 ft from the locust. Planting black locust between the rows of a black walnut plantation of which the mean annual growth had been 0.29 ft increased the current annual growth in the fourth and fifth years to an average of 0.95 ft per year, compared with 0.31 ft for the control area.

Increased vigor and growth of pines and beachgrass are evident in the blow sands in Oregon where they have been planted near Scotch broom and purple beach pea. The seaside lupine has no such stimulating influence but other lupines are reported to have it (30). In North Carolina, shortleaf pine benefits from the association with *Lespedeza sericea,* which continues to grow well under the trees.

The nitrogen content of the soil decreased at an average rate of 0.001 per cent per foot of distance away from locust plantations in the range of nitrogen content from 0.2 to 0.1 per cent (80). The beneficial effects were found in increased growth of catalpa, white ash, yellow poplar, black oak, and elm and extended to distances of 50 to 90 ft from the locust. In the range of from 0 to 65 ft the average decrease in height growth for all species was at the rate of ½ ft per ft of distance. Combining the nitrogen and growth relations makes it appear that a decrease of ½ ft in height is associated with a decrease of 0.001 per cent nitrogen content of the soil.

101. Soil Deterioration.—A striking example of the detrimental effect of the repeated removal of the forest litter is reported by Cope (100) from loblolly pine 55 years old in Maryland. Adjacent areas of the same stand with the same number of trees per acre and the same average diameter were under different ownership. On the one side the litter had been removed each year for mulch on a truck farm, and on the other it had been left to accumulate. The latter area with the litter had an average dominant height of 78 ft or 10 ft higher than the other, and the board-foot volume per acre was 24,800 or 6,200 in excess of the stand where the litter had been removed. As Cope points out, the litter that remained had added over $1 per acre per year to the value of the stand, but that removed had added considerably more than that to the values derived from the truck farm.

The deterioration of the site quality as a result of the annual collection of the litter is generally recognized in the European countries where that practice has been prevalent for many years. Other forms of deterioration associated with the development of forest stands have also been reported. In Idaho, Gibbs and others (137) found in the soil toxic effects that they attributed to the residual products from coniferous forest.

Much has been written in the European literature of the deterioration and actual stagnation of forest stands as a result of the excessive accumulation of an increasingly acid forest floor, called "raw humus," which intensifies the process of podsolization with the resulting leaching of the A_2 horizon and the formation of a cemented *B* horizon. This development goes on both in the northern countries where conditions for decomposition are unfavorable, and also in the central countries where pure stands of pine and spruce have been planted outside of and below their natural ranges. Similar results have been suggested as probable in some of the older plantings of pine in the northeastern United States. On the whole, however, examples of soil deterioration associated with the development of forest stands are rare in this country.

102. Litter and Forest Floor: Summary.—1. The oven-dry weight of the annual accumulation of forest litter is a function of the stand and varies from over 3.5 to less than 0.5 metric ton per acre in moderately to well-stocked stands.

2. The annual fall varies widely in the same stand in different years, to such a degree that the maximum in one year may be as much as three times the minimum in another.

3. Differences between species and types, between deciduous and coniferous, or between light and heavy-crowned species are not well-defined.

4. The annual fall is smaller on poor than on good site qualities.

5. The heaviest annual fall in well-stocked stands occurs about the age of culmination of the current annual increment and is less at older and at younger ages.

6. The total oven-dry weight of forest floor, including litter, duff, and leaf mold, varies from over 100 to less than 1 metric ton per acre in moderately to well-stocked stands.

7. It is a function not only of the forest stand but of all the environmental factors as they affect the fauna and flora and the biological and chemical processes of decomposition in the forest floor.

8. The amount of forest floor tends to be greatest in the forests of cool climates of high latitudes and altitudes, and lower as the climate becomes warmer. Exceptions are the heavy accumulations in warm dry climates, as under chaparral in southern California, under pine in southern France, and under spruce planted out of its natural range in the German lowlands.

9. The amount of forest floor in a given region and type increases with age for 30 to 80 years and thereafter varies with age, site, and density, but often without systematic relation to any one of these factors.

10. Forest floor tends to keep frozen soil porous, loose, and permeable when bare soil becomes solid and impermeable.

11. The annual fall of litter varies from more than one-third to less than one-thirtieth of the total forest floor.

12. The chemical role of litter and forest floor in maintaining or changing soil productivity depends chiefly upon their content of N, Ca, P, and K.

13. The amounts of these constituents, extracted from the mineral soil by the vegetation and returned to the surface in the litter, vary with species and with the amounts available in the root zone of the soil.

14. The percentage contents of CaO and N in litter and floor are usually correlated, and both tend to be less under conifers than under deciduous species, less on poor than on good sites, and less in pioneer than in climax stages of succession. CaO varies from almost 5 per cent to less than 0.5 per cent, and N, from 2 per cent to less than 0.5 per cent.

15. The percentage contents of CaO, P_2O_5 and K_2O on the basis of volatile matter increase progressively with depth and degree of decomposition from litter through duff to leaf mold.

16. The percentage contents of P_2O_5 and K_2O vary from about 0.4 to 0.1 without evident relation to condition of stand or environment.

17. Addition of forest floor to soils of poor sites increases forest growth and removal of forest floor or its excessive accumulation as peat or raw humus deteriorates site quality and decreases growth.

18. Fires, in destroying the forest floor, cause a total loss of the N, which volatilizes, and a conversion of the portions of Ca, P, and K bound in organic complexes to more available and more soluble forms subject to rapid washing and leaching.

19. Fires may increase the content of nitrate nitrogen in the surface layer of the mineral soil for 1 or 2 years after the burn.

20. Dry weight of roots will usually equal or exceed the dry weight of leaves of stands of trees or of nonwoody vegetation.

21. Species of the pea family, notably black locust among the trees, which fix atmospheric nitrogen through their root nodules, increase the nitrogen content and productivity of the soil.

22. As a stand increases in age, at least up to a certain point, the A_1 horizon increases in content of organic matter and of clay fraction, while the volume weight decreases.

23. The removal of forest floor results in reduction of degree of aggregation, and, conversely, reforestation gradually increases it.

CHAPTER XVIII

SOIL MOISTURE

In previous chapters the water cycle has been followed from the atmosphere through the vegetative canopy to the forest floor and soil. The next step is to follow the process of movement and retention in the layers of the soil, beginning with the forest floor at the surface.

103. Field Capacity of Forest Floor.—The rain that passes through the forest canopy first wets the forest floor, to the extent that it will retain water against the pull of gravity, before any of the water passes down into the mineral soil. The amount of water thus retained, usually expressed as a percentage of the oven-dry weight of the material, has been called the "normal moisture capacity" or "field moisture capacity," and defined as the minimum amount of water retained by absorption and film forces when the water is free to move downward through a mass of soil (316). For forest-floor materials these amounts of water have been reported to be from one to five or more times the dry weight. Fresh pine litter in Minnesota and in the southern states retained from 150 to 350 per cent (266). In the Appalachian and Central states percentages for conifers have been reported as 340, for hardwood 360 to 460, for moss 890, and in plantations 12 to 20 years old, from 100 to 250. In New England for spruce and northern hardwoods the values ranged from 300 to 900 per cent. The methods of determining these values have not always been the same and some of them were called "water holding or water absorptive capacity." Probably some gravitational water was included, and hence the values are higher than the field capacity as previously defined.

A series of values for field moisture capacity of the forest floor of different forest types, all determined by a slight modification of Shaw's method of determining normal moisture capacity, are given in Table 49. The range from 130 to 225 per cent of the oven-dry weight is not large, considering the different types and conditions represented. In general the A_0 horizon in these types contains only a small proportion of the decomposed leaf mold or humus. The white and red fir and Monterey pine, which have higher field moisture capacities, tended to have a deeper forest floor with a larger proportion of the humus layer. In the case of California buckwheat, the litter was very light, and it is probably the consequent low value of the denominator of the percentage fraction that results in the high figure. Deciduous species of more rapid decomposition that form a mull profile with larger proportions of decomposing humus material have high values, ranging up to 500 per cent. The higher moisture capacities of the duff and leaf mold in comparison with the undecomposed litter have been shown

TABLE 49.—FIELD MOISTURE CAPACITY OF FOREST FLOOR OF DIFFERENT FOREST TYPES

Forest type	Field moisture capacity, per cent	Ref.
Manzanita–*Ceanothus leucodermis*	130	(200)
Chamise–*Ceanothus crassifolius*	139	(200)
Virginia and table-mountain pines, 30 years	144	(232)
Douglas fir	150	
Interior live oak	158	(200)
Ponderosa pine	160	(51)
Ponderosa pine–sugar pine–fir	161	(225)
Christmasberry	164	(200)
Bigcone spruce	167	(200)
Pitch pine, 22 years	168	(232)
Canyon live oak	173	(200)
Red pine, 22 years	174	(232)
White fir	178	(51)
Monterey pine	186	
California buckwheat	190	(200)
Coulter pine, 25 years	207	
Oak and red maple, 40 years	208	(232)
White pine, 22 years	216	(232)
Red fir–white fir	225	

both in this country and in Europe. For Norway spruce in France the litter absorbed 214 per cent, and the duff and leaf mold from 530 to 630 per cent of the oven-dry weight.

For forest-floor materials the field moisture capacity is appreciably higher than the moisture equivalent. Most of the moisture equivalent determinations on such materials range from 100 to 120 per cent with occasional values from 80 to 150, whereas the field capacities usually exceed 120 per cent. The maximum water-holding capacity of Hilgard includes considerable gravitational water and hence is always higher than the field capacity.

Field capacity is ordinarily expressed as a percentage of the oven-dry weight, which provides a standard through which different determinations are comparable. It should be kept in mind, however, that what seems to be dry forest floor under field conditions rarely contains less than 10 or 15 per cent of moisture. Thus when it becomes desirable to estimate the amount of moisture or precipitation that will be retained by the forest floor it is essential to start with the difference between the field capacity and the actual moisture content just prior to the time of the rain. Several factors influence the actual moisture content of the forest floor in the field, and Stickel (332) has indicated their relative importance by a correlation analysis. He found the following correlation indexes for relations to moisture content of the litter:

Evaporation rate	0.74
Temperature of the surface floor	0.67
Hours since last rain	0.67
Dew-point depression	0.59
Air temperature	0.58
Relative humidity	0.56

For many purposes it is convenient to express the moisture retained by the forest floor in inches depth of water comparable with precipitation. This conversion from the percentage moisture content may readily be made if determinations of dry weight per unit area or of volume weight are also available.

In the case of a heavy forest floor with a dry weight of 45 metric tons per acre and a retention of five times the dry weight, the water retained would amount to 2.0 in. depth. On the other hand more nearly average figures of 5.5 metric tons per acre dry weight and 200 per cent retention give a depth of 0.1 in. of water. For actual retention under field conditions these figures would have to be further reduced to make allowance for the amount by which the actual moisture content exceeded the oven-dry condition.

Even at the maximum the retention of precipitation by the forest floor does not account for all the water of heavy rains or melting snow. The analogy to a sponge of almost unlimited absorptive capacity, formerly emphasized, can hardly be justified. Light showers or rains of 0.1 or 0.2 in. depth may be wholly retained in the forest floor, but large amounts of precipitation are likely to contribute water to the mineral soil beneath.

104. Moisture Retention in the Mineral Soil.—There are two important phases of the moisture in the mineral soil: one the retention or storage, and the other the infiltration and percolation. Both affect, although in different ways, the amount of water that passes down to the water table and reappears in stream flow. From the viewpoint of retention and storage there are three so-called moisture constants that are important: One is the field moisture capacity or normal moisture capacity of Shaw (316) representing the moisture content at which gravitational and lateral movement becomes extremely slow. In irrigation the field capacity is used to represent the uniform moisture content throughout the depth to which soil is wetted by a surface application of water, after water and soil have nearly attained equilibrium in 2 to 5 days. Although the foregoing definition mentions a "uniform moisture content," actually in forested soils the distribution is rarely uniform, first, because of the variable additions of water at different locations resulting from stem flow and interception and, second, because large pores may carry gravitational water locally below the level of the layer wetted completely to field capacity. The same constant has also been called "capillary saturation," "field capillary moisture capacity," and "specific retention," the latter usually expressed as a percentage by volume rather than by weight. The field capacity has been shown to correspond closely to the moisture equivalent for medium-textured soils but not for sands or clays or soils with large proportions of organic matter. The moisture-equivalent values for sands may be as much as 4 per cent lower than the field capacity and as much as 30 per cent higher for soils high in colloidal material.

A second important constant is the moisture content at which moisture ceases to be available to the roots or at which losses by transpiration stop. This is the wilting point or permanent wilting percentage as determined by actually growing plants in the soil. The wilting point determined in this way, however, does not always correspond to the wilting coefficient obtained by dividing the moisture equivalent by the factor 1.84. Depending upon various characteristics of the soil this factor has been shown to vary from 1.4 to 3.8. The wilting point is considered to be independent of the kind of plant or of the environmental conditions and thus a property of the soil itself.

A third constant, the saturation capacity, has significance to complete the outline of water storage in floor and soil. As indicated for field capacity, an appreciable period of time, sometimes as much as 5 days, may elapse after saturation before the gravitational water drains away. This water above field capacity constitutes a form of temporary storage with an upper limit at saturation. The volume of water in a soil at saturation is almost always somewhat less than the pore space because of entrapped air that remains even after thorough wetting. Thus the pore space can be used only as a rarely attained upper limit of saturation. Actual saturation in some 200 samples in West Virginia in a range from near 0 to 22 per cent, averaged about 9 per cent by volume less than the pore space (327). Moreover the unsaturation varied with the cover. It was low (5 per cent) for sod of grass or clover and high (15 per cent) for an area of hardwood forest. The high figure for the forest is probably related in part to the resistance to wetting of organic materials commonly observed in the surface layers of forested soils.

The distribution of percentage of water and space by volume in a saturated soil of fine texture might be somewhat as follows: solid particles, 43; capillary moisture, 29; noncapillary or gravitational water, 18; and pore space resisting saturation, 10 per cent. A column of soil of unit cross-sectional area, 12 in. deep and containing 5.6 in. of water at saturation would have, according to this distribution, 3.5 in. of capillary and 2.1 in. of gravitational water. The capillary moisture might include 1.2 in. below the wilting point and 2.3 in. of available moisture. From measurements of discharge and of water levels in ground-water wells in a forested cove of the Appalachian mountains, the storage in the macropores of the soil between depths of 3 and 9 ft was computed by Hursh and Fletcher as 4.1 in. or 0.68 in. per foot of subsoil (182).

By way of summary, a certain amount of rain reaches the forest floor, which retains the difference between the field capacity and the actual moisture content, and the excess of gravitational water passes down to the mineral soil. The mineral soil retains in each layer the difference between the field capacity and the actual moisture content, and if the rain is sufficient, is wetted to field capacity. The excess water again passes downward until it reaches the water table.

For many purposes it is convenient to express the amounts of water in inches depth, which usually involves a conversion from percentage moisture content to depth. This can be computed readily from the formula

$$d = \frac{M \rho_a D}{100}$$

where d equals inches depth of water, M is the moisture content in percentage by weight, ρ_a is the volume weight, and D is the inches depth of soil. When the depth of water is known, the same formula can be transposed and solved for the depth of soil that will be wetted to field capacity by that depth of water. The percentage moisture content can be used either as a single figure or as the difference between two percentages, as for example between the field capacity and the actual moisture content or between the moisture contents at successive times of sampling. By making certain assumptions an idea may be obtained of the range of depths of water which may be retained in 1 ft depth of soils of different characteristics. At one extreme in a fine-textured soil with a field capacity of 37 per cent, a wilting point of 20 per cent and a volume weight of 1.4, 1 ft of soil would retain 2.86 in. of water. On the other hand a sandy soil with a field capacity of 7 per cent, a wilting point of 4 per cent, and a volume weight of 1.8 would retain a little more than 0.6 in.

The moisture contents in percentage of the dry weight have been reported in many studies, but most of them do not give also the volume weights, which are necessary to determine the depths of water contained in the given depths of soil. For some purposes the percentages are useful, as for example in the effect of changes in vegetation on the moisture content of the soil. In the Santa Monica Mountains at an elevation of 2,300 ft on a south aspect, Bauer (40) sampled the soil moisture throughout the dry summer in the chaparral and in an adjacent area that had been burned clean 2 years previously. At a depth of 10 cm he found that the moisture content was below the permanent wilting percentage of 7.7 in May in both areas. At 30 cm depth on July 30 the moisture content in the chaparral was 6.9 per cent or below the wilting point, whereas it did not go below a minimum of 13 per cent in the 2-year burn. This illustrates the effectiveness of the roots of vegetation in the deeper layers of the soil in removing moisture down to or even a little below the wilting point.

The distribution of the roots of different species and types of vegetation will obviously affect the distribution of moisture in any dry period. Evaporation by itself is not effective in reducing soil moisture appreciably below a depth of 1 ft. Most grasses have shallow root systems, and thus in grassland the moisture is strongly reduced in the upper 1 or 2 ft but not much below that depth. In a forest with roots distributed to a depth of 4 or 5 ft the moisture content at the lower depths is likely to be decidedly lower than that under grassland or exposed soil, whereas nearer the surface the moisture content is lower under the grass. In Russia, the influence of the forest in reducing soil moisture has been reported to extend as deep as 50 ft.

Moisture determinations in the successive layers of the soil in Shasta and Tehama counties of California in plots covered with large dense brush, compared with adjacent plots where the brush had been cut and burned the year before and replaced by annual vegetation or sprouts, showed the same moisture content during the winter rains, a greater reduction in the upper 6 in. of the burned plots during the growing season, and a greater reduction below 12 in. in the brush-covered plots. In the latter case the whole 42 in. sampled was reduced to the wilting percentage, which was not true of the burned plots below the upper foot (359).

Regionally, there are differences in penetration associated with the distribution of the precipitation. For example, in southern California where the precipitation is concentrated during the winter, when the use of water by plants is small, a rainfall of 15 to 20 in. may penetrate below the root zone. By contrast, 30 in. of rainfall in Kansas and Nebraska during the growing season did not penetrate below the root zone, and there was no replenishment of ground water in 15 years.

The subject may be summed up by the statement that after a rain or the melting of snow stops, the downward movement of water in the soil almost stops within 2 to 5 days after its disappearance from the surface, and losses from the first foot by both evaporation and transpiration and from the remaining depth of the root zone by transpiration continue. Those losses must be replenished by subsequent rains before additional water will pass downward to a greater depth or to the water table.

105. Infiltration.—Infiltration is defined as the process by which liquid water enters the surface soil or zone of aeration. It includes both wetting to the field moisture capacity and the subsequent progressive downward movement of free water by gravity flow. The term "percolation" has also been used in the same sense, but to avoid ambiguity it should be reserved for the movement of water below the surface layer of the soil. In ground-water hydrology, percolation is defined as the movement of ground water by laminar flow through capillary pores produced by hydrostatic pressure caused by elevation of the water table. Infiltration is the most important part of the process by which water reaches the soil below the root zone and is added to the water table and thence to stream flow.

Infiltration is influenced by several factors, and through them some of the probable effects of vegetation on the process can be anticipated.

1. The importance of soil texture is evident when the difference in the rates of infiltration between sands and clays is considered. Soil texture is essentially independent of vegetation and hence also the effect on infiltration.

2. Soil structure includes the number and sizes of pores and the amount of incorporated coarse organic matter. All three are likely to be high in forest areas where the roots and floor and abundant organisms all contribute.

3. The moisture content of the soil will affect infiltration through the contracting or swelling of colloids and the consequent changes in pore space. The lower moisture content in the root zone of forest or grassland is therefore likely to result in higher infiltration, at least temporarily.

4. The turbidity of the water affects infiltration. This in turn is affected by the exposure of the soil and the size and velocity of fall of the raindrops. The water after passing through the forest floor is notably clear.

5. The degree and kind of cation saturation becomes important in alkali soils but is not likely to be significant in forest areas.

6. The soil temperature will be higher in the open, and therefore the viscosity of the water will be lower, and infiltration will be correspondingly increased. Thus, infiltration will also vary with the season of the year.

7. The depth and degree of freezing of the ground is a factor which affects permeability and infiltration. The soil freezes less deeply and less tightly under forest cover.

8. Finally, the duration of the rain affects the rate of infiltration.

106. Trends of Infiltration with Time.—The change of infiltration with time serves as a basis for a rational analysis and understanding of the process. Infiltration capacity during rainfall has been shown in numerous experiments to have an initial high rate that decreases rapidly at first and then more slowly until it approaches a constant rate in a period from ½ to 1½ hours or longer. This trend may be attributed to a group of causes in the soil including changes in structure, dispersion of aggregates, filling of pores, puddling of the surface layer by the impact of raindrops, swelling of colloids, closing of cracks, and decreasing hydraulic gradient. As rain continues the effects of these changes become progressively less so that the resulting infiltration may be considered in the nature of an exhaustion process, that is, one in which the rate of doing work is proportional to the amount of work remaining to be done. Such processes may be expressed by the inverse exponential law. In the case of infiltration the work remaining to be done is that of changing infiltration capacity from its initial high to its lower ultimate constant value f_c. If infiltration capacity is represented by f at any time T, Horton (161) has derived the following equation from theoretical considerations,

$$f = f_c + (f_0 - f_c)e^{-KT}$$

where f_0 is the original infiltration capacity, e is the base of natural logarithms, K is a constant, and T is time. Empirical data conform well to this trend.

If the equation is transposed to the form

$$f - f_c = (f_0 - f_c)e^{-KT}$$

observed values of f give a linear trend on semilogarithmic paper when plotted over T. Then the value of K can be determined from this trend line.

In the foregoing equation the initial infiltration capacity f_0 may be given any desired value, and if the time T is measured from the point where this value occurs on the curve, the form of the equation or the value of the constant K will not be changed. In this way, curves starting from any assigned value of f_0 and the corresponding time, for different kinds of cover, will start at the same point, and the differences in trend should reflect the differences in vegetation.

Frequently, it is helpful to be able to determine the critical time T_c at which infiltration capacity becomes constant. If the other data for infiltration are known,

$$T_c = \frac{1}{K} \ln \frac{100\,(f_0 - f_c)}{f_c}$$

This critical time is not constant for a given soil or cover because it varies with the initial soil moisture and the amount of rain before the infiltration capacity is exceeded.

The relations so far have been in terms of rate of infiltration. By integrating for any time interval from 0 to T, the total infiltration F may be obtained by the following equation:

$$F = f_cT + \frac{(f_0 - f_c)}{K}(1 - e^{-KT})$$

In a large series of tests on 68 different sites (129) the cumulative infiltration in inches depth of water F could be expressed as a function of the time of application in minutes T by the equation

$$F = bT^a$$

The constant b varied from 1 to 0.009 and a from 0.04 to 0.82 for initial tests. The graph of the equation is linear on log-log paper, from which the values of a and b may be obtained. By differentiation of this equation, the rate of infiltration f becomes

$$f = baT^{a-1}$$

If, as many tests have shown, the rate of infiltration becomes constant after sufficiently prolonged applications of water, then the equations do not represent that condition because, according to them, infiltration continues to change with time indefinitely. In this respect and because of its theoretical justification, Horton's formula is preferable wherever appropriate values of the initial and constant infiltration rates are available or can be determined.

107. Methods of Determining Infiltration.—The use or reproduction of the natural conditions under which rain or snow water enters the soil is the desideratum of determinations of infiltration. If natural rainfall is not used, the accurate simulation of the rain and its application without disturbing the soil are desirable. Both of these requisites involve difficulties so serious that they are only partially obviated in the methods that have been devised. In general, there are four classes of methods: (*a*) the tube or ring methods, (*b*) the rainfall-simulator methods, (*c*) methods involving the analysis of rainfall and runoff records, and (*d*) flood irrigations.

The tubes have been of different cross-sectional areas from a pipe of 1⅝ in. inside diameter (20) to a 36-in. diameter ring (249) or a frame 1 ft square (19). The infiltration is likely to vary with the area within the tube up to a diameter of 9 in. The tubes are inserted into the soil either just enough to prevent leakage to the surface outside (20) (usually about 1 in.) or to a depth of 16 to 22 in. (129). In operation a given volume of water may be added, and the time until its disappearance from the surface recorded (331). This involves a variable head of water. Others maintain a constant head and obtain the infiltration as the amount of water added in any desired period of time. The head has been as much as 9 in. (20) or as little as ¼ in., in the latter case maintained by a self-dispensing calibrated burette (129). Obviously, the greater the hydrostatic pressure, the greater will be the infiltration and the less the results are likely to correspond to those from natural rainfall. The errors caused by lateral movement of water in the soil below the tubes have been corrected by the use of buffer rings or compartments surrounding the central area where infiltration is measured (205). Water is applied to the buffer area as it is to the central area. Statistical analyses indicated that from 6 to 22 replicates would be required with tubes 9 in. in diameter and 18 in. long to obtain an average of satisfactory accuracy for a small area of a single soil type (129). Comparisons of the ratios of infiltration between two soil types corresponded closely with the reciprocals of surface runoff in the same types, indicating that relative infiltration between different areas should be reliable by the tube method (269).

The rain-simulator method consists essentially in the determination of infiltration as the difference between rates of artificial rainfall and surface runoff within small portable plot frames where both can be measured at short intervals. The plots are protected from the wind and measurements are also made of the sediment in the runoff water, the soil temperature, slope, and other conditions of site and vegetation. Several types of instruments have been devised and used as infiltrometers. Pearse used a frame 1 ft square to which water was applied at the upper edge from a graduated container through a perforated pipe (281). The North Fork (299), Rocky Mountain (368), and type FA [Soil Conservation Service (365)] infiltrometers use frames enclosing areas of 2.5 to 8 sq ft to which the water is applied through sprinklers under pressure. The type F infiltrometer has a plot frame 6.6 by 12 ft (365). Otherwise the instruments differ chiefly in the details of applying water and measuring rainfall and runoff. Improvements are being studied currently.

A comparison of the Pearse, North Fork, Rocky Mountain, and type F instruments led to the conclusion that satisfactory but only relative estimates of true infiltration could be obtained with any of them (365). The type F gave somewhat higher infiltration than the other three, which were in close agreement. The largest part of the wide variation in infiltration rates was associated with variations of site within a single vegetation type. Variations in rainfall intensity and soil temperature produced significant differences between instruments, which became nonsignificant when their influences were evaluated.

Infiltration determined on the same 13 sites both by infiltrometer and by tube gave a highly significant correlation ($r = 0.76$). When the tube determinations were made with turbid water, the difference between the means by the two methods was not significant. On the whole the foregoing methods seem to give reasonably consistent estimates of relative infiltration.

Flood irrigation of basins or small nearly level areas where the rates of application and outflow can be measured gives a direct measure of infiltration. It is useful in the infrequent areas of wild land where topography, water supply, and facilities for the necessary installations make it feasible.

The difference between the rainfall in excess of the infiltration capacity P_e and the surface runoff Q_s on a drainage basin is approximately equal to the total infiltration F in the time of rainfall excess T_e

$$F = P_e - Q_s$$

If this difference in amount of water is divided by the time of rainfall excess, the mean infiltration capacity f is obtained

$$f = \frac{P_e - Q_s}{T_e}$$

This method requires good records of rainfall with deductions for interception by vegetation. At least one record must be from a recording gage. It also necessitates the separation of surface runoff from subsurface flow in the records from the stream-gaging station. Methods of doing this and of correcting for surface detention will be considered in Chap. XX. Details of this method of deriving infiltration may be found in the work of Horton (171) and Sherman (249).

The two latter methods give average infiltration for large areas of land, whereas those previously considered were based upon more or less numerous samples, each of a small area. Comparisons between the two groups of methods are therefore difficult to make and are not available.

108. Causes and Magnitudes of Variations in Infiltration.—Determinations of relative infiltration together with 10 of the soil properties likely to influence infiltration on 68 soil profiles provided the basis for a correlation analysis (129). The following list of soil properties for surface or subsoil in descending order of the simple correlation coefficients r with infiltration roughly indicates their relative importance.

	r
Noncapillary porosity, subsoil	0.54
Organic matter, surface	0.50
Clay content, subsoil	−0.42
Organic matter, subsoil	0.40
Noncapillary porosity, surface	0.36
Total porosity, subsoil	0.36
Volume weight, subsoil	−0.33
Aggregation, surface	0.30
Moisture equivalent, subsoil	−0.30
Suspension, surface	−0.29
Total porosity, surface	0.24
Silt + clay, subsoil	−0.24
Volume weight, surface	−0.24

The regression of infiltration f in inches per hour over percentage of noncapillary pore space of the subsoil p was approximately

$$f = 0.078p - 0.23$$

When factors were combined in multiple correlations the highest multiple coefficient of 0.71 was obtained with noncapillary porosity and organic matter in both surface and subsoil, and clay content in the subsoil.

Examples may be used to illustrate the magnitudes and some of the sources of variation in infiltration capacity. Computed values for whole watersheds vary from 0.64 in. per hour in the highly cultivated Wabash River drainage to 0.16 in. as an average for several New England rivers and the French Broad River in North Carolina, which drain steep shallow-soiled forested areas (162). If the vegetation were the only factor involved in these instances its effect could not be said to be favorable.

The infiltration in the same area may vary widely from season to season. Averages by months for several streams including those just mentioned range from 0.09 in. per hour for April, increasing to a maximum of 0.72 in. in August and again decreasing to 0.17 in. in November.

Infiltration varies so much even within a soil type that the statement has been made that specific values do not exist, and therefore comparative rates between soils must be used (269). Determinations by the cylinder method on wet soils in Georgia gave the following results.

Soil type	Depth of *A* horizon, in.	Infiltration, in./hr	Erosion
Cecil sandy loam	11	0.55	Little
Cecil sandy clay loam	3	0.42	Considerable
Cecil clay loam	1	0.09	Excessive
Iredell loam	6	0.01	
Ruston sandy loam*	8	2.06	
Ruston sandy loam†	8	0.82	

* Clear water.
† Turbid water, 2 hours.

The three soils of the Cecil series represent progressive stages in the removal of the permeable sandy *A* horizon by erosion and the consequent nearer exposure of the impermeable red clay loam of the *B* horizon. Infiltration in the clay loam is only one-sixth of that in the sandy loam. The Iredell loam is an example of a soil of very low infiltration capacity because of its structure rather than its texture. Ruston sandy loam has a high infiltration capacity, which was reduced to less than one-third of its original value by the addition of turbid water for 2 hours. The infiltration of the Cecil clay loam increased from 0.09 to 0.22 in. per hour after 40 years' occupation by a forest of pine.

The effect of texture of the soil on the rate of infiltration is indicated by the results of 126 runs with the North Fork infiltrometer in the Pecos River drainage. When 1.42 in. per hour on the coarsest texture, the sandy loams, was taken as 100 per cent, the relatives decreased, as the texture became finer, to 86 on sandy clay loams, to 75 on clay loams, and to 61 on clays (325).

An inverse relation between rate of infiltration and the size and velocity of fall of raindrops has been indicated by Laws (214). The rainfall was characterized by a rain-erosivity index E/A derived as the ratio of the kinetic energy E of the falling drops to their cross-sectional area A. Infiltration f at the forty-fifth minute when plotted over the E/A index gave a linear trend that is represented approximately by the equation

$$f = 1.11 - 0.00216\frac{E}{A}$$

when f is in inches per hour, and E/A is in ergs per square millimeter. These tests were on the same soil, and the constants would doubtless be different for other soils. Within the range of drops tested, the difference in infiltration was as much as 300 per cent.

The impact of the raindrops when they strike the soil causes a splash and turbulence by which fine soil particles are set in motion and tend to form a seal over the pores in the soil. Thus this relation to raindrops is one aspect of the demonstrated phenomenon of decreasing infiltration with the addition of turbid or muddy water to the surface. In Lowdermilk's experiments (225), after passing clear water through tubes of fine sandy loam, muddy water containing 1.8 per cent by weight of sediment was applied for 1 week. The rate of infiltration rapidly decreased to one-tenth of the rate with clear water. The decrease was ascribed to the filtering of the muddy suspension at the surface of the soil, where a thin layer of fine-textured and relatively impervious sediment was deposited. The results have been confirmed by Hendrickson (150), who found that silt increased the time for the infiltration of a given amount of water three to four times, that clay increased it seven to eight times, and that the fine materials settled as a film 1/16 in. thick. Subsequent application of 27 in. of clear water caused no downward translocation of the materials in the film.

Part of the reason for the decrease of infiltration rate with time is of course the increasing moisture content of the soil, so that a similar trend of infiltration with soil moisture might be anticipated. On an annually burned plot of sandy clay loam in the foothills of the Sierra Nevada, a 5-year series of records gave a well-defined trend of infiltration rate decreasing at a decelerating rate as the relative wetness increased from below the wilting point to above field capacity (301). The following figures define the relation:

Relative wetness, per cent of field capacity	20	40	60	80	100	120
Infiltration, in./hr	0.91	0.43	0.26	0.17	0.11	0.08

109. Influence of Vegetation on Infiltration.—Numerous comparisons illustrate the effects of vegetation or of denudation. Tests for 20 minutes with tubes were made in the Kennett area north of Redding, California, where all vegetation was destroyed by the smelter fumes on 67,000 acres between 1905 and 1910. The soils are stony clay loams badly eroded in the denuded area but covered by ponderosa pine and Douglas fir in the surrounding undamaged area. The forested soils had an average infiltration rate 13.2 times that of those denuded, with a range in the ratios from 2.2 to 32.8 (302).

In Arkansas on cherty silt loam, Auten obtained the following comparisons expressed as relatives of volume of infiltration (19). Earlier,

Cover	Relative Infiltration
Oak	100
Old field pine	95
Oak, burned	25
Open pasture	22

in the Central states, where old-growth hardwoods showed relative infiltration of 100 at 1 in. depth, adjacent fields gave 2. At 3 in. depth the relatives were 100 to 7 and at 8 in., 100 to 45. Comparing the three depths under the forest, if the relative for 1 in. is 100, that at 3 in. is 35, and that at 8 in., 4 (20).

Forest plantations, when compared with open areas presumably representing conditions at the time of planting, showed relatives of 100 to 7 at 1 in. and 100 to 23 at 8 in. depth. Increases of 1.3 to 2 times those in the open were found in 12-year plantations, and, in those over 14 years, increases were from 3.8 to 113 times. Subsequently determinations with an infiltrometer in the same region in plantations of black locust and of sassafras of different ages have indicated the progressive improvement of infiltration as forest cover grows older (32).

Simulated rainfall was applied at the rate of 5.08 in. per hr to paired locations in the plantations and in the abandoned fields like those on which the trees were planted. The resulting infiltration rates in the open fields and the ratios of forest to open are given in Table 50. The differences between the means of forest and open, based on 24 determinations in each age class of locust paired with an equal number in the open, were statistically significant except in the 0–10-year class. All three differences were significant in sassafras with 36 replications. The improved infiltration is doubtless associated directly with the increase in coverage and depth of the forest floor as the plantations grow older.

TABLE 50.—INCREASE OF INFILTRATION WITH AGE OF BLACK LOCUST AND SASSAFRAS PLANTATIONS IN THE CENTRAL STATES

Age, years	Black locust		Sassafras	
	Open, in./hr	Ratio, forest/open	Open, in./hr	Ratio, forest/open
0–10	1.84	0.89	1.07	1.67
11–20	1.36	1.57	1.34	1.83
21–40	1.02	1.66	1.26	2.29

Using the tube method in upland hardwoods on seven soil types in Missouri, Arend (15) found that the infiltration in annually burned areas was from 38 to 82 per cent of that from areas undisturbed for several years. The corresponding percentages for unimproved pasture ranged from 27 to 56.

In the chaparral of southern California the ratios of burned to unburned areas have been found to vary from 50 to 70 per cent. Moreover if these ratios are plotted over rain intensity they form a descending trend becoming less steep as the intensity is greater. This suggests the possibility of predicting the effect of burning on infiltration from the intensity of the rain.

A comparison on the same soil in Mississippi gave infiltration of 0.35 in. per hr in pine pole stands and 0.14 on cropped and abandoned lands (249). From the same source the infiltration was 1.17 in. per hr on lands properly grazed, whereas under excessive grazing on the same soil the rate was only 0.15 in. per hr. The higher rates of infiltration were ascribed to good land use, which results in more organic matter, better aggregation, and more large pores.

Tests on 6-in. soil samples from adjacent forested and cultivated areas on 10 different soil types in Pennsylvania gave higher infiltration in every type on the forested soils, as an average, by 6.5 times the cultivated. This difference was associated with higher contents of organic matter, 4.0 for forests as compared with 2.3 per cent, and lower volume weights, 1.12 for forests, compared with 1.31 (4).

Infiltration in 24 coniferous plantations on loam soils in New York was 2.7 times as great in those with mull as in those with mor humus layers (108).

TABLE 51.—AVERAGE INFILTRATION CAPACITIES AFTER 30-MINUTE APPLICATIONS TO MOIST SOIL FOR PLANT-SOIL COMPLEXES IN THE SEVIER LAKE BASIN, UTAH

Cover type	Brush		Herbaceous	
Cover density, %	5	25	5	25
Geological formations	In./hr			
Limestone	1.9	3.3	0.4	1.0
Igneous	1.3	3.9	0.4	0.9
Sandstone	1.1	1.8	0.2	0.3

Differences in infiltration capacity associated with different types and densities of vegetation and with soils of different geological origin in the Sevier Lake basin in Utah were derived from type F infiltrometer runs by Woodward (378). Figures in Table 51 read from his curves indicate higher infiltration capacities for brush than for herbaceous types, higher for high than for low densities of cover, and a descending sequence from limestone through igneous to sandstone in geological formation.

Increases in infiltration from 30 to 160 ml per min in Auten's tubes were associated with increases from 26 to 68 in site index for old-field shortleaf pine, 40 to 110 years old, near Durham, North Carolina (49). However, this relation cannot be claimed as an influence of the forest in any direct way. It merely serves to throw light on another factor that may underlie variations in infiltration.

Comparisons of infiltration based on many samples in three different types and the relation to the densities of vegetation appear in Table 52. Although the actual rates of infiltration and content of organic matter are highest in the grassland and lowest in the creosote bush type, the increase in infiltration associated with increase in the density of the vegetation is greatest for the piñon–juniper and least for the grassland. Thus changes in density in the woody types seem to be more influential than those in grassland.

The effect of the removal of the 3 in. of forest floor under ponderosa pine in Colorado was tested with an infiltrometer on 18 paired plots. The average infiltration rate after it became constant on the litter-covered area was 1.52 and on the bare area, 0.92 in. per hour (190). The difference of 0.60 in. was highly significant statistically. It was also noted that the rate of infiltration increased with the thickness of the A_1 horizon of the soil.

TABLE 52.—INFILTRATION* IN RELATION TO TYPES AND DENSITIES OF VEGETATION AND ORGANIC MATTER IN THE UPPER ¼ IN. OF SOIL IN THE PECOS RIVER DRAINAGE (325)

Type	Median increase in infiltration per 10% increase in vegetal density, in./hr	Mean rates of infiltration, in./hr	Organic matter, %	Number of plots
Piñon–juniper	0.41	1.24	2.98	39
Creosote bush	0.38	0.71	1.77	63
Grassland.........	0.16	1.55	3.69	24

* Measured with North Fork infiltrometer.

Various similar tests have been made in Switzerland, which tend to show that undisturbed mature forests show maximum infiltration, and the heavily grazed pastures a minimum, whereas the establishment of forest plantations tends to increase the infiltration progressively, as the trees grow older, for a period of at least 50 years. Grazing in the woods or disturbance by thinning reduced infiltration (65).

Freezing of the soil reduces infiltration. On Volusia gravelly silt loam in the headwaters of the Susquehanna River, a frozen layer of 3 to 4 in. prevented infiltration on areas of pasture and corn stubble but had little effect in the forest (18).

The addition of ¼ to ½ in. of grass litter to an "in-place" lysimeter with a 10 per cent cover of grass and shrubs in Arizona resulted in an increase in infiltration for the year of 4.39 in. or 19.2 per cent (149).

An interesting and important practical example of the value of vegetation in promoting infiltration has come from the experiments on the spreading of water in southern California (257). In general, the practice of spreading water involves diverting the flow from the canyon mouths into shallow basins or low-gradient ditches on the alluvial gravels where it has time to infiltrate and thereby replenish the groundwater reservoirs from which a large part of the supply is pumped. At the mouth of the San Gabriel River, shallow basins were laid out and subjected to different treatments to determine comparative infiltration. Measured applications of water at different times were continued for 82 and 105 days, with the following results:

Cover and Treatment	Relative Infiltration
Grass and mulefat, undisturbed	100
Small basins, cleared	77
Vegetation removed, plowed and harrowed once	69
Furrowed	59
Plowed 2 ft deep, fallowed 1 year, and replowed and harrowed	39

The maximum infiltration of the vegetated basin was at the rate of 5.5 acre feet per acre per day.

From evidence derived from records of flood runoff it appears that soils tend to reach a minimum infiltration capacity after prolonged wetting, and it is this minimum that is important in major floods, where it is likely to be reached. Under these circumstances it also seems to be true that the influence of vegetation may become small as judged from the flow in channels. This may be related to the fact that forest soils with many pores of greater than capillary size carry water rapidly to the stream channels by subsurface flow.

110. Percolation in Rock Strata.—Before leaving soil moisture, something should be said about deep seepage, storage, and percolation of water into the rock strata, where it may travel long distances into other watersheds or into the ocean. This is not directly a matter of the influence of vegetation, but it becomes important in the completion of the equation of the water cycle. The amount of the loss or gain from this source varies with the kind of rock and the slope of the strata. In general, granites and similar igneous rocks carry little water, whereas lava, sandstones, and limestones may carry considerable amounts. Montecito, California, gets part of its water supply from tunnels in the sandstone above the city. In Wisconsin the run off from areas of crystalline rocks was 13.1, from sandstones 10.3, and from limestones 9.4 in. The differences of 2.8 and 3.7 in. were attributed to deep seepage into the sedimentary strata (126). The average annual precipitation on the drainage area of Fall River in California is 24 in., although the stream flow averages 30 in., an excess of 6 in. or 25 per cent. The basin is underlain by porous lava formations, and apparently water in subterranean channels enters from outside the headwaters area and comes to the surface within the basin as part of the yield of large springs.

Temporary storage in the rock strata of drainage basins in the mountains of southern California has been determined by Rowe (300) as the residual water after deducting from the cumulative precipitation the cumulative runoff, evaporation, transpiration, interception, and soil moisture deficit at the beginning of the season. In the fractured and disintegrating igneous formations the storage frequently amounted to 20 to 40 in. depth of water.

All ground water in rock or soil can be considered as temporary storage. It may be the subsurface flow in transit through noncapillary pores of the soil on a slope or a perched water table above an impervious layer and barrier or the water in the alluvial deposits of a stream valley or the water in crevices or pores of the rock. In any case, it is likely to be an important source of water for stream flow, when it drains into the channels. The level of the water table, where it occurs, and changes in level can be measured in ground-water wells periodically or continuously by water-stage recorders. If the water level is within the root zone of the vegetation, the record will usually show diurnal fluctuations reflecting the use of water for transpiration. Recurrent minor fluctuations may also be caused by changes in atmospheric pressure or geophysical phenomena. After heavy rains the ground water in a well in porous alluvial soil may rise and recede to form a trace on the chart like that of the hydrograph of the stream. The water table rises to the surface when surface runoff begins, and on well-drained slopes recedes as promptly after the rain stops. The decelerating recession after rain may be a useful measure of the rate of drainage of ground-water storage. In wells that penetrate alternating impervious and pervious soil or rock layers where perched water tables and seepage from higher levels on the slope may be involved, the interpretation of the changes in water levels becomes difficult, and for this reason some wells so located have been abandoned. In semiarid regions or where there is a long dry season, there may be no water table except in the alluvial bottoms or temporarily in the rainy season or during the melting of the snow. Even in the humid region near Norris and Lexington, Tennessee, wells in the upland did not reach ground water within 50 ft of the surface. On the other hand, near Ithaca, New York, Coshocton, Ohio, and high on the Appalachian ridges of the Coweeta Experimental Forest in western North Carolina ground water was found usually within 15 ft of the surface. Although more work needs to be done on the analysis and interpretation of the records, ground-water wells are likely to be a valuable supplement to other methods of hydrological study of percolation and subsurface storage.

111. Influences on Soil Moisture: Summary.—1. The forest floor will retain, after saturation and draining, from one to five times its dry weight of water. Needles retain less than leaves. Litter retains one-half to one-third as much as leaf mold. The maximum retention of precipitation is rarely over 1 in., and the average for dry floor would be about 0.2 in.

2. The water retained is also affected by the initial wetness of the floor, which will rarely be less than air dry or about 15 per cent of its oven-dry weight.

3. Only the water in excess of the field moisture capacity of the floor passes down into the mineral soil.

4. The soil water that may become available for stream flow is only that in excess of the field moisture capacity of the mineral soil, which may range from 10 to 50 per cent, according to its texture and other properties.

5. The amount by which the actual moisture content is less than the field moisture capacity, as well as current evaporation and transpiration losses must be made up by precipitation before any excess becomes available for stream flow.

6. Downward movement of water above the *C* horizon of the soil almost stops within 2 to 5 days after its disappearance from the surface.

7. Forest floor increases the percentage of unfree moisture retained by the soil to the extent that it reduces evaporation.

8. High infiltration is maintained at a maximum by undisturbed natural forest canopy and floor. Soil under second growth may have infiltration as high as under old growth. Under plantations more than 25 years old it may approach that under natural forest.

9. The infiltration capacities of soils under forest floor are usually greater than those of the same soils bare.

10. Infiltration is greater in dense than in open plantations; greater in old than in young plantations; greater in ungrazed than in pastured woods; greater in unthinned or lightly thinned than in heavily thinned stands and lowest in overgrazed pastures and in fields.

11. The difference in infiltration between forested and disturbed soils, which may be in the ratio of 100 to 2 at a depth of 1 in., tends to

decrease rapidly with depth and may be only 100 to 40 at 8 in. below the surface.

12. The influence of vegetation on infiltration is least and may be negligible on sands or other highly permeable soils.

13. Percolation may be important in permeable soils and in porous rocks such as sandstone, limestone, and lava but negligible in granites and other igneous rocks.

14. Ground-water wells may provide useful information as to the amount of subsurface water in transit or in storage.

CHAPTER XIX

TRANSPIRATION

The process and certain aspects of the significance of transpiration are familiar from work in plant physiology, ecology, and silviculture. From the point of view of forest influences it is one of the sources of loss of water of precipitation along with interception and evaporation before the remainder appears in stream flow. Thus the amount of water transpired per unit of land area should be expressed in inches depth so that it is comparable with precipitation and stream flow. For this purpose the transpiration from all the trees and vegetation of whatever size on the unit area or watershed should be included. Without reviewing thoroughly the findings as to the variations of transpiration with species and stand characteristics, with site quality, and with the climatic factors, only certain useful relations will be considered.

112. Relation to Temperature.—On the basis of the relation of transpiration to air temperature, Meyer (251) has devised a graph from which the inches depth of transpiration may be read directly from Weather Bureau records of mean monthly air temperatures. The scheme has the advantages of ease and simplicity but obviously neglects physiology and other factors except as they may be associated with temperature. The use of the graph, however, seems to give reasonable results. The trends are not quite linear but they may be closely approximated by two equations, first, for March to July,

$$T = 0.084t - 3.38$$

second, for August to November,

$$T = 0.087t - 3.72$$

where T is transpiration in inches per month, and t is the mean monthly temperature in degrees Fahrenheit. The transpiration plus evaporation from the data of Thornthwaite and Holzman (342) at Arlington, Virginia, for April, May, and June, when increased by 19 per cent (341), may be represented roughly as

$$T + E = 0.083t - 3.35$$

This trend for a field of grass with the inclusion of evaporation in a locality with ample rainfall is close enough to Meyer's trend to afford some confirmation.

The linearity of this relation is also confirmed to the extent that plottings of transpiration plus evaporation E for natural vegetation in the Santa Ana River Valley and for willows and saltgrass in tanks in southern California also give straight lines but with different constants. Thus for the Santa Ana River bottom,

$$T + E = 0.20t - 8.1$$

The relation to temperature has also been shown (339) by records of stream flow and ground-water levels where willows or tules were growing in the same locality. The daily fluctuations of transpiration and temperature corresponded closely, and the seasonal decrease in stream flow coincided with the increase in transpiration rate.

In general, Meyer concluded for the North Central States that the normal annual transpiration of deciduous forests is 8 to 12 in.; of small trees and brush, 6 to 8 in.; and of conifers, 4 to 6 in.; subject to adjustments for deficient precipitation or available supplies of ground water.

113. Relation to Soil Moisture.—The statements that transpiration varies with soil moisture may be questioned in view of Veihmeyer's work, which seems to indicate that such variation is found only near or below the wilting point and that much of the earlier evidence that apparently showed a progressive decrease of transpiration with decrease in soil moisture was really caused by unequal distribution of moisture in containers, so that only part of the root system had enough to function.

The usual methods of determining or estimating transpiration and some of the results are well outlined by Baker (30), Burger (67), and Raber (291). These methods depend in part upon the application of a factor determined from trees grown in pots or tanks to trees growing under natural conditions. The reliability of this application must be seriously questioned in view of the large series of experiments by Veihmeyer (unpublished) in which alfalfa grown in tanks used almost three times as much water as did the same crop growing in the same soil in the field. This is probably related to the finding that any disturbance of the normal life processes of a plant increases its use of water (340). If pots or tanks or lysimeters constitute physiological disturbances for parts of the roots, as they probably do, then the factors obtained from plants so grown and the estimates of transpiration are likely to be seriously high.

114. Relation to Evaporation from a Free Water Surface.—Comparisons of transpiration of plants growing in containers have not led to usable relations to the evaporation from free water surfaces in standard Weather Bureau or other deep pans. When the depth of the water is reduced to about 0.5 in. in a shallow black pan, the diurnal fluctuations and sometimes the seasonal totals of evaporation correspond closely to those of transpiration. Thus, near San Bernardino, California, the evaporation from such a pan from May through October was 47.8 in. while the transpiration plus soil evaporation from the alluvial bottomland was 53.9 in. (339).

115. Relation to Foliage.—Most of the methods also involve a problem in sampling to apply results from seedlings or branches to all the foliage of all the trees on an acre or larger area. This problem will be greatly simplified if converting factors can be worked out by which the large amount of data on growth and yield of trees and stands can be translated into terms of leaf area or weight and use of water. For individual prune trees in large tanks of soil the correlation coefficient between water use and leaf area was 0.97 and between water use and twig elongation, 0.995 (358). The regression of water use on leaf area is linear, so that leaf area or weight could be used to estimate transpiration whenever sufficient data could be obtained to define the trend. The relations of leaf area or weight to diameter and annual growth, mentioned in Chap. IV, would be useful in this connection.

In methods where transpiration is determined as the volume of water used by individual trees, those volumes to be useful must be converted to depth of water over an area of land. For a single tree the depth of transpiration (T_1) on the projectional area a, of its crown, derived from the volume V of water transpired, is

$$T_1 = \frac{V}{a}$$

Similarly for all the trees in a stand, the depth of transpiration on the projectional area of the crowns T_P is

$$T_P = \frac{\Sigma V}{\Sigma a}$$

If it is desired to express the depth of transpiration T_A on the total area occupied by the stand A, including space between the crowns, then

$$T_A = \frac{\Sigma V}{A}$$

The relation between T_A and T_P becomes

$$\frac{T_A}{T_P} = \frac{\Sigma a}{A}$$

But the ratio of the sum of the projectional areas of the crowns to the total ground area occupied by the stand is the crown closure C, or crown density. Therefore the conversion between the two expressions for transpiration can be made when the crown closure is known, because

$$T_A = T_P C$$

A method by which transpiration could be estimated from forest yield-table data was suggested by Horton in 1923 (172). He derived the formula

$$T = 2.76 \times 10^{-7} \, WLdhN \times \frac{E}{45}$$

where T is transpiration in inches, W is the weight of water used per unit of dry-leaf weight (von Höhnel), L is the dry-leaf weight in ounces per inch-foot of the product of the diameter and height of the experimental trees at Voorheesville, New York, d is average stand diameter from yield table, h is average height from yield table, N is number of trees per acre from yield table, and $E/45$ is the annual evaporation from a free water surface at the locality of application compared with 45 in. in Austria, where von Höhnel made his determinations of W.

As a result of more recent findings, estimates by this method could probably be improved by certain modifications. Determinations of W for small plants in pots are not difficult for any desired species, although the results will probably be higher than they should be for larger trees growing under natural conditions. Dry-leaf weight can be estimated directly for trees and stands from the annual growth or from the diameters and numbers of trees per acre in stand tables (199). The conversion for the difference in climate between the locality where W is determined and where the transpiration is to be estimated, represented by $E/45$, would probably be improved for the United States by using the ratio of losses by transpiration, interception, and evaporation for the two localities from the map of those losses (202) rather than the ratio of evaporation from free water surfaces.

In a comprehensive and suggestive study of transpiration in Java in 1937 (102), Coster determined the average annual transpiration use of water in inches depth over the areas of 11 different types of grass and shrubby and forest vegetation as the product of the rate of transpiration in grams per hour per gram of green weight of foliage separately during hours of sun and hours without sun in the period from 8:00 A.M. to 4:00 P.M., multiplied respectively by the average percentages of sun and cloudiness, multiplied by the green weight of foliage per hectare, multiplied by 8 hours per day and 365 days per year. The rates of transpiration were determined by repeated weighings of cut leaves, cut plants, and potted plants, checking one against the other.

Recent suggested methods of approach include the relation of transpiration to the diurnal fluctuations of dendrograph traces. Another possibility is the adaptation of the tensiometer principle to measurements in the conducting elements of trees. A third approach is being tried by Cummings at San Bernardino by the use of the ratio of the total radiant energy received by a horizontal projection of the foliage to the latent heat of water. Thus in cgs units radiant energy in calories per square centimeter per minute divided by latent heat in calories per gram would give transpiration in grams per square centimeter per minute. The method of measuring temperatures, vapor pressures, and wind velocities at two levels above the vegetation has been mentioned in connection with evaporation (342). The separation of concurrent evaporation and transpiration would still be necessary.

116. Relation to Radiation, Temperature, and Humidity.—Experiments under controlled conditions have indicated quantitative relations with some of the factors. Thus, with tobacco plants in pots under controlled temperature, humidity, and radiation, the rate of transpiration doubled with an increase of 2.3 times the radiant energy at temperatures between 73 and 78°F (16). The relation was independent of relative humidity between 50 and 88 per cent. The increase of transpiration of sunflower leaves with visible radiation was linear also in Martin's experiments (241).

The total transpiration T in cubic centimeters per 24 hours for 12 potted seedlings of six Rocky Mountain conifers plotted over saturation deficit H_d in inches of mercury by Bates (35) gave a linear trend in the range from 0.1 to 0.95 in. of Hg, which can be represented by the equation

$$T = 235.1H_d + 28.1$$

Although there is considerable dispersion of the points about the line, the regression of increasing transpiration with increasing saturation deficit is unquestionably significant.

The transpiration rate of *Quercus robur* at constant temperature was proportional to the difference between the saturation vapor pressure at the temperature of the leaves and the actual vapor pressure of the surrounding air. When the vapor pressure difference was held constant the transpiration at 104°F was five times that at 68°. Similarly, at constant relative humidity the transpiration increased seven to ten times with the same difference in temperature. At air temperatures of 68 and 86°F in darkness the relation of transpiration to relative humidity in the range from 97 to 10 per cent was nearly linear. At 104°F the relation was linear from 90 to 50 per cent (240). Corresponding relations to relative humidity were obtained with lemon cuttings and leaves by Bialoglowski (48). However, the relation to relative humidity is not always linear over so wide a range as indicated in the following findings:

TABLE 53.—RELATIVE TRANSPIRATION RATES OF DIFFERENT CONIFERS PER 100 SQUARE IN. OF LEAF SURFACE AT CONSTANT TEMPERATURE OF 75°F FOR DIFFERENT HUMIDITIES RELATIVE TO 90 PER CENT AS 1, IN SOIL AND IN WATER CULTURE

Species	Culture medium	Relative humidity, %				
		90 Grams/24 hr	90 = 1	70	50	30
Ponderosa pine	Water	29.9	1	1.88	2.12	2.30
	Soil	20.3	1	1.53	1.72	2.04
Bigtree	Water	26.3	1	1.51	1.74	1.84
	Soil	25.1	1	1.55	1.67	1.82
Port Orford cedar	Water	17.2	1	1.50	1.62	1.71
	Soil	22.5	1	1.50	1.65	1.71
Monterey pine	Water	27.0	1	1.41	1.43	1.52
	Soil	23.0	1	1.55	1.58	1.82
Redwood	Water	25.1	1	1.44	1.62	1.59
	Soil	26.0	1	1.30	1.37	1.59
Vapor-pressure difference		0.225* =	1	1.78	2.54	3.30
Atmometer loss		110 =	1	2.40	3.70	5.00

* In inches of mercury between temperatures of air and of the interior of the leaf, assumed to be higher than air by 5°F.

Similar controlled experiments with five tree species by Daniel[1] in which the seedlings were grown both in water culture and in continually moist soil showed the variation of transpiration with humidity

[1] From T. W. DANIEL, "The comparative transpiration rates of several western conifers," Ph.D. thesis, University of California, with the kind permission of the author.

when the air temperature was kept at 75°. The relative amounts of water transpired per 100 sq in. of leaf surface, if the average of the initial and terminal transpiration at 90 per cent relative humidity is 1, for relative humidities of 70, 50, and 30 per cent are shown in Table 53. The relatives for vapor-pressure difference and losses from an atmometer are also shown, and the constant differences in these relatives indicate a uniform humidity gradient corresponding to the gradient of relative humidity at a constant temperature. The transpiration rates, however, are not proportional to the relative humidities. They increase at a decreasing rate as the relative humidity is reduced. The only instance in which the relative of transpiration exceeds the relative of vapor-pressure difference is that of ponderosa pine in water culture at 70 per cent relative humidity. In soil, redwood relatives are the lowest. The other four species have closely corresponding relatives of about 1.5 for 70, 1.7 for 50, and 1.8 for 30 per cent relative humidity. When soil and water cultures are compared, the relatives for ponderosa pine are considerably higher in water. Those for Monterey pine are higher in soil. For the other species the differences are small. Only Port Orford cedar used appreciably less water when growing in the water culture than in soil.

117. Estimation from Changes in Ground-water Levels.—The diurnal fluctuations of ground-water levels may also be utilized in estimating transpiration where a continuous record of water levels in wells below the surface is available. When the water levels at 1- or 2-hour intervals are plotted over time and the points connected, the curve represents changes in the volume of water stored in the soil resulting from the effects of withdrawals by transpiration and of inflow from the river or from higher ground. When transpiration is the greater, the trend is downward, and when recharge is greater, the trend is upward.

Two graphic methods have been suggested to estimate transpiration (347). In Method 1 it is assumed that transpiration is at a minimum and negligible between 2 and 4 A.M., so that the trend at those hours should be the rate of recharge r exclusively. If there were no transpiration loss, this trend extended for the 24 hours, or $24r$ would be the total recharge for the day. Actually, after 24 hours the water level may be a little above or a little below where it was 24 hours previously. Letting s represent this difference, the total change in ground-water level in 24 hours attributable to transpiration will be $24r \pm s$. The change in ground-water level must be converted to equivalent depth of water by multiplying by the specific yield y of the soil in order to get transpiration loss. Finally, therefore,

$$T = y \frac{(24r \pm s)}{100}$$

Specific yield y is defined as the ratio of depth or volume of water yielded from the soil to depth or volume of soil expressed as a percentage. If a volume or depth of soil just above the water table is at field capacity (capillary saturation), then the addition of a volume or depth of water d_w will cause the water table to rise by a depth D_s. Then, by definition,

$$\frac{100d_w}{D_s} = y$$

Then the depth of water involved in a change in the level of the water table is

$$d_w = \frac{D_s y}{100}$$

The depth of water corresponding to the difference between moisture percentages by weight between saturation M_s and field capacity M_f is

$$d_w = \frac{(M_s - M_f)\rho_a D_s}{100}$$

and moisture percentage by weight M_w times volume weight ρ_a gives moisture percentage by volume M_v.

$$M_w\rho_a = M_v$$

Then,

$$100\frac{d_w}{D_s} = (M_s - M_f)\rho_a = M_{sv} - M_{fv} \qquad \text{(both by volume)}$$

But M_{sv} is also pore space as a percentage by volume. So the specific yield is also pore space minus field capacity, both by volume.

$$y = M_{sv} - M_{fv}$$

Actually the specific yield is usually less than would be indicated by this equation, because the total pore space in a soil is rarely more than 90 per cent occupied by water. Actual values for specific yield vary from 25 per cent for a silt loam, 12 per cent for a clay, 5 to 20 per cent for sandstone, 1 to 8 per cent for limestone, and from 0.03 to 0.8 per cent for granite.

The foregoing method assumed that the rate of recharge did not change during the 24 hours. Actually, it will certainly be more rapid as the water level is drawn down by transpiration in the afternoon, becoming a maximum at the trough of the diurnal cycle and a minimum at the crest. The following method makes an approximate allowance for the change in rate of recharge.

In this second method the hourly changes in ground water level are plotted over time, plus above and minus below a horizontal line of 0 change to form a histogram. During the hours when transpiration is at or near 0 the trend of the histogram will represent rate of recharge. As indicated above, the times of maximum and minimum are fixed. Thus a curve may be sketched to represent rate of recharge for the whole period. Then the sum of the hourly areas below this curve will be the transpiration loss as indicated by rate of change in ground-water level. This multiplied by the specific yield as before gives depth of transpiration use.

Ordinarily the value of transpiration will be lower and more accurate by Method 2. Evaporation will be a factor insofar as it results in part of the loss from the upper foot of soil. The methods would not be applicable where the water table is below the depth of root penetration. For the growing season the total transpiration would be the sum of the daily amounts. If transpiration is not 0 between 2 and 4 A.M. as assumed, the effect of the inaccuracy in that assumption on the estimate of transpiration would be small. In Method 2 the error that might result from the fact that the trend of the recharge line is only partially determinate is also small.

In support of the assumption that transpiration is negligible from midnight to 4 A.M. the transpiration of 2-year-old yellow poplars on Aug. 11 in Ohio was 0.4 g per plant for those hours or 1.6 per cent of the 25.4 g for the 24 hours (252). The transpiration from 1 to 4 A.M. by five conifers in a controlled temperature of 75°F in Daniel's data varied from 2 to 4 per cent of the 24-hour amounts. This was probably higher than would occur under natural conditions where the night temperatures would be lower.

Changes in ground-water levels in 24-hour periods in shallow wells under four forest types have been recorded in the table on page 224. The specific yield for the alluvium under the willow thicket was 12 per cent, so that the transpiration plus evaporation was 0.19 in. of water in the 24 hours. In the yellow poplar cove at 3,000 ft elevation the specific yield was about 1.5 per cent, and the consequent transpiration, 0.10 in. per day.

Complete removal of the riparian vegetation along a small creek on the Coweeta Experimental Forest in North Carolina eliminated the diurnal fluctations and increased the base flow by about 20 per cent as a maximum for a single day.

Forest type	Location	Date	Depth to water level, ft	Change in water level, in./day
Willow thicket...........	Riverside, Calif.	Aug. 19–20	3	1.60
Cottonwood grove........	Virden, N. M.	July 8–9	8	2.04
French tamarisk thicket ..	Safford, Ariz.	Sept. 9–10	6	2.17
Yellow poplar...........	Coweeta Exp. For., N. C.	July 14–15	6	6.7

118. Use of the Storage Equation.—The combined loss by transpiration and evaporation can also be determined by periodic sampling of soil moisture, at different depths to the bottom of the root zone, and application of the storage equation. If M_1 and M_2 are the moisture contents at the beginning and end of a period, P is precipitation, I is interception, E is evaporation, T is transpiration, and X is seepage below the root zone, all in inches depth,

$$M_1 + P - I - (E + T) - X = M_2$$

This equation can be transposed and solved for $E + T$. If there is no rain during the period, P, I, and X become zero. The difficulty remains of separating T and E on the same area.

A similar application of the storage equation can be made from stream-flow measurements in and out of a basin where the leakage is negligible. In this case inflow and outflow are substituted for the soil-moisture contents, and X becomes zero.

119. Amounts of Water Transpired in Relation to Species, Age, and Site.—The actual data by which transpiration from unit land area can be estimated are woefully fragmentary and unsatisfactory. Many of the estimates depend on von Höhnel's old data from potted seedlings in Austria (381). Some figures are available for water requirement or transpiration ratio that vary from about 100 to over 1,400. For western conifers, determinations range from 982 for limber pine to 359 for Engelmann spruce (35, 283). There is a suggestion in these figures of a decrease in transpiration ratio as the tolerance of the species increases and as the succession approaches the climax.

Green ash phytometers in Nebraska during the summer gave relative transpiration per unit of leaf area of 100 in the open prairie, compared with 44 in a sumac thicket (156). In the hills south of Palo Alto, California, the following values of cubic centimeters transpired per square decimeter of leaf surface per 24 hours were obtained (98):

Species	In chaparral	In forest
Chamise	2.7	1.4
Manzanita	1.7	1.0

The difference between the two species as well as the contrast in the environment is strikingly indicated.

Relative transpiration indexes from a large number of tests in the Santa Monica Mountains gave, for *Ceanothus cuneatus*, 0.12; for California scrub oak, 0.15; and for chamise, 0.28 (314).

Tests with $CoCl_2$ paper at 5,000 ft elevation in Butte County, California, indicated that the transpiration rate of the two species of ceanothus was more than double that of California chinquapin or California black oak (101).

The variation of transpiration with age is indicated by some old figures for beech in Germany growing in well-stocked stands, for the period from June to November (381).

Age, years	35	55	115
Transpiration, in.	2	9	15

On the other hand, more recent figures from Switzerland by Burger, still using von Höhnel's old pot experiments for rates, are 10.0 in. annually for 35-year-old and 7.9 in. for 98-year-old spruce. These reflect in part the decrease in the ratio of leaf area to ground area from 19.2 to 18.0 in the same period, following the culmination of the current annual growth. By comparison with the 10.1 in. for pure spruce, the transpiration of a mixed stand of spruce and fir, 37 years old with a ratio of leaf area to ground area of 22.0, was 10.0 in. (66).

Daily rates of transpiration in July, August, and September were determined by a thermoelectric compensation method from readings of a thermocouple in the sap stream (311) as shown in Table 54. By comparison with spruce the transpiration of stands of larch is double, and of pine, one-half annually. These figures for annual depths transpired by well-stocked stands are somewhat higher than those heretofore available. The relations for daily transpiration per unit of green-leaf weight are not the same as for stands, because the amounts of foliage in stands of the different species are quite different.

An estimate by Hursh and Fletcher by an analysis of the water cycle in a small drainage basin in western North Carolina where the precipitation was 67 in. gave 15.42 in. annual transpiration (182). Complete removal of the forest in a neighboring basin indicated the transpiration had been 17 in. annually, according to Hoover (158).

The figures for European species commonly quoted give water use for the vegetative season for beech, 10.8 in.; for spruce, 8.3 in.; for oak, 4.8 in; and for pine, 2.9 in.

TABLE 54.—TRANSPIRATION OF EUROPEAN SPECIES

Species	Daily transpiration, g/g of green leaves		Annual transpiration of stands 60 to 100 years old, in.
	July 19 to Aug. 6	Sept. 21–28	
European larch.......	4.6	2.2	26.8
Norway spruce.......	1.0	0.5	12.6
Scotch pine..........	1.9	0.7	6.3
White pine...........	2.1	0.7	
European beech......	4.1	2.3	

A comparison of the water use by plants on the *A*, *B*, and *C* horizons of the same soil by Sinclair and Sampson (324) indicated that, although the greater fertility of the *A* horizon resulted in a decrease in water requirement, the greater total growth more than compensated for that decrease to give an increased use of water per unit of soil area. Doubtless the same relations hold for different site qualities. The water requirement is less on the good sites but the total transpiration is greater.

The attempt by Horton (172) to apply such data as were available on water requirement, leaf weight, and stand characteristics, to estimate transpiration losses for different species, sites, and ages from yield tables, gave figures ranging from 1.8 in. annually for ponderosa pine in Arizona to 27 in. for red spruce on site I in New England. Some such method to convert the large amount of yield-table data into terms of water use would be extremely useful and should be satisfactory if and when reliable figures for water use per unit of leaf weight applicable to natural stands are made available.

Transpiration of a 55-year beech–maple forest in New York was estimated by Minckler as 4 to 5 in. annually (255).

In general, annual transpiration for most forests in the United States is probably between 5 and 15 in., and may approach 35 in. for large dense stands on the best sites. In the tropics, where the climate is favorable and growth is rapid, transpiration may be as much as 122 in. annually (102).

120. Combined Losses by Transpiration, Evaporation, and Interception.—Another approach to transpiration may be made by considering the combined losses from transpiration, evaporation, and interception. The sum of these losses on upland areas where the water table is below the reach of the roots cannot exceed the annual precipitation. For example, in southern California 19 in. of rainfall may be completely consumed in transpiration, evaporation, and interception without increment to ground water. If the precipitation in some years is only 12 in. the sum of the losses will not exceed 12 in.

If the roots of the vegetation reach the water table, as in canyon bottoms, then the losses are not limited by the precipitation. In Temescal Canyon, near Corona, California, losses of 12.9 in. were recorded in 30 days, from Apr. 28 to May 27 (339). In Cold Water Canyon, the loss for 6 months, from May to October, totaled 53.9 in. Along the Santa Ana River bottom, the losses for 2 years averaged 50 in. a year. These are large losses, chiefly by transpiration, but it must be remembered that the canyon bottoms occupy only a small fraction, perhaps 5 per cent of the land area, and when they are prorated over the whole drainage area they become much less impressive.

For some purposes the combined losses by transpiration, interception, and evaporation serve the needs. In forestry, however, it is important to separate the three because they do not change at the same rate or in the same direction with changes in the vegetation. As the vegetation increases in foliage area or volume, transpiration and interception increase but evaporation decreases. T and I are at a maximum for dense stands on the best sites probably near the culmination of the current annual growth when evaporation is at a minimum. E is at a maximum on a clean burn when T and I equal zero (201).

The separation of concurrent transpiration and evaporation from the same soil is a difficult matter. In a humid region a rough approximation may be made, however, by remembering that evaporation is mostly from the upper foot of soil and that transpiration is taking water both from the upper foot and from the lower layers to the depth of the root zone. Thus if the root zone extends to a depth of 3 ft and evaporation and transpiration each take one-half the moisture from the upper foot, then transpiration for the whole layer will be five times as large as evaporation. The roots almost always extend more than 1 ft deep so that it is likely that transpiration will exceed evaporation wherever the vegetation is at all dense.

121. Transpiration: Summary.—1. Water loss by transpiration in open-grown trees of the same age is closely proportional to twig elongation and to leaf area or weight.

2. The rate of water use is independent of the moisture content of the soil as long as that content is above the wilting point throughout the soil mass contiguous to the roots.

3. The rate of transpiration tends to rise with increase in saturation deficit of the atmosphere.

4. Winter losses by transpiration for deciduous species are only from 0.4 to 4 per cent and for coniferous species 10 to 18 per cent of those during the rest of the year. Ordinarily, the minimum transpiration is during the month of minimum temperature.

5. The water requirement of tree seedlings and nonwoody plants in weight of water used per unit of dry matter or transpiration ratio, under constant environment, varies with species from less than 100 to over 1,400. For conifers or broad-leaved evergreens it is usually less than that of deciduous trees. Relations to crown density, rate of growth, or habitat factors are not consistent in existing data.

6. The transpiration per unit of leaf area varies with environmental conditions, being less in full shade than in half shade and either greater or less in half shade than in full sunlight. Thus, it is greater in grassland than in brush thicket and greater in chaparral than in forest. Ratios of leaf area to ground area vary from 3.5 to 20 and probably more widely.

7. The water loss by transpiration per acre is the product of water use per unit of foliage for the species concerned and the amount of foliage on the acre. The amount of foliage on an acre, total or annual, may be approximated from leaf-weight determinations on sample trees or from relations between foliage and current annual increment of wood and diameters of stems or crowns.

8. The more productive the soil, the lower is the transpiration ratio but the higher is the total use of water, at least for equally stocked stands of the same species.

9. Insofar as transpiration losses are proportional to leaf area and weight, the minimum use of water by vegetation and the maximum

yield in stream flow will be obtained from vegetation of small size and low density.

10. If rainless periods within the growing season are sufficiently prolonged, the available moisture in the root zone may be used by transpiration, irrespective of the kind, size, or density of the vegetation.

11. Any disturbance of normal life processes increases the water requirement. Transpiration by plants in tanks where the roots are restricted by the walls is greater than in natural sites.

12. The transpiration of well-stocked forests of the same age and site quality is greatest in the climax types and progressively less as the types are further removed from the climax, probably because of differences in amount of foliage and in soil productivity, site II for a pioneer type being less productive than site II for a climax type. The annual transpiration may vary from less than 1 to more than 30 in.

13. The total transpiration of vegetation in canyon bottoms with subirrigation is greater during dry seasons than on mountain slopes where the soil moisture in portions of the root zone is at or below the wilting point.

CHAPTER XX

RUNOFF AND STREAM FLOW

122. Definitions, Units, and Sources of Information.—The term "runoff" may be defined as the total recoverable portion of precipitation water received on a drainage basin, whether it moves on the surface or by subsurface flow. It is usually expressed in acre-feet or in inches depth. The part of the runoff that passes over the surface is called "surface runoff." The subsurface flow is sometimes called "seepage."

"Stream flow" is the term commonly used for the water flow in stream channels and is usually expressed in cubic feet per second (cfs) abbreviated to second-feet. Discharge is the rate of flow of a stream at any point expressed in second-feet. The stage of a stream is the depth measured on a gage as the height above some datum level. Stage readings are converted to discharge by means of a rating curve. Units and equivalents for the different expressions in the measurement of water may be found in publications on hydrology and hydraulics (196).

There is a large amount of printed information on stream flow principally in the following publications: U.S. Geological Survey Water Supply Papers; U.S. Weather Bureau gage and flood records; Transactions of the American Geophysical Union, Hydrology Section; hydraulic engineering and hydrology books, and publications of the engineering departments of various states. Descriptions, formulas, and methods of rating weirs and flumes for the measurement of flowing water may be found in these publications and also in those on irrigation (183).

Almost all the surface and subsurface conditions in a drainage basin, of which Mead (243) gives a comprehensive list, affect stream flow.

123. Surface-runoff Analysis.—Surface runoff, as distinct from flow in channels, ordinarily persists only a short time after rain stops and includes only the sheet flow or that in incipient surface channels or small gulleys. In his analysis of surface runoff, Horton (171) makes use of the following factors.

1. Total rainfall P.
2. Total infiltration F.
3. Average depth of surface detention, including depression storage, δ_a.

4. An index I of turbulence of flow.
5. Length of overland flow l_o.
6. Slope s of surface in feet per foot.
7. Coefficient of surface roughness n.

In his "infiltration theory of surface runoff," Horton has developed a series of equations by which surface runoff may be predicted when these factors are known.

Surface runoff occurs only when the intensity of the rain exceeds the infiltration capacity of the soil, whereupon the excess precipitation first fills the depressions of the surface as storage including the interception storage of vegetation. When this depression storage is filled the depressions overflow, and the depth of flow plus the depth of depression storage is called "surface detention" δ_a. On plots when the total infiltration, precipitation excess, and surface runoff are known,

$$\delta_a = P_e - (F + Q_s)$$

When these quantities are expressed by lower-case symbols in inches per hour, the relation becomes

$$q_s = i - (f + \delta)$$

i being the intensity of precipitation. From Rowe's data for runoff plots at North Fork, California, when rates of surface runoff were plotted over intensities of rainfall, the linear relation between them was confirmed (301). The regression equation was

$$q_s = 1.015i - 0.15$$

If the infiltration rate was also determined this would provide another means of estimating the average surface detention. When values of the rate of surface runoff q_s are plotted over surface detention on log-log paper the resulting trend is a straight line that may be expressed by the equation $q_s = K_a\delta_a{}^M$. The slope of line M may be obtained from the coordinates of any two points and becomes the exponent of surface detention in the runoff equation

$$M = \frac{\log q_2 - \log q_1}{\log \delta_2 - \log \delta_1}$$

K_a is the value of q_s at the point on the trend line where δ_a equals 1. The value of M depends upon the degree of turbulence of flow and can be converted to an index of turbulence I.

$$I = 3/4\ (3 - M)$$

When the flow is fully turbulent,

$$M = 5/3 \text{ and } I = 1$$

If the flow is wholly laminar,

$$M = 3 \text{ and } I = 0$$

In order to evaluate the factor n for surface roughness, which is not obtainable from the published values for flow in channels, it is desirable to transform the average depth of surface detention δ_a to the depth at the foot of the slope δ by the relation,

$$\delta = \frac{M + 1}{M} \delta_a$$

This involves also a change in the constant from K_a to K_s by means of the relation,

$$K_s = \left(\frac{M}{M + 1}\right)^M K_a$$

The surface roughness factor may then be determined from the equation

$$n = \frac{1020\, s^{1/2}}{I\, K_s l_o}$$

Transposing,

$$K_s = \frac{1020\, s^{1/2}}{I n l_o}$$

If the value of K_s is substituted in the runoff equation it becomes

$$q_s = \frac{1020\, \delta^M s^{1/2}}{I n l_o}$$

For fully turbulent flow, I equals 1 and

$$q_s = \frac{1020\, \delta^{5/3} s^{1/2}}{n l_o}$$

The use of this formula permits the application of results from plots or small areas to larger areas and also makes possible the estimation of runoff when the other factors are known. It also permits the analysis of changes in surface runoff when any one factor is varied and the others are held constant.

If measurements of surface runoff are not available but records of intensity of rainfall i, infiltration capacity f, and duration of rainfall excess t have been made and K_s can be estimated, then the surface runoff can be computed by the equation (164),

$$q_s = (i - f) \tanh^M \frac{M + 1}{M} [(i - f)K_s]^{\frac{1}{M}} t$$

If the flow is three-fourths turbulent, M becomes 2, and the equation may be simplified to

$$q_s = (i - f) \tanh^2 \left[\frac{3}{2} \sqrt{(i - f)K_s}\, t\right]$$

As more data for values of K_s become available it should be possible to estimate its value for similar conditions in the same way that values of the roughness factor n are estimated for stream channels.

The two foregoing equations are applicable only when the $i - f$ factor is constant. When that factor is changing, a method of applying them to successive small intervals of time during which $i - f$ would be constant has been outlined (160).

These formulas indicate that surface runoff will increase as (*a*) length of overland flow decreases, (*b*) surface roughness decreases, (*c*) slope increases, (*d*) surface detention or depression storage increases, and (*e*) infiltration capacity decreases.

If the effect of vegetation on the different factors is known its influence on surface runoff can be evaluated. It has previously been shown that the effect of vegetation is to increase the infiltration capacity and hence to reduce surface runoff. It is unlikely that vegetation has any appreciable effect on slope or length of overland flow as they affect surface runoff. The forest floor probably does increase depression storage. A grass cover may cause "subdivided" flow and an increase in surface detention (170). For completely subdivided flow for depths not too small relative to the width between the stems, the flow would vary as the first power of the depth. However, experiments have indicated that the subdivided flow through grass is more nearly 75 per cent turbulent so that the flow is proportional to the square of the depth. It may be that the coarser fragments in the forest floor have a similar effect. The influence of forest or other vegetation on surface runoff generally will be to keep it at a minimum just as infiltration is kept at a maximum.

124. Influence of Vegetation on Surface Runoff.—Numerous examples of this influence have been recorded. The maximum surface runoff as a percentage of the precipitation in oak forest in southwestern Wisconsin was less than 3 compared with 26 in cornfields and seeded pastures (39). In post oak plots in Oklahoma (46) when undisturbed, the surface runoff was less than 0.01 per cent of the precipitation, and when the litter was burned, almost 2½ per cent. In striking examples in northern Mississippi (245), natural oak forest and broomsedge grassland yielded about 1 per cent of the precipitation as surface runoff, while abandoned lands and cotton fields yielded 47 per cent or more. In the same locality occupation for 22 years by a dense plantation of black locust and osage orange had reduced the surface runoff to 2 per cent, when it was presumably over 40 per cent on the old field before the planting. Burning of the litter in oak forest in Texas and

Oklahoma (23) increased the percentage of surface runoff from three to forty times. Similarly, in the Appalachian Mountains (266), burning the litter in an old stand of pine and hardwoods increased the surface runoff by ten times on the average and up to thirty-two times in single storms. Tests by artificial sprinkling in central Idaho showed 0.4 per cent runoff in the undisturbed wheat grass, whereas in the annual-weed type of the badly overgrazed ranges 61 per cent of the precipitation appeared as runoff (104). In tanks, by burning the litter of pine, white fir, and incense cedar the surface runoff was increased by nine times, and burning the litter of oak increased surface runoff twenty-two times (225).

TABLE 55.—SURFACE RUNOFF IN PERCENTAGE OF PRECIPITATION IN HEAVY STORMS ON ANNUALLY BURNED, ONCE BURNED, AND UNBURNED PLOTS IN THE BRUSH TYPE AT NORTH FORK, CALIFORNIA

Date	Rain, in.	Unburned	Once burned, 1930	Burned 1929, 1930, 1932, 1933
		Surface runoff, % of precipitation		
Feb. 19–27, 1930	4.98	0.022	....	0.73
Dec. 31, 1930–Jan. 2, 1931	2.23	0.025	0.77	0.07
Feb. 4–10, 1932	7.00	0.031	0.36	1.05
May 10–12, 1933	2.02	0.27	1.28	2.76
Dec. 12–15, 1933	3.28	0.05	0.74	5.36
Dec. 31, 1933–Jan. 1, 1934	3.02	0.036	0.51	16.9

The cumulative effect of repeated burning may be illustrated by the data for the runoff of single storms on plots in the brush type of ceanothus, buckeye, mountain mahogany, and *Rhamnus* on a 32 per cent slope (226), as given in Table 55. The unburned and once-burned plots show no progressive increase insurface runoff while the repeatedly burned plots show an increase at an accelerating rate. The maximum rate of runoff for the storm of January, 1934, represented a flow of 358 sec-ft per square mile, comparable with amounts recorded elsewhere in destructive floods.

When an area is covered with litter of pine, and probably the same thing happens with litter of other species, there is lateral movement of water along the needles down the slope before the soil mantle is saturated and probably before the infiltration capacity is exceeded. This kind of flow may take place under forest when none would be recorded from soil without the litter. Thus on runoff plots at Berkeley, California, this apparent surface runoff was recorded under a plantation of Canary pine with heavy litter when the cumulative seasonal rainfall reached 1 or 2 in., whereas 8 to 10 in. had to fall on the adjacent grassland plots before surface runoff began. The amounts of this concealed surface runoff are small. For example, in 1935, when 9 in. of rain had fallen and true surface runoff began in the grassland, the concealed runoff from the pine plots had reached a total of only 0.09 in. or about 1 per cent of the precipitation. In any case it would be a small factor in the flow in stream channels.

In the usual installation of a plot for the measurement of surface runoff, which is intercepted in a trough inserted in the soil at the lower end, the measured runoff is likely to be too high by an indeterminable and perhaps serious amount. This error results from the fact that a part of the subsurface flow in the soil of the plot is forced to the surface above the intercepting trough because the cross-sectional area of soil through which the subsurface water would otherwise pass is reduced by the area of the trough constituting a partial barrier below the surface. This water, which is thus forced to the surface just above the trough, cannot be distinguished from the true surface runoff with which it mixes and is recorded as surface runoff. A collecting device in the nature of a flat plate at the foot of a plot should obviate this source of error.

In general, the evidence is quite conclusive for most of the regions of the United States that surface runoff is very small or negligible from areas of undisturbed vegetation and may amount to over 50 per cent of the precipitation where the vegetation has been seriously disturbed by overgrazing, cultivation, burning, or other causes.

Not only does the removal of the forest floor increase the surface runoff but conversely the addition of a protecting, decomposing layer of litter to a denuded soil reduces thé surface runoff. This has been demonstrated experimentally on newly cleared land in North Carolina by the comparison of areas of cotton plantation with those to which litter was added. The results for the first 3 years are given in Table 56. In the first year the percentage of surface runoff from the cotton plantation was greater than from the litter-covered area, but in the 2 succeeding years it was only about half as great. The percentage of surface runoff from the cotton planting more than doubled in the next 2 years, whereas that from the litter-covered areas decreased. Hardwood litter was more effective than pine needles, and the effects became greater each year. The 23 tons to the acre of litter that was added each year was of course much greater than the annual accumulation in a natural forest would have been. In fact that amount is equivalent to a moderately heavy total forest floor. It is probable that the addition of smaller amounts of litter would have been almost equally effective. The minimum layer of litter that will keep surface runoff at a low point is one of the things that needs to be determined.

125. Stream Flow.—There are marked variations in stream flow as in precipitation, annually, seasonally, monthly, and, in the case of small streams, diurnally. These variations are often analyzed graphically by plotting stream flow over time to form what is known as a "hydrograph." The forms of these hydrographs vary with the factors

TABLE 56.—EFFECT OF UNDECOMPOSED FOREST LITTER IN 2-IN. LAYERS, 23 TONS TO THE ACRE APPLIED EACH FALL, ON RUNOFF OF NEWLY CLEARED CECIL SANDY CLAY LOAM ON 10 PER CENT SLOPE AT STATESVILLE, NORTH CAROLINA (47)

Cover		1934	1935	1936
	Rainfall, in.	49.34	42.80	60.00
	Runoff as per cent of rainfall			
Pine needles		11.0	11.3	9.0
Hardwood litter		10.9	8.6	6.1
Cotton		8.6	22.4	17.4

that affect stream flow and illustrate strikingly such differences, for example, as the difference between snow-fed and rain-fed streams.

The flow of water varies with the size of the drainage basin and consequently has little value for comparisons between basins of different areas. If however, discharge is divided by the area, an expression per unit of area is obtained that is more useful. Thus in California the Smith River in the redwood region yields 7.5 sec-ft per square mile; the Stanislaus River in the pine–fir region, 1.9; the small streams from the coast ranges near San Francisco Bay, 0.6; and in the chaparral regions of the south the Santa Maria River yields 0.2 and the Santa Margarita River less than 0.1 sec-ft per square mile annually. The same unit can be used as a standard of comparison for high and low water flows from the same basin at different times.

In large watersheds the combinations of variable conditions often confuse the effect of any one factor, such as vegetation. For this reason it is usually desirable to study and work first with small uniform drainage areas or even runoff plots that can be replicated within a larger topographic unit. Records from these small units can then be combined in application to larger units, preferably after a comparison with stream-flow records of such larger units.

126. Relation to Precipitation and Losses.—The most obvious and important causative factor in stream flow is the precipitation. Much work has been done in studying this relation, principally with the aim of predicting the stream flow from the precipitation. Actually stream flow is the residual after the losses of water are deducted from the precipitation. Thus stream flow is rarely equal to or greater than the precipitation, and it is often expressed as a percentage, ordinarily being less than 100 per cent of the precipitation. In California over a period of years these percentages have varied from 0 to 60. In the Middle West the ratio of stream flow to precipitation is usually between 20 and 30 per cent, and in the East about 50 per cent.

Locally in regions where fog drop is sufficient, it is possible that the stream flow might exceed the precipitation as recorded in normally exposed gages. This relationship has actually been reported from a portion of the Pyrenees Mountains in France. Similarly, below perennial snow fields in years of low precipitation and high snow melt the stream flow might exceed the precipitation in those years, although the balance would be reestablished in other years. Condensation of atmospheric moisture on snow or ice or bodies of cold water might also contribute to the same result. Similarly percolating water from other drainage basins may increase the stream flow above the amount of the precipitation falling on the basin. As mentioned previously, Fall River, California, has an average annual flow of 30 in., but the annual precipitation falling on its basin is only 24 in. Water comes into the drainage basin in subterranean channels through the porous lava formations from other basins at higher elevations. If most precipitation records are 5 to 10 per cent too low, this error might cause an apparent excess of stream flow over precipitation and in any case introduces a discrepancy in the relation between them.

On the whole it is more fruitful to analyze stream flow as the difference between the precipitation and the losses than as a percentage of the precipitation. If the stream flow Q accounts for all the precipitation P, then the losses L are equal to 0 and

$$Q = P$$

This is represented graphically in Fig. 19 by the straight line of 100 per cent runoff and 45-deg slope. Actually there are losses by evaporation, etc., from a drainage basin. If those losses remain constant then the stream flow becomes

$$Q = P - L$$

This may be represented by a straight line parallel to $Q = P$ and at a distance L below it. The vertical distance between the two lines will represent the losses for that drainage basin. Ordinarily measurements of stream flow and precipitation are more readily available than those of the losses, and thus the losses may be determined by subtracting the stream flow from the precipitation on the same areas. The conditions under which the losses become constant are that the pre-

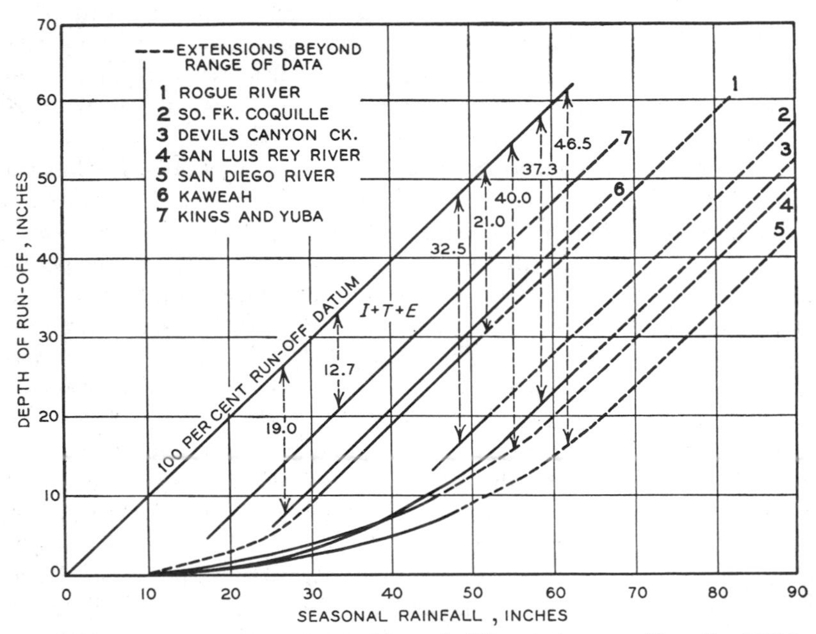

FIG. 19.—Seasonal runoff, rainfall, and losses in different streams. [*From Lee* (216).]

cipitation exceeds the losses and occurs mostly in the winter months. In the eastern part of the United States where much of the precipitation falls in the summer months, the trends with increasing seasonal precipitation continue to diverge slightly from the 100 per cent runoff line. Even in this case the average seasonal losses may be derived as the difference between the precipitation and runoff on the ordinate of the average seasonal precipitation. Several such trends for streams in different parts of the United States as plotted by Lee (216, 217) are reproduced in Fig. 19. The southern California streams like the San Diego River in the period of record have never had sufficient rainfall to reach the point where the trend becomes parallel to that of 100 per cent runoff. However, the curves can be extrapolated for approximate estimates of what the losses might be if the rainfall were sufficient to supply that amount of water. The losses derived in this way vary in the West Coast streams from 12.7 in. for the Kings River to 46.5 in. for the San Diego River. In the East they vary from 20 in. for the Red River of the North, and 21 in. for the Merrimac in New England, to 32 in. for the Chattahoochee River in Georgia. These losses are often termed "evapo-transpiration losses" or "consumptive use" or "retention."

If the losses are expressed as a fraction of the mean annual evaporation from a free water surface in the same locality the resulting ratio is known as the "evaporation opportunity." For some of the streams in Fig. 19 the evaporation opportunity varies from 0.90 for the south fork of the Coquille River, where there is maximum interception and transpiration by large dense forest cover, to 0.47 for the Rogue River, where the lava formation carries large amounts of precipitation below the root zone and where the vegetation is much less dense and less luxuriant.

127. The Water Cycle.—Already in considering the relation between stream flow, precipitation, and losses, a simplified form of the water cycle has been used. When the losses of water are distinguished as interception, transpiration, evaporation, and deep seepage, the water cycle equation becomes

$$Q = P - (I + T + E + X) \pm S$$

where Q is the total amount of water flowing in the channel, X is the unknown amount of deep seepage that passes out of the drainage basin through the rock strata, and S is the change in soil moisture in storage between the beginning and end of the period. It has been concluded previously that the precipitation P does not vary more than 1 per cent with differences in forest cover, and therefore it can be assumed to be constant. The variations of deep seepage X with the vegetative cover are unknown and probably small. Therefore X can also be assumed to remain constant, and in formations of igneous rocks it will be nearly zero. The change in storage S on the average over a period of years and also for a single year in a region where the soil is dry at the end of the water year will approach zero. Hence the stream flow Q will vary inversely as the sum of interception, transpiration, and evaporation $(I + T + E)$. It has been shown that interception and transpiration vary concurrently and that evaporation varies in the opposite direction with changes in the forest cover. Therefore it is the balance between the change of I and T on the one hand as compared with the opposite change of E on the other that determines how the stream flow will be influenced by the changes in the vegetation. The relations between different conditions of cover can be analyzed by a series of inequalities to derive logical relations to stream flow. Take, for example, the comparison between forest and bare land, letting the subscript f represent forest and b represent bare land. It is evident that

$$I_f > I_b = 0$$

$$T_f > T_b = 0$$

Similarly,

$$E_f < E_b$$

Therefore, if

$$I_f + T_f > E_b - E_f$$

$$Q_f < Q_b$$

Even in the absence of figures to substitute for the symbols, there can be little question about the conclusion that the stream flow from a forested area during the water year will be less than that from an otherwise similar area of bare land.

A similar analysis may be made for the comparison between two well-stocked stands, one at the age of maximum annual increment, assumed to be 50 years, and the other 15 years old just after the canopy has closed. For that comparison,

$$I_{50} > I_{15}$$

and

$$T_{50} > T_{15}$$

Very nearly

$$E_{50} = E_{15}$$

Then

$$(I_{50} - I_{15}) + (T_{50} - T_{15}) > E_{50} - E_{15}$$

Hence

$$Q_{50} < Q_{15}$$

In other words the stream flow from the area of the 50-year stand at the time of its maximum growth is less than that of the 15-year stand.

The relations in the equation of the water cycle as a forest stand develops from the seedling stage to old age are graphically represented in Fig. 20. In the absence of actual data, figures are assumed, but it is not unreasonable to suppose that they may represent rather closely the changes that might occur with increasing age. At zero age, transpiration and interception would obviously be zero. Both would increase slowly at first and most rapidly during the period of most rapid growth to a maximum at the culmination of the annual growth, which is assumed at 40 years. Thereafter the trends would be slightly downward, becoming more definitely so with the mortality and opening of the stand in old age. Transpiration is likely to be higher than interception in a forest stand. Evaporation would start at a maximum at zero age and would decrease to a minimum at the age of maximum development of canopy at 40 years, when it would be less than inter-

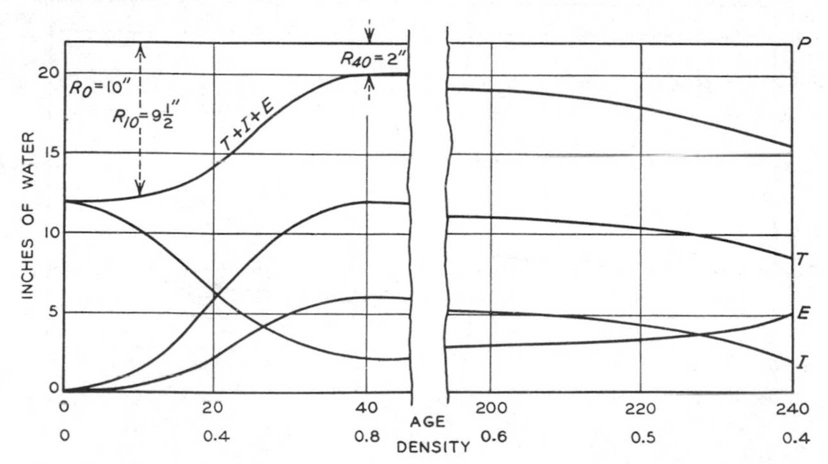

FIG. 20.—Transpiration, interception, evaporation, and runoff in relation to age and density.

ception. Thereafter there would be a gradual increase somewhat accelerated in old age.

The values for the three forms of water loss are summed at different ages and plotted as a summation curve for transpiration, interception, and evaporation. Because transpiration and interception after 20 years are decidedly greater than evaporation, the trend of the summation curve starting from the value of evaporation at zero age follows somewhat the trend of the transpiration and interception curves; reaching a peak at the 40-year age. An assumed value of 22 in. for average precipitation has also been plotted as a straight line. If now the values at different points on the summation curve for transpiration, interception, and evaporation are read and deducted from the value for precipitation, the difference will represent the stream flow at the different ages. These values in the figure are the vertical distances at different ages between the trend for $T + I + E$ and that for P. At 10 years the stream flow would be almost 10 in. and at 40 years it would be only 2 in.

Similar analyses may be made for variations in density or site quality or stages in natural succession, with considerable assurance that they represent qualitatively the relations that would be found. In the case of site quality it is certain that transpiration and interception will increase, while evaporation changes but little, as the site quality is

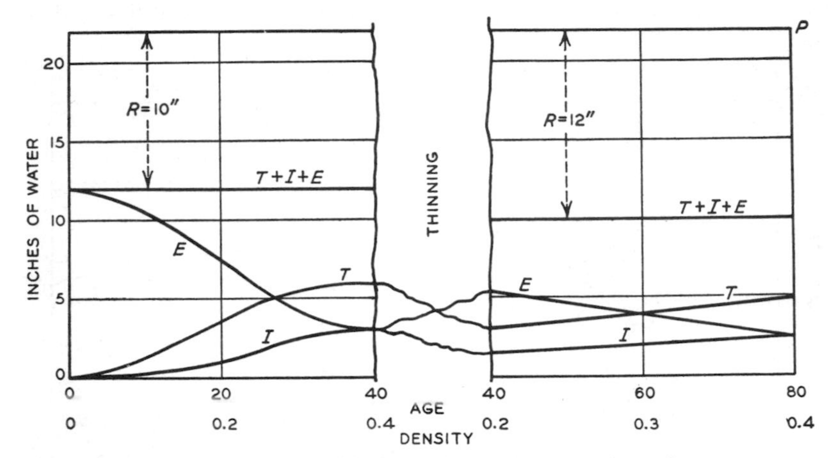

FIG. 21.—Effect of thinning on losses of water and runoff.

better. Thus the sum of $T + I + E$ will increase on the better site, more water will be lost, and less will remain for stream flow. Similarly, as natural succession proceeds toward the climax, the vegetation becomes more stratified and more dense, the transpiration and interception will increase, and evaporation may decrease slightly. The sum of the three factors will increase, and therefore the losses of water become greater in the later stages, and the stream flow is correspondingly reduced.

Another graphical analysis has been made in Fig. 21 of the changes that might result from thinning a stand from a crown closure of 0.4 to one of 0.2 at about the age of the culmination of growth. Trends are assumed for the factors of loss, such that the sum of the losses remains constant up to the time of the thinning. By the thinning the transpiration and interception might be approximately halved and the evaporation something less than doubled, because the forest floor would still protect the soil from evaporation after the thinning as before. After thinning the stand might regain its former crown closure of 0.4 at an age of 80 years, and, during that period, transpiration and interception would again increase and evaporation decrease. Immediately after the thinning the summation of the three losses would be less than before, and the trends have been drawn so that the summation remains constant during the 40 years at a value of 10 in. as compared with 12 in. before the thinning. The stream flow following the thinning would consequently be 2 in. more than before. There are no actual data to prove that this increased yield of water could be obtained by thinning or partial cutting, but neither is there any proof that such a result cannot be obtained. The relations seem plausible and suggest an interesting possibility of silvicultural treatment for the purpose of increasing water yield.

Obviously the trends of the curves and the relations between transpiration, interception, and evaporation will change not only with differences in the vegetation but also with regional differences in climate. The amount of the annual precipitation becomes limiting when it is not more than enough to equal the losses. The distribution of the precipitation by seasons, by size of storm, and by intervals between storms is also important. Thus on the Pacific slope in California where the precipitation is concentrated in frequent heavy storms during the winter, transpiration might be 8 in., interception 4, and evaporation 3 in. annually. But the same type of forest with the same annual precipitation in Arizona, where the storms are smaller and less frequent but more widely distributed through the year, the transpiration might not exceed 5 in., interception 4, and evaporation 8 in. The curves in Fig. 20 were drawn as if transpiration and interception changed with age as do growth and foliage. In a region where rainfall is deficient during part or all of the growing season, however, it might be the root development rather than that of the crowns which limited transpiration. For individual trees the root spread is greater than the crown spread. Therefore, root closure probably comes earlier than crown closure in a stand. Thus the curve of transpiration might rise earlier than shown and reach a lower culmination. Admittedly in the present state of knowledge, these are speculations only justifiable as suggestions for study.

128. Numerical Examples of the Water Cycle.—Values for the different factors in the equation of the water cycle have been derived for certain drainage areas, but in no case has there been a rigorous determination of the separate values for transpiration, evaporation, and deep seepage. The following figures from various sources illustrate the influences of changes in the vegetation on the factors in the water cycle and on the resulting stream flow. All values are in inches depth of water.

Equation of the water cycle	Q	$= P$	$-(I$	$+ T$	$+ E$	$+ X)$	$\pm S$
Swiss stream-flow experiment (118):							
Forested	37.1	62.6	9.1	11.8	4.6	0	+ 0
Two-thirds grass	40.4	65.1	7.7	5.3	11.7	0	+ 0
Wagon Wheel Gap experiment, Colo. (37):							
Forested	6.2	21.2	3	5	7	0	+ 0
Deforested	7.3	20.8	0.5	4	9	0	+ 0
Coweeta Experimental Forest, N. C. (158):							
Drainage basin 17:							
Before cutting:							
1937–1938	19.3	63.3	8.1	20.3	15.2	0	− 0.4
1938–1939	42.0	88.2	7.8	17.4	18.9	0	− 2.1
1939–1940	15.1	51.0	5.9	21.5	12.4	0	+ 3.9
1940–1941	21.0	62.2	6.5	19.2	15.6	0	+ 0.1
After cutting:							
1941–1942, sprouts cut	36.6	62.4	0	0	21.9	0	− 3.9
1942–1943, some sprout growth	46.9	78.4	1	6.0	24.7	0	+ 0.2

In all three examples the yield of water was greater when the forest cover was less. In the high rainfall area of North Carolina the removal of the forest increased the annual runoff by more than 15 in. whereas above 9,000 ft in Colorado the average increase was only a little over 1 in. The marked variation of runoff with precipiation, illustrated by the figures of the 4 years before cutting at Coweeta, is the corollary of the fact, previously mentioned, that the sum of the losses is nearly independent of or increases only slightly with precipitation where rainfall is ample. The slight correlation is chiefly in the factor of evaporation from the soil, which does increase with precipitation and may even exceed transpiration. Transpiration is not only independent of but varies inversely with precipitation. After complete cutting the evaporation from soil and slash, all left after felling and lopping, increased about 6 in.

Over a period of time in a drainage basin underlain by granite or other impervious rock, the sum of the losses by interception, transpiration, and evaporation will equal the difference between the precipitation falling upon the drainage basin and the stream flow from the basin. On the Stanislaus River drainage in California, for example, the stream flow averages 26 in. annually and the precipitation is about 43 in. leaving 17 in. of precipitation that does not appear as stream flow. Under existing conditions this 17 in. may be estimated to include 6 in. intercepted, 8 in. transpired by the forest, and 3 in. evaporated from the soil.

If the value of the water from that river were sufficiently high and the need for more water were great enough, would it be possible to modify the forest to increase the stream flow? Unquestionably, by removing some of the trees, particularly the larger ones, and by thinning the denser stands, the interception could be reduced from 6 to 4 in. and the transpiration from 8 to 6 in. On the other hand, by thus opening the forest and exposing more of the soil surface to the sun and wind the evaporation would be increased, perhaps from 3 to 4 in. Thus the combined loss from interception, transpiration, and

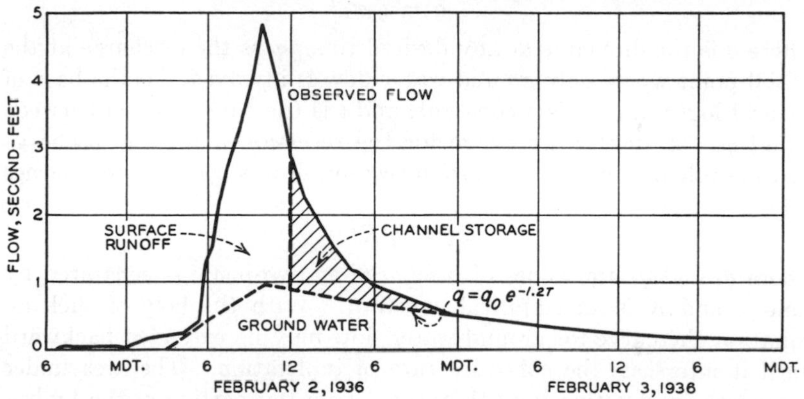

FIG. 22.—**Surface runoff, channel storage, and ground-water flow as constituent parts of a hydrograph.**

evaporation would be reduced by 3 in., and this 3 in. would appear in the stream flow, making it 29 instead of 26 in. annually. That saving of 3 in. of water could probably be made without seriously impairing the other influences of the remaining forest on floods and erosion. If, however, in order to obtain a maximum yield of water the drainage basin were wholly denuded of its forest and other vegetation, it is reasonably certain that damaging floods and erosion would result. In any drainage basin the gain in yield of water that may result from removing part or all of the vegetation must be balanced against the damage that increased floods and erosion may cause as a result of exposure of bare soil.

129. Analysis of Stream Flow.—Total stream flow, at least during high water, is composed of three parts, namely, surface runoff, channel storage, and ground-water flow. The channel storage is of course that water during and following a flood which is in excess of the normal ground-water flow after surface runoff has stopped. An assumed hydrograph of flood flow is indicated in Fig. 22. The total area under the curve represents the total stream flow including all three component elements. The rise of the stream and the early part of the falling stage include the surface runoff. The termination of surface runoff, according to Horton (171), comes at the point on the time scale where the descending limb of the hydrograph changes from convex to concave upward. The portion of the hydrograph below the curve at the extreme right where the rate of decrease is very slow and where the trends after different storms coincide represents ground-water flow. Curves of this type may be represented by an equation of the form

$$q = q_0 e^{-ct}.$$

where q is the discharge at any desired time, q_0 is the discharge at the initial point where only ground-water flow is involved, e is the base of natural logarithms, c is a constant, and t is the time interval between q and q_0. A similar expression for the recession limb of a discharge hydrograph of surface and ground-water flow suggested by Barnes (31) is

$$q_2 = q_1 K^t$$

where q_1 and q_2 are values of discharge at two instants separated by time t, and K is an empirical constant. With the help of such an equation the curve for ground-water flow may be extended backward until it intersects the curve for rate of infiltration. The area under these two intersecting lines then represents the portion of the hydrograph assignable to ground-water flow. The portion above these two lines and to the left of the point of inflection on the descending limb of the hydrograph represents surface runoff, and the remaining wedge, crosshatched in the figure, is the channel storage. In this way surface runoff, which is often the matter of chief interest, may be determined from a flood hydrograph including also the other types of flow. Other methods of segregating these components of flow have been suggested, but they involve larger elements of personal judgment as to where the lines of demarcation should come.

Engineering study of flow in open channels has led to a useful group of functional relations between the factors in stream flow, which are already, and may become more, useful in evaluating the influences of vegetation. First, the discharge of a stream may be represented by the equation

$$Q = av$$

where Q is the discharge in second-feet, a is the cross-sectional area of the water in square feet, and v is the mean velocity in feet per second.

Extension of this formula is sometimes useful in deriving relations for different points along the same channel. In this case,

$$Q = av = a_1v_1, \text{ etc.}$$

The cross-sectional area at any point may be obtained by a survey with levels and soundings at a sufficient number of points across the stream. The mean velocity v is about nine-tenths of the surface velocity, and when a current meter is available for measuring velocity below the surface the mean velocity is ordinarily found at about six-tenths the depth; that is, if the stream is 10 ft deep, the mean velocity will be at 6 ft below the surface.

If the velocity is not known, as after the passage of a flood crest when no record was obtained, or for use in predicting floods, it can be computed by Manning's equation

$$v = \frac{1.49\, r^{2/3} s^{1/2}}{n}$$

where r is the hydraulic radius, s is the slope in feet per foot, and n is a roughness coefficient. The hydraulic radius is defined by the equation,

$$r = \frac{a}{p}$$

where p is the wetted perimeter, defined as the perimeter of the wetted channel excluding the water surface. In natural channels the wetted perimeter is approximately equal to the width plus the mean depth. The $2/3$ power of the hydraulic radius is based on the assumption of turbulent flow. The $1/2$ power of the slope is derived from the law of falling bodies

$$v = \sqrt{2\,gs}$$

In this case the s is the vertical distance through which a body falls, and thus corresponds to the vertical component of the slope in Manning's equation. Values of the roughness coefficient n have been compiled for many types of natural stream channels and are available in most reference works on hydraulics (23, 28, 251, 354). The values for use in Manning's equation are the same as those used earlier in the Kutter formula. The roughness coefficient n according to Strickler is a function of the average diameter D of the particles forming the bed and may be computed as

$$n = \frac{D^{1/6}}{21.3}$$

within the limits that $4 < r/D < 2{,}000$, where r is the hydraulic radius expressed in the same units as D. The numerical constant has also been given a value of 24 in some cases. When velocities and discharge are computed after the recession of a flood, the cross-sectional area a should be that at the time of the flood and not after the channel has been scoured out by the receding water.

For surface runoff on a strip of unit width ($w = 1$), because the depth d is very small,

$$p = 1$$

and

$$a = dw = d$$

Therefore

$$r = \frac{a}{p} = d = \delta$$

The hydraulic radius, δ, being the depth of surface detention as in the earlier analysis of surface runoff, is also approximately equal to the depth in wide rivers.

Since

$$Q = av$$

by substitution in Manning's formula

$$Q = \frac{1.49\, ar^{2/3}s^{1/2}}{n}$$

and for surface runoff,

$$Q_s = \frac{1.49\, \delta^{5/3}s^{1/2}}{n}$$

The reasons for some of the foregoing relations may be seen from a consideration of the following factors on which the velocity depends.

1. Slope of the bed; steepest and velocity highest in the upper reaches.
2. Volume of flow; greatest velocity during high water.
3. The amount of load; addition of load reduces velocity.
4. Straightness of channel; velocity lower in crooked channels.
5. Smoothness of channel; velocity lower in rough channels.
6. Depth of channel; velocity is lower at the top and bottom where there is friction with the air and bed.
7. Width of channel; velocity is lower at the margins where the shallower water is more affected by friction.

130. The Rational runoff Formula.—Another approach to the prediction of stream discharges for small drainage areas of less than 1,000 acres has been made by the so-called "rational runoff formula"

$$Q = CiA$$

where C is a runoff coefficient, the ratio of the rate of runoff to the rate of rainfall; i is the rate of rainfall in inches per hour, and also very nearly in cubic feet per second per acre; and A is the area of the drainage basin in acres. Approximate values for C have been suggested for a variety of conditions. For example, on the western range for brush and timber, C equals 0.4; for areas of sod, 0.6; and for bare or cultivated land, 0.9 (350). Ayres (23) gives the following:

Type of land	Slope, %	C
Cultivated	5–10	0.60
Cultivated	10–30	0.72
Pasture	5–10	0.36
Pasture	10–30	0.42
Timber	5–10	0.18
Timber	10–30	0.21

The coefficient also varies with the area, and it has been proposed that for areas of 2 to 5 square miles values between 0.2 and 0.3 should be used; from 6 to 50 square miles, from 0.1 to 0.2; and from 50 to 200 square miles, from 0.05 to 0.1. For any given drainage area the runoff coefficient for individual storms will tend to increase with the intensity of the rainfall. Thus the values will start from 0 at the point where the intensity just equals the infiltration capacity, rise steeply at first and then more slowly, approaching a constant as the soil mantle becomes saturated at the higher intensities.

The rate of rainfall i is the equivalent rate per hour of the maximum fall to be expected in an interval equal to the time of concentration. This rate may be obtained from rainfall intensity maps (23, 350). The time of concentration is the interval of time required for water to flow from the head to the outlet of a drainage basin by the longest channel. It may be approximated by determining from a topographic map the length of the longest channel and the slope so that velocity may be computed by Manning's formula. Refinements and aids in the use of the rational method have been suggested by Bernard (354).

131. Unit Hydrograph and Pluviagraph.—Another approach to the analysis of runoff and its relation to precipitation and other factors has been developed under the terms "unit hydrograph," "distribution graph," and "pluviagraph" (52, 176). A unit hydrograph is a hydrograph of surface runoff resulting from a storm occurring in a single definite interval of time, such as an hour or a day. For the same drainage area, unit hydrographs tend to coincide as to width of base. The ordinates of unit hydrographs when converted to percentages of the total precipitation for the same unit storm also tend to coincide irrespective of the amount of rain. When several such graphs for a given drainage basin have been plotted on the same set of coordinates by days or hours or shorter units, a single best-fitting trend line can be drawn to generalize the individual unit storms and can be adjusted so that the sum of the percentages for the whole graph is 100. Such a graph is known as a "distribution graph."

The pluviagraph is derived from the distribution graph by applying the percentages from the latter to each unit of an actual storm and then summing for all units. The depths of rainfall can then be converted to second-feet for each unit of time and compared with actual records of stream flow for the drainage basin. The trend of the pluviagraph should closely parallel the actual hydrograph at a nearly constant percentage of it. The percentage of hydrograph to pluviagraph may thus be computed and represents the ratio of surface runoff to precipitation and thus affords a means of obtaining the runoff coefficient previously mentioned in connection with the rational runoff formula.

Inasmuch as surface runoff is the complement of infiltration, and the rate of infiltration is reduced as the duration and amount of rainfall increases, the surface runoff represented by the hydrograph of actual stream flow would be expected to approach the trend of the pluviagraph in the heavier and more prolonged storms. This has been shown to be the case, and this lack of parallelism in the trends, although small, should be taken into account in deriving the percentage relation. The method offers the possibility of predicting the surface runoff from rainfall of any amount and intensity. It can be used in this way also to predict flood crests either from rainfall records or for an assumed maximum storm greater than any in the record to date. For this purpose obviously the percentage relation of the hydrograph to the pluviagraph should be based upon a comparison of the two trend lines toward the end of the storm when infiltration has reached a minimum and surface runoff a maximum.

132. Influence of Vegetation on Flow of Water.—Differences in the condition of the vegetation on a drainage area may be strongly reflected in the form of the distribution graphs. Excellent examples are provided by Brater (52) from small drainage basins in the southern Appalachian Mountains, where the distribution graphs from a variety of watersheds, using a 2½-minute interval as a basis for the percentages, were compared. At one extreme in a small, completely denuded watershed in the Copper Basin in Tennessee, the distribution graph had a base with a width of about 20 minutes and a peak of 50 per cent. Not far away a well-forested drainage basin in the Coweeta Experimental Forest had a base more than 100 minutes wide and a peak of only 7 per cent. A drainage area representing chiefly abandoned pasture was intermediate with a base about 60 minutes wide and a peak at 26 per cent. The forms of the distribution graphs are also affected by large differences in the areas of the drainage basins, so

TABLE 57.—ROUGHNESS COEFFICIENTS AND SAFE VELOCITIES FOR SMALL CHANNELS LINED WITH VEGETATION

Channel lining	n	Probable safe velocity, ft/sec	Slope, %
Bare	0.02	1	6
Centipede grass	0.04	8	10
Lespedeza sericea	0.05	3	6
Bermuda grass, 15 in.	0.045	6.5	20
Bermuda grass, 15 in.	0.05	8	10
Bermuda grass, 15 in.	0.09	...	3
Bermuda grass, 15 in.	0.11	...	0.25
Bermuda grass, 4 in.	0.06	...	0.25
Kikuyu grass	0.07	...	18
Kudzu, February	0.08	2.5	3
Kudzu, August	0.09	4	3

that, for comparisons of the influence of different conditions of vegetation on different basins, it is important that the areas be not too different.

The effect of vegetation in channels is chiefly reflected in the values of the roughness coefficient n. In recent work on erosion control, test flows in small channels lined with different species of vegetation gave results indicated in Table 57 (96, 292).

Several useful comparisons may be derived from Table 57. First, the larger and coarser the vegetation, the larger is the roughness coefficient. The Kudzu is an outstanding example of a large coarse-leaved species. Any vegetation increases the roughness coefficient compared with bare soil. Clipped and probably also grazed grass give a lower coefficient than unclipped or ungrazed grass of the same species on the same slope. The roughness coefficient for the same vegetation tends to increase as the slope decreases. The safe velocity, meaning the velocity below which no scouring of the channel bed will occur, tends to increase with the roughness coefficient when the slope is constant. These values of n for small waterways and probably also for surface runoff are not the same as the published values applicable to larger stream channels by the use of Manning's formula. For these small channels, n varies inversely with the hydraulic radius and depth of flow, increasing rapidly in the range of hydraulic radius below 0.4 ft.

Median values of n for the range lands of the Concho River basin in Texas were 0.06 for the sites with poor cover of density less than 2.5 per cent and 0.45 for sites with good cover of over 5 per cent density. The difference of 0.39 between the two medians was highly significant.

In larger stream channels the values of Manning's n also increase with the size and density of the vegetation. Thus in small sluggish streams in Illinois the following comparisons of values of the roughness coefficient with and without willows and other riparian growth on the banks are reported (286).

Condition of Bank Slopes	n
Cleared	0.029
Low vegetation	0.036
Not cleared	0.050
Dense growth	0.10

In another instance in Missouri the roughness coefficient in a ditch after the development of a thick heavy growth of willows was 0.073 and the velocity was 0.73 ft per second, whereas before the willows grew up, the velocity was 1.29 and n was 0.031.

133. Forests and Stream Flow.—Stream flow includes both surface runoff and the water that infiltrates and contributes to subsurface flow. Surface runoff is the way in which water reaches the stream channels most rapidly, as compared with ground-water flow. Rapid concentration in the stream channels is one cause of floods. Deforestation and denudation increase surface runoff and hence may be expected to increase the height of floods. It is also true that deforestation and denudation decrease interception and transpiration more than evaporation is increased and especially during the growing season. Thus the stream flow will be augmented whether at flood or at minimum stages. The increased runoff will further increase the height of flood crests. In the case of minimal flows the low-water discharge may also be raised.

The chief source of water for stream flow during dry seasons is the ground-water flow resulting from infiltration. Denudation reduces infiltration as it increases surface runoff, and therefore denudation will tend to reduce the minimal flows. Thus, there is a conflict in the effects of removal of the vegetation on the minimal flow of streams.

If the transpiration is large, as in early summer, the net effect by deforestation is likely to be an increase in the minimal flow. If the transpiration is small, as in late fall or winter, the net effect may be a reduction in the minimal flow. Unquestionably it is this conflict and the fact that observations made in different places and at different times have sometimes represented the one and sometimes the other condition that has led to some of the controversy over the effects of forests on stream flow. Attempts at generalization for all seasons and conditions are obviously unsafe and under some circumstances will certainly be contrary to the facts.

134. Runoff and Stream Flow: Summary.—1. Runoff or stream flow is the residual difference between precipitation and losses due to interception, transpiration, evaporation, and deep seepage. $Q = P - (I + T + E + X) \pm S$. Since P and X vary only to a negligible extent with changes in cover, runoff varies inversely as the sum of interception, transpiration, and evaporation. Hence maximum runoff should be obtained under a cover that intercepts and transpires a minimum amount of water and at the same time reduces evaporation from snow or soil.

2. Surface runoff is reduced by a layer of forest floor to nearly the same extent that infiltration is increased. The surface runoff is maintained at a minimum by a complete cover of vegetation with the accompanying floor, under which soil permeability is at a maximum. Channels formed by roots and soil fauna contribute to this influence.

3. Surface runoff from burned or otherwise bare soil, due to the sealing of the soil surface by a clayey film, exceeds that from litter-covered soil from two to seventy fold, or sometimes the equivalent of 3 in. per hr of rainfall. The excess is greater for clay loam than for coarser-textured soils and increases with rainfall intensity to a maximum with intensities of over 1 in. per hr. The excess is greatest on the steepest slopes.

4. Grass or brush or the vegetation following cutting of the forest may be as effective as uncut forest in minimizing surface runoff, provided the A_0 and A_1 horizons of the soil are well developed and essentially undisturbed.

5. Surface runoff from areas of well-established and undisturbed vegetation, if any, occurs only in heavy rains and is less than 3 per cent of the precipitation, whereas it starts after lighter rains and may be over 60 per cent from denuded and cultivated lands.

6. Insofar as vegetation reduces surface runoff and promotes infiltration, and also by offering mechanical obstructions, it retards the move-

ment of water to the drainage channels and thus tends to retard and lower flood crests and prolong increased flow in low-water periods.

7. Comparisons of stream flow before and after deforestation or denudation by fire, and between forested and deforested watersheds, indicate the following effects of deforestation on stream flow:

a. The annual flow is increased and to the greatest extent the second or third year after deforestation, assuming the vegetation is allowed to reestablish itself. Most of the increase occurs during the rising stages of floods.

b. The stream flow during dry summer or winter periods may be increased or decreased and the date of the summer minimum flow may be later as a result of deforestation.

c. During snow melting in spring or in the period of reduced evaporation and transpiration in the fall, discharge from forested areas averages higher than from grassland or deforested areas.

CHAPTER XXI

FLOODS

135. Definitions.—Floods are important and easily recognized in their extreme forms, but the definition of a flood, involving a definite limit between what is and what is not to be considered flood flow, has been subject to a variety of interpretations. Recently, flood has been defined by the National Resources Committee as "any relatively high stream flow overtopping the natural or artificial banks in any reach of a stream." The United States Geological Survey (186) gives two definitions, the first essentially like the forcgoing, namely, a flood is a "relatively high flow as measured by either gage height or discharge quantity." The second defines a flood as "any flow equal to or greater than a designated basic flow. Ordinarily, the item of record is the average discharge for a 24-hour period." In the latter sense the technical flood has been used by Bates and Henry (37) as the flow from the initial date, when stream discharge first exceeded a certain arbitrary volume, to the last day in which it had another arbitrary amount of discharge. Thus the periods of flood flow varied in different years. They also used the concept of an arbitrary flood that was defined as the discharge between two dates, as for example Mar. 1 and July 10, which included the average period of spring high water. Others have defined flood as an arbitrary high-water stage such that there were from two to six floods and from 4 to 23 days of flood flow above these stages in a year. The flood period was defined by Hoyt and Troxell (177) as that period when the daily runoff exceeded about one-half the average maximum daily runoff.

The following series of definitions have been proposed by Hazen (186). The maximum 1-day flood is the highest average rate through any consecutive 24 hours that can be found in the record period. The annual 1-day flood is the highest corresponding rate in any 1 day in the year. The average 1-day flood is the average of the annual 1-day floods, one for each year in the record period. The same definitions may be applied to any other period of time, such as an hour instead of a day. Hazen considered that the mean or average 1-day flood, when the day is used as the unit of time, is the best standard of comparison, and the best available index of the flood-producing capacity of different drainage basins. However, it requires a long period of record for its most satisfactory use. A period of more than 10 years is desirable, although a 10-year record can be used if nothing longer is available.

136. Estimation.—Prediction of floods and flood damage involves both magnitudes and frequencies of occurrence of flood-producing storms. For this purpose, several different criteria from storm records have been used. One is the total storm rainfall, which is related to saturation of the soil mantle and to flood runoff. A second is the maximum day or 24-hour rainfall, which is a measure of intensity over a period sufficiently long to be related to maximum high-water flow. Third, the maximum 1-hour intensity of rainfall may be closely related to peak flows in small drainage areas subject to short heavy downpours of the thunderstorm type. For any given storm, these three criteria may correspond to very different frequencies. For example, in the storm of March, 1938, in Los Angeles County, the 1-hour intensity gave a frequency of 2 years; the maximum 24-hour intensity, 21 years; and the total storm rainfall, 76 years. One-hour, 24-hour, and other rainfalls to be expected in different periods of years have been compiled by Yarnell in a series of maps (379).

The necessity of estimating maximum floods in the design of dams has stimulated proposals of different methods of forecasting by many workers. Several of them have been classified, analyzed, and illustrated by Jarvis and others (186). In general, flood flows tend to increase as the area of the drainage basin increases. This is the basis for several formulas of the general form

$$Q = CA^n$$

where Q is the flood discharge in second-feet, A is the area of the basin, and C and n are constants. The values of n are usually between 0.5 and 0.8. C is a coefficient representing the conditions of climate, topography, soil, and vegetation and is therefore subject to wide variation and difficulty of estimation. In some formulas the area factor is treated more specifically by separating it into its components of width and length.

By incorporating the percentage relation of any given flood to an assumed maximum flood as 100 per cent, the Myers rating was developed from the formula (186)

$$Q = 100\% \, A^{1/2}$$

in which % is the percentage of the assumed maximum. In this case the % reflects the influence of vegetation along with other conditions.

Transposing,

$$\% = \frac{Q}{100A^{1/2}}$$

If the flood discharge is expressed in second-feet per square mile q, then since

$$Q = qA$$

$$\% = \frac{qA^{1/2}}{100}$$

This percentage may be computed for any recorded flood and may be compared with the percentage ratings for other floods in the same or other basins. For areas of less than 4 square miles, if the formula is used the first power rather than the square root of the area should be substituted. These Myers ratings for maximum flood have been compiled on a map by Jarvis (186). Sometimes, as in Texas, the ratings may exceed 100 per cent. The forested parts of the United States rarely show ratings in excess of 50 per cent and are usually under 30 per cent.

Several methods of estimating maximum floods use graphic analyses of existing flood records with extrapolations for the maxima. Either annual 1-day floods, or peak flows, or arbitrary floods may be used. For any given stream, when the flood flows, however defined, are arranged in descending order of magnitude and plotted on logarithmic probability paper over a percentage scale representing the proportion of time or of flood events in which given limits of flow are equaled or exceeded, the trend tends to be linear, at least toward the upper end. Thus this straight line may be extrapolated to provide an estimate of greater floods of less frequent or likely occurrence than any in the record.

For example, from a daily record of stream flow covering 100 years, the maximum 1-day floods, one in each year, may be arranged in descending order of magnitude and plotted on the logarithmic scale of probability paper of which the time scale is graduated so that a cumulative frequency distribution in percentages becomes a straight line. The sample of 100 annual 1-day floods, lacking other 100-year records for the same stream, may be considered a random sample of a much larger series. Actually, each annual 1-day flood represents a single sample in a class interval of one year or 1 per cent of the 100-year total. Thus the largest flood of record occurred in only 1 per cent of the time or once in 100 years. A flood as large as or larger than the second in the series occurred in only 2 per cent of the time or once in 50 years. Similar statements can be made for each flood magnitude in the series and for momentary peak flows as for 1-day floods. However, because it is the floods which will be equaled or exceeded in any percentage of time, the values are plotted at the mid-points of the class intervals. These mid-points may be computed as

$$\% = \frac{(m - 0.5)100}{n}$$

or

$$\% = \frac{(2m - 1)100}{2n}$$

where $\%$ is the percentage of time, m is the rank of the item in the series, and n is the total number of years in the series.

A similar graphic method using logarithmic probability paper may be used by selecting from the record the 1-day or momentary peak floods above some arbitrary limit whenever they occurred without regard to the time interval. Thus there may be two or more notable floods in one year and none in others. The cumulative percentage scale in plotting, therefore, represents proportions of flood events equal to or exceeding certain magnitudes rather than percentage of time. The interpretation accordingly becomes a statement that a flood of given magnitude will be equaled or exceeded only once in 100 flood events or another flood, 10 times in 100 corresponding to a 10 per cent chance flood. For plotting such curves only a certain number, perhaps 100, of the largest floods need be used. The trend at the upper end on logarithmic probability paper is likely to be nearly linear and therefore suitable for limited extrapolation for purposes of estimation of maximum floods. If the floods selected in this method are all those above the designated limit within a given number of years of record, then the percentages also apply to the proportion of flood events in that period of time and can be converted to a frequency of occurrence per 100 years. In this, as in the former method, the points should be plotted at the mid-points of the class intervals on the cumulative percentage scale.

The "rational" runoff formula of Chap. XX, $Q = CiA$, may also be used to estimate floods by substituting the estimated maximum value of rainfall intensity for i. However, the reliability of applying the formula to areas of more than 1,000 acres is doubtful. For use on larger areas and in making allowance for shape or slope, the following methods are preferable.

Obviously, the shape of a drainage basin will affect the degree of synchronization of flood flows from the tributaries and consequently the magnitudes of floods in the main stream. In streams of the same length, the width of the basin may be used as an index of shape. On this basis, a shape factor has been proposed as the ratio of the square root of the area A in square miles to the distance l_c in miles from the center of gravity of the basin to its outlet. A formula for maximum discharge in second-feet per square mile using this shape factor becomes

$$Q = \frac{CiA^{1/2}}{l_c}$$

where C is a flood coefficient derived from recorded floods, and i is the average of the precipitation in inches per hour during the time of concentration and during the hour of maximum intensity. For the San Gabriel Mountains of southern California the flood coefficient C was about 235.[1]

In the preceding and in the "rational runoff formula" the time of concentration used in deriving the intensity of precipitation only partially represents the slope of the basin, another factor affecting floods. The slope-factor formula, also known as the Burkli-Ziegler formula, attempts to allow for the slope by inserting a factor $(s/A)^{1/4}$ to give the equation

$$Q = CiA\left(\frac{s}{A}\right)^{1/4}$$

where Q is the maximum rate of discharge in second-feet, C has values from 0.2 to 0.75 depending on the imperviousness of the area, i is the average precipitation in inches per hour during the time of concentration and during the hour of maximum precipitation, A is the area in acres, and s is the average slope of the area in feet per 1,000.

The average slope of an area may be obtained from a contour map as follows: First, determine the areas A by planimeter of altitudinal zones delimited by contours at convenient intervals, often 100 or 200 ft. Second, measure the lengths l_1 and l_2 of the delimiting contours. Then the mean horizontal distance w between contours is

$$w = \frac{A}{\frac{1}{2}(l_1 + l_2)}$$

The slope s between two contours bounding any altitudinal zone is the contour interval or vertical distance h divided by the horizontal w, or

$$s = \frac{h}{w}$$

[1] BELL, J. A., and others, "Report on maximum flood discharges in the San Gabriel River watershed," to Los Angeles County Flood Control District.

Converting slope to per cent and horizontal distance from miles to feet, this becomes

$$s = \frac{100h}{5{,}280\ w} = \frac{h}{52.8w}$$

Finally, the slopes for the different altitudinal zones may be averaged to give the mean slope of the drainage area.

The pluviagraph has been suggested in Chap. XX as a means of estimating floods, through the use of the flood coefficient or the ratio of the maximum ordinate of the flood hydrograph to that of the pluviagraph. For the estimation of a maximum flood, a further refinement may be made by superposing the isohyetal pattern of the maximum regional storm upon a map of the drainage basin in such a way as to give the maximum rainfall on the basin. The pluviagraph for the transposed storm is then computed, and the greatest flood coefficient is applied to the ordinates of this pluviagraph to obtain the hydrograph of flood flow for the maximum regional storm in a location where it would produce the maximum rainfall.

These and other methods of estimating maximum floods have been analyzed and illustrated in Water Supply Paper 771 (186). None of them makes any specific evaluation of the influence of vegetation or provides for predicting the effect of a change in vegetation except in some instances by estimating the change in a coefficient.

A method has been proposed by Rowe (300) in which all the factors in the water cycle are evaluated as a means of predicting flood flows or the changes in peak flows following changes in vegetation or soil. First, hydrologic complexes were delineated and successively subdivided according to isohyetal zones, types of vegetation, soil and geologic formations, and land use. Second, for each complex a composite infiltration-capacity curve was derived. Third, these infiltration-capacity curves were adjusted to measured relations of precipitation and runoff. This step involved the construction of rainfall-intensity-duration histograms; separation of surface runoff, channel storage, and ground-water flow in storm hydrographs; computation of soil and rock water-storage capacities; determination of depression storage; and evaluation of losses by interception, evaporation, and transpiration. Fourth, the relations established in the first three steps were applied to existing or modified conditions on the drainage basin. This method has been used in the analyses of some of the Flood Control Surveys and is preferable to others where changes in flood flows resulting from changes in vegetation such as fires or reforestation are to be evaluated. At the same time in most drainage basins it requires considerable preliminary field work to obtain the hydrologic information necessary for such an analysis.

Another approach similar in some respects was used in the Flood Control Survey of the Connecticut River basin and has been outlined by Morey (261). Vegetation-soil complexes were defined and areas determined. Effects of grazing, fire, and past use were considered. Relations between age of stand and depth of humus were used to estimate depth of humus for whole complexes. Soil-moisture sampling for each horizon of the profile in each of six humus types afforded a basis for the construction of soil-moisture-depletion curves. From these, by adding the depletion of the humus to that of the *B* horizon, soil-moisture-depletion curves were derived for each complex. The depletion curve for the whole drainage basin was then synthesized by prorating the curves for the different complexes according to their areas. The synthesized depletion curve was finally checked with the normal depletion curve of stream flow. This method stresses the importance of the moisture above field capacity detained by the humus on and in the soil in retarding the release of water over a period of several days with a corresponding reduction of flood crest. It is simpler than the preceding method, but the greater simplicity is obtained largely by neglecting differences in infiltration and water storage in the rock formations, which may not always be adequately reflected by data on soil moisture.

Still another method was developed by the Tennessee Valley Authority to determine what reductions in flood crests might be obtained by changes in vegetation and land use (72). The land area of the Upper French Broad basin was classified according to use as forest, pasture, or cultivated. Each class was subdivided into three or four conditions of cover and soil corresponding to differences in average rates of infiltration for large storms. By using available experimental determinations and checking against observed infiltration rates derived from storm hydrographs of surface runoff, adjusted average rates of infiltration, separately for summer and for winter seasons, were assigned to each soil-cover class. Hydrographs of the largest floods with good records were prepared and separated into surface-runoff and ground-water components. The effects of improved land use of two degrees were computed as reductions in surface runoff proportional to the changes in land use. These changes also caused changes in ground-water runoff, and it was assumed that the reductions in surface runoff would be compensated by gains divided equally between ground water and losses. The possible reductions in peak gage heights for different subdrainages in the 1940 floods resulting from improved land use varied from 0.6 to 5.2 ft. The figures for the Ivy River for the storm of late August, 1940, are given as an example in Table 58.

These last three methods, in contrast to those which preceded them, depend upon analysis of factors like infiltration, organic matter, and soil moisture which closely reflect the influences of vegetation or the natural or artificial changes that the vegetation may undergo.

137. Influence of Forest on Floods.—Evidence is accumulating from different sources and in general is consistent in showing that seri-

TABLE 58.—HYDROLOGICAL DATA FOR THE FLOOD OF LATE AUGUST, 1940, ON THE IVY RIVER AND THE CHANGES ANTICIPATED AS RESULTS OF A 20-YEAR INTENSIVE PROGRAM

	Precipitation, in.	Surface runoff, in.	Ground-water flow, in.	Losses, in.	Ave. rate of infiltration in./hr	Peak flow, sec-ft	Peak gage height, ft
Actual........	4.92	= 1.27	+ 0.86	+ 2.79	0.26	8,880	12.67
20-year possible departures from actual..	0	= −0.45	+ 0.22	+ 0.23	+0.09	−2,860	−2.24

ous disturbance or removal of the forest may and usually does have definite effects on the magnitude of flood flows and the resulting damages. However, the influences are expressed in different ways, and consequently it is often difficult to make comparisons.

In a comparison of flood flows from areas of young growth, coniferous, and broad-leaved species in Japan, Hirata (154) obtained conflicting results in the ratios of annual runoff to annual precipitation. However, differences in the total hours of discharge above normal stage for storms of more than 1.18 in. were distinct: for the young wood, 103; for the conifers, 121; and for the broad-leaved forest, 144 hours. The maximum hourly discharge for the precipitation class from 1.97 to 3.94 in. for the broad-leaved forest exceeded that of the coniferous, and in another district the nonforested area exceeded the broad-leaved area. He concluded that the coniferous forest provided a larger retention of precipitation than the broad-leaved forest, which in turn retained more than nonforested land.

Using the ratio of the number of days of high-water stage to the annual precipitation as an index of the regimen of streams, Leighton (219) found in the Allegheny region that the values varied from

0.24 to 1.48, and that they were higher following logging or fires or clearing.

The influence of forest on floods and the factors that may be involved are indicated by Spaeth and Diebold (330) at the Arnot Forest within the headwaters area of the Susquehanna River in New York for the disastrous flood of March, 1936. Precipitation of 6.35 in. was recorded between Mar. 10 and 19, and the peak flood flow occurred on Mar. 18. At open-field stations the snow disappeared several days before the peak, whereas at hardwood-forest stations, 3 in. water equivalent of snow remained on Mar. 18. Only the bare soils of open fields were frozen during the flood period as a result of insufficient (7 in.) snow cover to prevent deep freezing earlier in the winter. The soils under the 17 to 31 in. of snow in the forest did not freeze at any time that winter. The surface runoff from the frozen bare soils was one of the principal sources of the flood flow. One such area yielded 7.9 in. of runoff between Mar. 10 and 19.

In the Wagon Wheel Gap stream-flow experiment in Colorado (37) following deforestation, the beginning of the rise in the deforested basin was 12 days earlier than in the forested. The dates of the crests of the floods in the deforested watershed were 3 days earlier, and the height of the crests averaged 69 per cent higher. The maximum average daily discharge after deforestation was 46 per cent higher. Most of the increase resulting from deforestation occurred in the rising stages of the floods.

The comparison of the records before and after the fire in Fish Creek near Monrovia, California, by Hoyt and Troxell (177) indicated the maximum gain in flow occurred in the second year after the fire and amounted to 3.12 in. or 26 per cent. In the first year after the burn the gain was 2.47 in. or 231 per cent. The latter figure is a good example of the exaggeration in percentage resulting when the denominator of the fraction is small. In this case, the comparison between the 26 per cent in the second year and the 231 per cent increase in the first year after burning is quite misleading. The effect of the denudation is also reflected by a change in the Myers rating for peak flows. The first year after the fire the rating was 8.5 per cent, compared with 1.3 before the burn.

Several recent examples of serious floods have provided comparable figures expressed as the maximum flow in second-feet per square mile of drainage area. In the Appalachian Mountains comparisons of 23 small watersheds with different conditions of cover indicated maximum runoff from the forested drainages of 13 to 56 sec-ft per square mile; from the areas of abandoned fields and partial broomsedge cover, 332 to 900 sec-ft per square mile; from overgrazed pasture 579; and from the completely denuded drainages of the Copper Basin from 1,263 to 1,434 sec-ft per square mile (52, 266). These figures correspond to the percentages for the crests of the distribution graphs.

In southern California in the major floods of 1914 and 1916 the peak runoff in Sawpit Canyon, area 7 square miles, averaged 550 sec-ft per square mile (115). In the flood from the recently burned drainage basins above Montrose, California, in 1934 the peak flow was estimated at from 500 to 1,000 sec-ft per square mile. In the same storm the flows from the adjacent unburned Arroyo Seco and San Dimas basins, where the rain was even heavier, amounted to only 58 and 51 sec-ft per square mile, respectively (210, 266). This is another striking example of the magnitude of the differences in flood flows that may result from denudation by fire.

More recently, in March, 1938, an even greater storm with 4-day precipitation of 20 to 23 in. occurred in the same San Gabriel Mountain area of California (68). That storm, during a preliminary period from Feb. 27 to Mar. 1, produced from 5 to 10 in. of precipitation, of which from 5 to 8 in. was retained in the soil mantle. A maximum precipitation of 15.28 in. or about 50 per cent of the total fell in the 24 hours of Mar. 2, with a maximum 1-hour intensity of 1.8 in. On the 1/40-acre runoff plots in the San Dimas Experimental Forest only 1 to 2 per cent of the precipitation appeared as surface runoff. In drainages less than 20 square miles in area the peak flows were between 1,000 and 2,000 sec-ft per square mile. The maximum on one small drainage of 0.8 square mile was over 3,500 sec-ft per square mile. From the more than 20 square miles of San Dimas Canyon the total runoff was over 50 per cent of the precipitation, and in the San Gabriel River for Mar. 2, from an area of 202 square miles, 64 per cent of the 24-hour rainfall appeared as runoff in the succeeding 24 hours. These records were from areas of well-established chaparral or forest cover that, except locally, had not been burned for several years.

Here then is a striking example of the fact that, if a storm is sufficiently intense and prolonged beyond the period when the soil mantle is saturated, maximum flood flows may result, even when the vegetation is essentially undisturbed. In such storms when the cover is intact, it is not the surface runoff but rather the wet-weather seepage or subsurface storm flow that causes the floods. This storm, in comparison with the Montrose flood 4 years earlier from the same mountains, provides a striking contrast. In the one case a major flood occurred in well-vegetated drainage basins, whereas in the earlier storm the serious floods issued only from the burned areas. The comparison indicates the danger of attempting to generalize as to the results of deforestation on floods. Unquestionably, it is the opposition of well-authenticated examples of these two different conditions, without discriminating between them, which has resulted in the controversies in past years as to the effect of forests on floods.

138. Indexes of Water Regime.—Because denudation usually causes higher floods and may also result in lower minimal flows, the ratio of maximum to minimum flow in any stream should be larger for disturbed or denuded areas. This ratio of maximum to minimum discharge has been used as a measure of the failure of a drainage area to exercise its full storage function, and consequently also the need for increasing the protective influence of vegetation. In the Wagon Wheel Gap experiment, before denudation the ratio was 12:1, and after, 17:1. The critical ratio representing the line between satisfactory and disturbed conditions was suggested as about 25:1. For large rivers, Ashe (17) showed that in the Northeast, the Lake states, the northern Rocky Mountains, and the Pacific Northwest the ratios tend to be narrow, ranging from 25:1 to 205:1. On the other hand, in the Great Plains, Texas, and the Southwest the ratios are much wider, running up to 2,820:1 and in one case even to 18,000:1.

Not only do flood flows frequently discharge 300 to 1,500 times more water than low-water flows, but also the velocity or speed of travel of the water during floods is from twenty-five to seventy-five times greater than during low-water flows. The importance of this large difference in velocity, which is the principal factor in the cutting and carrying power of water, will appear in the chapter on erosion.

Another index of the water regime, "the water loss–infiltration ratio," was suggested by Horton (169). It is the difference between the precipitation P and the total runoff Q divided by the total infiltration F. Since F is nearly equal to the precipitation minus the surface runoff Q_s, this index might also be expressed as

$$\frac{P - Q}{P - Q_s}$$

When expressed as annual totals the ratio for the same vegetation and soil tends to be nearly constant and independent of variations in precipitation. In arid regions of torrential rainfall or where impervious soils are exposed, this index should be high and might approach the value of 1. In localities of coarse-textured permeable soils or in well-forested areas where the surface runoff is negligible, the ratio would be low. At Coshocton, Ohio, the value for cultivated soils was 0.82 and for a near-by woodland drainage was 0.64. As more information becomes available on infiltration this ratio may have considerable value in comparing different drainage areas with respect to the likelihood of floods and the need for protective measures. In this ratio both numerator and denominator increase as the vegetation becomes denser or larger. The ratio serves as an index because the infiltration responds in vegetation differently from the losses but it might not always change in the same direction with treatment of the vegetation. A more sensitive index would be obtained by substituting surface runoff for infiltration in the denominator. Surface runoff decreases while the losses increase as the vegetation increases, and thus the ratio would vary widely with changes in cover and probably also with variations in precipitation. Examples of the two ratios follow.

Drainage basin	$(P - Q)/Q_s$	$(P - Q)/F$
Coshocton, woodland, Ohio	207.0	0.64
Red River, Minn.	51.3	0.99
Neosho River, Kans.	6.9	0.97
Coshocton, cultivated, Ohio	4.0	0.82
Miami River, Ohio	3.3	0.86
Chattahoochee River, Ga.	3.2	0.75
Pomeraug River, Conn.	2.0	0.73

For the five larger drainage basins the rank order is the same for the two ratios. For the two small areas near Coshocton, where the major difference is in the vegetation, it is reversed. Both ratios reflect differences in soil as much or more than in vegetation and thus provide an unequivocal index of the influence of vegetation only where the soil is the same.

If deforestation increases the heights of floods and also the minimal flows, the yield of water after removal of the forest will be higher, and, conversely, reestablishment of good forest cover will reduce the yield. This is usually the case notwithstanding claims to the contrary. The significant question in evaluating the role of the forest involves the usableness of the water. In minimal flows, even from denuded areas, the water is clear and usable. Flood flows from forested areas are usable for most purposes, but those resulting from surface runoff are prolific sources of erosional sediment and debris. This material renders the flood flows from denuded areas unusable. It fills reservoirs, clogs spreading grounds, ruins turbines, and damages property. In comparing forested and deforested areas it is therefore the value of the usable water that must be balanced against the damage by floods and erosion.

In large drainage basins it may be difficult to detect the influence of vegetation or changes in the vegetation on stream flow or floods, because such differences are obscured by variations in the degree of synchronization of the flows from the tributaries. This results partly from the fact that most storms do not contribute equal or simultaneous amounts of precipitation to the different tributaries, so that only a part of them are in flood at any one time, and partly because the concentration of the flood flows in the main stream may be strongly influenced by the shape of the drainage area and the arrangement of the tributaries. To express this characteristic numerically, a shape factor derived by dividing the area by the square of the length has been suggested. Obviously the concentration of flood flows in a river where the tributaries are distributed along the length of the main stream will be quite different from that where they all converge near the same point from a fan-shaped system of tributaries. It is therefore difficult or impossible to draw conclusions as to the influences of vegetation from large river basins.

139. Flood Control.—The word "control," when major floods are concerned, is a misnomer because it is impracticable in large floods to do more than reduce the peaks by small amounts and protect lands that are subject to damage by overflow or cutting of banks. In general, for flood reduction or protection, five groups of methods are recognized:

1. The reduction of flood flows by storage and gradual release.
2. Increasing channel capacities.
3. Providing by-pass channels.
4. Maintenance of gradients and banks of channels by sills, revetments, levees, and other local protective measures.
5. Reducing synchronization of tributary floods by retarding flows in some and accelerating those in others.

In considering the part that vegetation may play in flood control, it will be helpful first to list the factors that affect flood flows as follows:

1. Precipitation, intensity, duration, areal extent, path of storm.
2. Arrangement of tributaries.
3. Slope.
4. Length of overland flow.
5. Channel storage.
6. Rock storage.

7. Soil storage.
8. Depression storage.
9. Surface roughness.
10. Infiltration capacity.
11. Freezing of soil.
12. Melting of snow.
13. Vegetation, interception, evaporation, and transpiration.
14. Accumulations of debris or sediment.

The first six of these factors are not affected by vegetation while the last eight do or may change with it. The factors are already familiar and the influences of vegetation upon them are known. It may be useful to notice, however, that interception in this list is included with vegetation, number 13, although it is sometimes considered a part of depression storage, number 8. Also, soil storage involves depth, texture, and organic content, of which only the latter changes with the vegetation. By way of summary, in the extreme case, as bare soil is covered by vegetation increasing in density, age, and size, the soil storage, depression storage, surface roughness, infiltration capacity, interception, and transpiration increase, while freezing of soil, snow melting, evaporation, and accumulation of eroded material decrease. With two groups of factors changing in opposite directions as the vegetation changes, it becomes necessary to know the quantitative relations and how they come into play in flood control.

Refer back to the five groups of methods of flood control. It is evident that, although vegetation might affect the synchronization of tributary floods, it is quite unlikely that its effect would be distinguishable. It would not be effective in providing by-pass channels. It will tend to decrease rather than increase channel capacities wherever riparian species encroach upon the channels. The maintenance of banks, as by willow mats, is primarily a form of erosion control and will be considered in a later chapter on that subject. Locally, vegetation may function in this method of flood control. There remain only the methods of storage and retarded release of flood flows, and it is in this connection that more detailed consideration of vegetation is required.

It will be convenient to distinguish between (*a*) floods resulting primarily from surface runoff, (*b*) those from melting snow, and (*c*) those from subsurface flow in and after major storms. In the first case, usually associated with areas of bare soil, reestablishment of the vegetation often may so increase the infiltration capacity, interception, depression storage, and surface roughness and reduce erosion, that effective flood control is obtained without other measures. In the second case, if the snow can be largely shaded by reforestation the rate of melting by itself will be sufficiently gradual and the infiltration high enough that serious floods will not occur. For example, in the Stanislaus River basin in California between 5,000 and 6,000 ft, in the period of rapid melting after Mar. 1, the average rate of drainage of water from the snow is about 0.5 in. in 24 hours. On occasional days it may reach 1.4 or rarely 2.0 in. In the major flood of Mar. 19, 1907, from the whole river basin above 300 ft altitude, the maximum flood runoff in 24 hours was 2.16 in., and the average daily runoff for the month of March was 0.41 in. The snow melt at the rate of 0.5 in. per day with occasional days at 1.4 in. or higher would produce large floods if the melting occurred over the whole drainage at the same time. Actually, however, in March for example the snow is gone on the lower third and has hardly begun to melt on the upper third so that, for the entire basin, the rate is not much more than one-third as great, and certainly not sufficient to cause a major flood. Incidentally, the flood of Mar. 19, 1907, must have been caused largely by rain, because there was 19 in. of rain recorded that March at Sonora, well below the snow line.

The third case, of major floods from subsurface flow resulting from prolonged storms, is the one for which the influence of vegetation requires careful evaluation. Such storms exceed the total storage capacity of the drainage basin, including surface, soil, and rock. If more and larger vegetation is to have any effect, it must be by increasing the storage or by otherwise decreasing the amount of water or rate at which water reaches the channels. Interception would increase with the vegetation and might reduce the rainfall reaching the ground by 5 or 10 per cent. This would probably include the small increase in transpiration. Infiltration capacity would be increased, and that would reduce the flood runoff to the extent that more water might move more slowly below the surface rather than as surface runoff. The same would be true of the reduced freezing of the soil and consequent increased infiltration. Surface roughness might be increased, but the effect on the total water reaching the channels to form the flood peak would be small. Rock storage would not change appreciably. Depression storage, even if it were increased by 0.1 in. would have little effect. There remains only the soil storage and the question of how much it may be increased by greater accumulations of organic matter on and in the soil. If the forest floor were increased by 2 in. in depth, had volume weight of 0.1, and retained temporarily 300 per cent of its dry weight of water, that is, more than field capacity, the increase would store only 0.6 in. depth of additional water. The content of organic matter in the upper foot of mineral soil under favorable conditions might be increased by vegetation over a sufficiently long period of time by 5 per cent. If such a soil retained 200 per cent of water and had a total weight of 80 lb per cu ft, the storage would be 1.54 in. of water. In a storm of 10 in. the decreases in depth of water available for flood flow might include interception, 1 in.; depression storage, 0.1 in.; forest floor, 0.6 in.; and organic matter in mineral soil, 1.5 in.; a total of 3.2 in. A reduction of 0.67 in. depth of runoff would reduce the peak gage height by 2.24 ft in a storm of 4.92 in. in 1 day in the Ivy River, North Carolina, as a result of improved land use for 20 years, according to estimates based on infiltration and surface runoff by hydrologists of the Tennessee Valley Authority (72). In the same storm with average rainfall of 7.25 in. over the much larger area of the French Broad River basin above Marshall, North Carolina, a reduction of 0.58 in. in runoff was estimated to lower the flood peak gage height by 5.2 ft. The figures would obviously be different for every drainage basin, but they would likely be of the same order of magnitude. Thus vegetation may reduce considerably the crests of floods resulting from major storms, if not always sufficiently to prevent major floods.

Usually, as in the foregoing example, it is convenient to evaluate the factors in the influence of vegetation on flood flows and their sum in inches depth of water. However, those figures must then be converted to gage heights as the useful expression of the level or change in level of flood peaks. This was done in the Flood Control Survey of the Little Tallahatchie Watershed in Mississippi by plotting recorded flood crests over volumes of surface runoff and using the curve fitted to the points as a means of estimating flood peaks corresponding to any desired runoff. The volumes of surface runoff were determined by applying infiltration-capacity curves to rainfall patterns of storms of different sizes and summing the amounts of rainfall excess.

If the flood flow in cubic feet per second Q is known, that unit may be converted to gage height G by the use of a rating or discharge curve. When values of G minus a constant e are plotted over Q, the trends are usually linear on log-log paper and may be expressed as

$$Q = p(G - e)^n$$

or

$$G = \left(\frac{Q}{P}\right)^{\frac{1}{n}} + e$$

where p, e, and n are constants that may be determined from the curve or from the coordinates of the three points necessary to define the trend. For the Mokelumne River in California,

$$G = \left(\frac{Q}{594}\right)^{\frac{1}{1.22}} + 3$$

For the Ivy River in North Carolina,

$$G = \left(\frac{Q}{74}\right)^{\frac{1}{1.92}} + 0.6$$

140. Floods: Summary.—1. The influence of forest in reducing floods is usually large where the floods result from surface runoff in storms of high intensity exceeding the infiltration capacity of denuded soil. It may be negligible in those major storms in which the rainfall exceeds the total storage capacity of the drainage basin.

2. The estimation of the effect of vegetation on floods by the use of formulas in which that effect is included with others in a coefficient is less satisfactory than by methods which attempt to evaluate infiltration, surface runoff, subsurface storage, interception, and other factors in the water cycle.

3. In floods resulting from surface runoff or melting snow the heights of the crests are reduced and the times of occurrence are retarded by reforestation. Peak flows from forested areas rarely exceed 60 sec-ft per square mile, whereas from eroded or denuded lands they may be 500 to 1,000 or more sec-ft per square mile.

4. In major storms the proportion of the rainfall appearing within 24 hours in the stream channels increases with the area of the drainage unit.

5. The ratio of maximum to minimum flow tends to be lower as the protective influence of vegetation is more effective.

6. The debris and sediment carried by floods may be more important than the volume of water as a source of flood damage.

7. In reduction of floods, vegetation may be effective in reducing surface runoff to a minimum and in retarding the rate of drainage of water from melting snow when these are the primary causes. In major storms when the total storage capacity of a drainage basin is far exceeded, vegetation, by increasing the storage capacity or retarding release of water, has only a limited influence.

CHAPTER XXII

EROSION

The phenomenon of erosion results from the movement of soil by water, either as flowing water or waves, or as semiliquid movement of soil in a saturated condition. Wind erosion will be considered separately. The soil, composed of particles of different sizes and weights with rough surfaces, as water flows over those surfaces is subject to (*a*) mechanical pressure on the projecting particles and (*b*) friction as it passes the particles. The particles thus cause eddies, rising currents, and turbulent flow. The loosely held and projecting particles tend to be moved by the water, the larger rolled and the smaller lifted. The greater the velocity, the greater the pressure and friction and the more and larger the particles that are moved. The particles rolled along the bottom are known as "bed load," and the smaller particles carried in suspension by the turbulent flow are termed "silt load" or "sediment." The distinction between bed load and suspended sediment has been placed at a critical diameter of particle of 0.35 mm in the textural grade of medium sand. The material in suspension is at a maximum in the early stages of a flood and is derived from erosion caused by the flood-producing storm. The bed load, however, is more nearly related to the discharge and is material in transit along the channel bottom resulting from a series of past storms. In the large tributaries of the Missouri River the suspended load, with one exception, exceeded the bed load. The ratio between them varied from 180:1 in the Milk River to 2.3:1 in the Platte and was 0.3:1 in the Niobrara, the only one whose course lay mostly in the sand hills.

141. Viscosity of Flow.—Water carrying large amounts of solid matter may become highly viscous as in mudflows, to mention an extreme example. For capillary tubes, the velocity of discharge varies inversely as the viscosity of the liquid. The magnitude of the variation in velocity that may result from this difference in viscosity may be realized from the fact that the coefficient of viscosity for clear water is 0.01, for light machine oil 1.0, and for heavy machine oil 6.0. Streams that may carry solid matter to the extent of two or three times the volume of water will thus have low velocities in comparison with clear water flowing in the same channel. A muddy, viscous flow for the same velocity will exert greater friction on the bed and banks of a channel, both because of the viscosity and also because of the higher density, mineral soil material having a density of about 2.65 compared with 1.0 for water. As the density increases, the buoyancy of the fluid also increases, so that particles are more easily moved or carried in suspension. Hence, the result of a large increase in the solid material carried by a stream may be greater erosion of its channel, notwithstanding the lower velocity. On the other hand, experiments by Vanoni (356) show that additions of sediment up to 0.6 per cent by weight increase rather than decrease the velocity of flow. The explanation lies in the effect of the sediment in damping the eddies that transmit the retarding force of friction from the bed or banks to the interior of the flow. The effect varies both with amount and size of the material in suspension.

142. Energy Relations.—Moving water expends the energy of its mass in abrading the resisting bed or bank material. The kinetic energy K of a mass m moving with a velocity v is

$$K = \tfrac{1}{2} mv^2$$

Thus the cutting or erosive power of water is proportional to v^2, that is, if the velocity is doubled, the cutting power is increased four times. The carrying power, or quantity of material of a given size that can be carried by the water, varies as v^5. In other words, as the velocity doubles, the carrying power is increased thirty-two fold. The size or volume of particles carried by pushing or rolling varies as v^6, that is, as the velocity doubles, the size increases sixty-four fold. If size is expressed as diameter that varies as the cube root of the volume, then the diameter of particles rolled along will vary as v^2. The justification for these relations may be seen quite simply. The force exerted by moving water (or wind) upon a fixed surface increases as the square of the velocity because, as v doubles, there will be twice as much water striking the surface with twice the velocity. If the velocity remains constant, the force increases with the surface area or the square of the diameter. Then the moving power will vary as the product, v^2d^2. But the work done or the weight of the object transported varies as the cube of the diameter. Hence the volume of particles carried varies as v^6. These relations must be considered as approximations that, because of factors other than velocity and slope, will vary considerably in specific applications.

143. Relation to Slope and Runoff.—Inasmuch as the velocity varies as the square root of the slope, the foregoing relations between velocity and erosion may be converted to relations between slope and erosion by taking the square root of the velocity quantities.

The rate of erosion as a function of slope and intensity of rain has been derived by Neal (270) from experiments on 0.001-acre plots of Putnam silt loam in Missouri as

$$W = 0.2s^{0.7}Ti^{2.2}$$

where W is weight of soil eroded in tons per acre, s is the slope in per cent, T is time in hours, and i is the intensity of rain in inches per hour.

From the same data and soil type, Horton (160) obtained the rate of erosion as a function of slope and rate of surface runoff as

$$D = s^{5/8}q_s^2$$

where D is depth of soil removed from the surface in inches per hour, s is slope in per cent, and q_s is the rate of surface runoff in inches depth per hour. The total weight eroded in a storm would be obtained by multiplying by the duration of the surface runoff in hours. Another factor would necessarily be introduced if the formula were to be generalized for more than the one soil type. Both these equations require further confirmation to test their applicability to other soils and the variation to which the coefficients and exponents may be subject.

More recently, Horton (163) suggested that erosion is a more complex function of slope and is proportioned to $(\sin \alpha)/(\tan^{0.3} \alpha)$ when α is the angle of slope in degrees. According to this function the erosion increases nearly in proportion to the slope up to 20 deg and reaches a maximum at 57 deg. However, Renner's data from the Boise River basin indicate that maximum erosion occurs on slopes of about 40 per cent or 22 deg (293).

If the angle of slope α, average depth of erosion D, length of slope l_o, intensity of surface runoff q_s, and critical length of the belt of no erosion x_c are obtained by measurements in the field, then it is possible to compare different areas quantitatively through Horton's erosion constants: R_i, the resistance to erosion; and K_e, the ratio of erosion depth to erosion force F'. The resistance to erosion would include the effects of vegetation and forest floor along with other factors, and hence a basis for evaluating the relative protective value of different types of cover. The critical length of the belt of no erosion is the distance measured along the slope from the divide to the point where sheet erosion starts. Then the resistance to erosion in pounds per square foot of slope area is

$$R_i = 0.082(x_c q_s n)^{3/5} \frac{\sin \alpha}{\tan^{0.3} \alpha}$$

The n is the surface roughness factor of which the derivation was given in Chap. XX as

$$n = \frac{1{,}020\, s^{1/2}}{I\, K_s\, l_o}$$

The other constant, K_e, requires for its computation the eroding force F', which is

$$F' = \frac{w}{12}\left(\frac{q_s n}{1020}\right)^{3/5} (l_o^{3/5} - x_c^{3/5})\, \frac{\sin \alpha}{\tan^{0.3} \alpha}$$

where w is the weight per cubic foot of the water plus its load of sediment. As the necessary data become available, tests of these measures of erosion can and should be made.

The usual observation that the muddiness of the water is associated with the stage of the stream would suggest that sediment carried in suspension might be a function of rate of discharge. For large streams the averages for periods of months have been found to be related by an expression of the form

$$W = kQ^n$$

where W is weight of sediment transported in unit time and Q is the rate of discharge, usually in second-feet (75). The exponent n for the Missouri River at Kansas City was 2.16 and for the Red River, 2.04, so that it may be true in other streams that the suspended sediment varies approximately as the square of the discharge.

In a channel of given slope formed by deposition from muddy water, clearing the water, as by control of erosion in the headwaters, will tend to increase the velocity of flow and hence its cutting power, so that beds of earlier deposition from muddy flood flows will again be eroded without change in slope. This frequently happens following storms, when the flood flows carried and deposited sediment in the lower reaches, and these deposits are deeply cut by the clear water of the receding flow after the flood.

The profile of a stream channel becomes flatter in the lower reaches. Theoretically, this profile can be expressed by the equation

$$S = S_o e^{-ax}$$

where S is the slope, S_o is the slope at the initial point, and x is the distance between S and S_o along the profile. If Z is the elevation at any point of slope x, and Z_o is the elevation at slope x_o, then, according to Shulits (320),

$$Z_o - Z = \frac{S_o}{a}(1 - e^{ax})$$

As the velocity decreases with the decreased slope the coarser particles are deposited and the bed of the stream is aggraded. This frequently happens on the alluvial fans of streams issuing from the mountains, and as the process continues the bed is built up to a point where flood flows can no longer be carried and the overflow finds a new channel down across the fan and the old one is abandoned. Similarly, the continual deposition of eroded material in the lower reaches of the Mississippi and other large streams gradually elevates the bottoms of the channels, and the levees must be correspondingly elevated to prevent overflow. A continuation of the process without dredging of the channel would eventually result in the stream flowing between levees on a bed elevated above the surrounding flood plain. Apparently this happened in some of the prehistoric canals in the southwest, of which the remaining beds stand above the level of the surrounding surface.

The relation of erosion to slope of the land is often obscured by variations in the soil, character of surface, presence of rock, cultivation, and vegetation. Within certain ranges on specific areas, correlations between slope and erosion have been found. Thus, between 6 and 23 per cent, Lowdermilk (224) found such correlation, but the relation tended to disappear with the formation of erosion pavement until such time as the pavement was undermined and gullying again accelerated. Erosion of clay soil was found (272) to vary with slope up to 12 per cent, and above that much more rapidly. On cleared lands, slopes as flat as 1½ per cent may erode, and it is generally considered that slopes above 15 per cent should not be cleared. On the other hand, cleared slopes up to 30 per cent may not erode seriously on certain soils.

As the land slopes become steeper the gravitational component becomes more important both in erosion by water and in mass movements of slips and slides. For different materials in artificial banks, observations have established the following critical slopes or angles of repose, above which movement may be expected.

Material	Angle of Repose, Deg
Damp clay	45
Gravel	39 to 48
Dry sand and mixed earth	21 to 37
Wet clay	17

Recent experiments indicate that the angle of repose increases with the average size of the fragments, with the roughness of the surfaces of the particles, with the compaction of the material, and, in the case of sand, with the moisture content up to saturation. Above the saturation point the angle of repose again decreased (355). Under natural conditions and particularly where the soil is anchored by the roots of vegetation, the angles of repose may be decidedly steeper locally and for certain periods of time than those indicated above. Talus slopes without vegetation stood at angles of 34 to 39 deg, whereas vegetated slopes derived from the same parent materials stood at steeper slopes, sometimes even as high as 60 deg (25). The supposition is that the vegetation could have started on the bare slopes only at the lower angle of repose and the oversteepening has resulted from erosion lower down since its establishment.

144. Safe Velocities.—As a result of experience with ditches and channels the values in Table 59 for the maximum velocities of flow that will not cause erosion have been compiled for various channel-bed materials (354).

TABLE 59.—SAFE VELOCITIES OF FLOW FOR BEDS OF CHANNELS OF DIFFERENT MATERIALS

Material	Safe Velocity, Ft/Sec
Fine sands or loose silt	0.5–1.0
Loamy sand, 15 per cent clay	1.2
Sandy loam, 40 per cent clay	1.8–2.0
Coarse sand	1.5–2.0
Loose gravelly soil	2.5
Loams, 65 per cent clay	3.0
Clay loams	4.0–7.0
Stiff clay	6.0
Stratified rocks	8.0
Small boulders	8.0–15.0
Hard rock	13.0

Such figures for safe velocities are intended to allow a margin of safety and therefore it is likely that appreciable erosion would occur only at velocities somewhat higher than those indicated.

145. Bed Load.—Several formulas derived to express the diameter of bed-load material that would be moved in channels as a function of their hydraulic characteristics are summarized below (321):

Formula	Author
$D = 4{,}770(Qs^{4/3}/w)$	Schoklitsch
$D = 1{,}200(Q^{2/3}s/w^{2/3}$	Swiss Institute of Technology
$D = 490(rs)^{2/3}$	Schoklitsch
$D = 3{,}080rs$	Krey
$D = 11{,}170rs$	Kramer
$D = 66{,}600{,}000(rs)^2$	U.S. Waterways Experiment Station

In these formulas, D is diameter of particle in millimeters, Q is discharge in cubic feet per second, s is the slope of the energy-gradient, nearly equal to the slope of the bed, in feet per foot, w is width of the stream in feet, and r is the hydraulic radius in feet. When the formulas are applied to actual data from large channels the results are less divergent than would be expected.

146. Relation to Impact of Raindrops.—Erosion losses from bare soil are also definitely related to the size and hence also to the rate of fall of raindrops. Erosion, expressed as percentage of soil in runoff water, may increase as much as twelvefold with increase in the size of drops according to Laws (214). He found accelerating trends of erosion with an index of rain erosivity or with the equivalent raindrop diameter. The index E/A was developed as the energy E per unit area A. The kinetic energy E of drops of mass m falling with velocity v is

$$E = \tfrac{1}{2}mv^2$$

The index E/A was in ergs per square millimeter. From the values of this index and the velocities of falling drops the equivalent raindrop diameter was derived. The relation between them is given by the equation

$$D = 0.157\left(\frac{E}{A}\right)^{0.45}$$

Some of the eroded soil is carried in the splash of the raindrops and this has been measured by Ellison (117). He found that the grams of soil in the splash W for a 30-minute rain could be expressed as a function of the velocity v, of fall of the drops in feet per second, their diameter D in millimeters, and the intensity i of the rain in inches per hour by the formula

$$W = Kv^{4.33}D^{1.07}i^{0.65}$$

This gives a series of accelerating trends when W is plotted over v for different values of D and i. The value of the constant K for eroded Muskingum silt loam was 0.0000766. Additional tests of these relations on other soils are needed before the variation in the constants and the extent of their application can be evaluated.

147. Forms and Classification of Erosion.—Erosion includes the continuing geological process termed "normal erosion," when soil and natural vegetation are undisturbed. This normal erosion is ordinarily very slow where the natural vegetation is dense and luxuriant. On the other hand it varies in different regions and climates and may become rapid and spectacular in semiarid regions where the vegetation is naturally sparse and the normal erosion results in phenomena like the Grand Canyon of the Colorado River.

In contradistinction to the normal erosion, abnormal or accelerated erosion is found in those areas where vegetation and soil have been disturbed by fire, grazing, logging, cultivation, or engineering works. The degree of disturbance is reflected in the degree of acceleration. Thus, disturbances like cutting timber, which affect chiefly the canopy, tend to be less serious than those like fire, which consume also the forest floor; and those like cultivation, which destroy vegetation, floor, and soil structure, tend to be the most serious. On the same area, accelerated erosion, as the term implies, will be more rapid than normal erosion. However, when different areas are compared, normal erosion in a semiarid region may be more rapid than accelerated erosion in parts of the humid region.

Evidence of accelerated erosion may be found in any one or more of a variety of forms. The texture of the layers in flood plain deposits in normal erosion becomes finer toward the surface whereas accelerated erosion reverses the sequence so that the coarser layers are above the finer. This may be observed in cut banks or by borings. The presence of erosion pavement indicates the removal of the finer particles from a former surface. On alluvial deposits the ages of the oldest trees will represent a period of years somewhat less than the time since the last deposition occurred. If the normal profile of the soil type is known, the reduced thickness of the *A* horizon, or absence of the *A* and removal of part of the *B* horizon, provides a measure of the depth of erosion. Historical records of dates when a piece of land now gullied was used for crops, pasture, or buildings give the time of beginning and the rate of accelerated erosion. If plants are standing with part of the root system exposed, the distance from the root crown to the present soil surface is the depth eroded since the plant became established. Plant indicators of deteriorated soil may afford qualitative evidence of erosion. Finally, increases in the silt load of a stream provide a sensitive index of accelerating erosion.

The different forms and degrees of erosion have been classified and given descriptive names, although, as in all classifications of natural phenomena, the boundaries between classes are often obscured by transitional examples. The following classes are useful in describing and referring to erosional phenomena.

Sheet erosion is the class in which the surface layers of the soil have been washed away more or less uniformly over the whole surface. The process may be hardly perceptible over short periods of time, but during a long period of years the removal of part or all of the *A* horizon and the exposure of roots of the vegetation afford convincing evidence.

Rill erosion is the incipient form of gullying and has often been referred to as "incipient" or "shoestring" erosion. Ordinarily it takes the form of a series of more or less parallel small gullies on a slope. The size of the gullies is not greater than can be plowed over in cultivation. Usually this would mean less than 1 foot deep or wide.

TABLE 60.—A CLASSIFICATION OF MASS MOVEMENTS ACCORDING TO PHYSIOGRAPHIC LOCATION AND PRINCIPAL CAUSE

Physiographic location	Original surface	Cause			
		Gravity, chiefly	Lubrication by water	Super-saturation	Under-cutting
		Class of movement			
Slopes	Partly intact Destroyed	Creep Slide	Earth flow Avalanche	Solifluction	Avalanche
Channels	Destroyed	....	....	Mudflow	
Banks	Partly intact Destroyed	....	....	....	Slump Caving

Gully erosion proper includes the gullies of larger size up to canyons and arroyos 100 or more feet deep and wide. Gully erosion may develop by the continuation of rill erosion when some of the rills converge and become large channels. Another form of gully erosion results from the headward cutting of gullies where concentrations of water drop over a bank of unconsolidated material. Frequently this process starts at a stream bank where a miniature waterfall undermines the bank and the sloughing of masses of soil forms a notch. As the water continues to fall over this notch a deep gully may cut headward quite rapidly. Lateral gullies may form in the same way as the main gully develops.

Stream-bed erosion is the obvious form where water running with sufficient velocity scours out the material of its bed to form a deeper channel or to change its course by cutting on the outside and aggrading on the inside of the banks.

Mass movements of soil that are usually not the direct result of flowing water include several forms of erosion. Along the banks of a stream where there is undercutting there may be either caving, where the undermined masses fall away from the bank, or slumping, where they slide down but lean back toward the bank, leaving part of the surface unbroken. Undercutting may be interpreted to include any dent in the natural arch of a slope that leaves a face or bank at a slope steeper than the angle of repose of the material.

In channels where soil material is mixed with water in such proportions that the mass becomes semiliquid, mudflows become a highly destructive form of erosion. They may carry large boulders and do great damage. Ordinarily mudflows are phenomena of dry regions. In arctic-alpine regions, with the melting of the surface layer of snow, ice, and soil on slopes, a somewhat similar phenomenon called "solifluction" occurs.

On slopes away from channels or banks, the term "slide" may be used to include a mass of soil and rock that breaks loose and moves rapidly, even when quite dry. The corresponding phenomenon when the mass of material is lubricated by water from rainfall may be called an "avalanche."

An earth flow like an avalanche occurs as a result of the reduction of the cohesiveness of the upper layers by water, so that a mass of soil becomes loosened and moves downslope, leaving large cracks at the upper edge and usually forming a slightly elevated roll at the lower edge. Earth flows frequently occur under a cover of vegetation. Obviously earth flows or other slips are more likely to occur in grassland where the roots are only 1 or 2 ft deep than in brush or forest where the anchorage is 4 or 5 ft deep. However, on deep clay subsoils like the "blue clays" in the redwood region, slips occur even under the old-growth forest, as evidenced by areas of leaning trees.

Finally, creep is a universal form of erosion on slopes, by which the surface soil moves downhill at an imperceptible rate. The evidence of it is clear in the downhill lean of fenceposts that have been in place for several years. It is a phenomenon that takes place either under or in the absence of vegetation. It is accelerated by the activities of livestock, deer, rodents, and birds. On the steep slopes under the chaparral in southern California, the annual creep was sufficient to override a 2-in. barrier on 17 per cent of the area (197).

A classification of erosion phenomena by mass movement on the basis of causes and physiographic location is suggested in Table 60. A classification on the basis of the kind and rate of movement and whether the material is dry or wet has been developed by Sharpe (315).

The proportion of different forms of erosion in different drainage basins is highly variable and somewhat difficult to estimate. In the

1-square-mile basin averaging 74 per cent slope of Brand Park Creek above Glendale, California, Garstka and Munson[1] estimated that, following the flood of January 1934, 2 per cent of the debris came from channel scour, 12 per cent from road banks, 13 per cent from slides, 15 per cent from rills, and the remaining 58 per cent from sheet erosion. In the flood-control survey of the Little Tallahatchie River basin in Mississippi, the sources of eroded material were estimated separately for the brown-loam area of loessal soil over sand and the flatwoods area underlain by clay, with the following contrasting results:

Source of sediment	Brown loam, %	Flatwoods, %
Cultivated and woodland	11	62
Abandoned fields	86	27
Roads	3	11

148. Erosion Pavement.—Erosion pavement, that is, a layer of gravel, pebbles, or stones on the surface, can be used as a measure of the depth of sheet erosion. It has been shown that the velocity of flow of water over a surface increases as the square root of the slope. On any given slope the velocity tends to be constant. The size of soil particles carried by water increases with the velocity. For any given velocity, particles below a certain size will be carried away, and those above will remain. It is this process in sheet erosion that results in a concentration of coarser particles on the surface to form the erosion pavement. Hence the degree of concentration of coarser particles and of the formation of erosion pavement should be a measure of the erosion.

It must be assumed that the original upper layer of the soil had the same proportion of coarse particles that the layer just below the erosion pavement now has. This assumption should be verified where possible. From samples of the pavement and of the underlying layer, the weights of the coarser particles above a certain size are determined, and a proportion may be made between weight and depth as indicated in Fig. 23, or, expressed in symbols,

$$\frac{x + d}{w_X} = \frac{D}{w_D}$$

where x is the depth eroded, d is the depth of erosion pavement, D is

[1] From "Investigation of reported flood, New Year's Day, 1934, Brand Park, Glendale, California," by W. U. Garstka and S. M. Munson, unpublished report of the California Forest and Range Experiment Station, U.S. Forest Service.

the depth sampled below the pavement, w_X is the weight of coarse particles in the pavement, and w_D is the weight of coarse particles in the depth sampled below the pavement. For the depth eroded

$$x = \frac{Dw_X}{w_D} - d$$

For such a determination it is only necessary to weigh the coarser particles remaining on a sieve with openings of a size suitable to separate the coarse from the fine in any given soil. Sieves of more than one size may be used, and presumably the one yielding the highest value of depth eroded gives the most reliable determination.

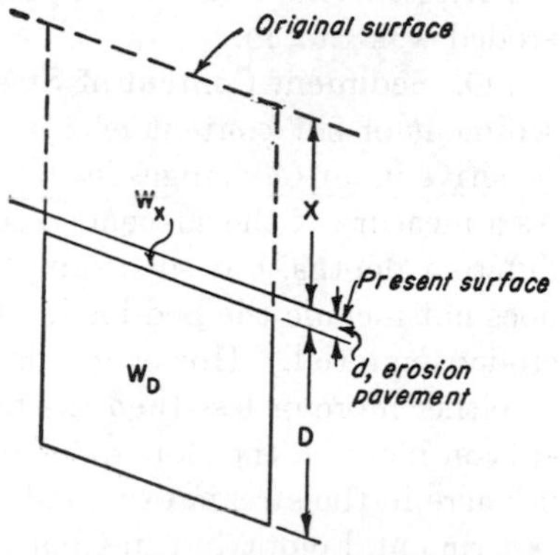

FIG. 23.—Erosion pavement as an index of depth of soil eroded.

For a pavement of stones and coarse material the same method can be used by collecting, in a bucket or box, the stones in the pavement from a marked area and marking the depth to which the container is filled. It is then emptied and stones above the same size from the same area are again collected from the soil layer below the pavement until an equal depth is filled in the container. Then the depth of the hole from which the stones came is equal to the depth of the layer eroded.

One of the difficulties in this determination is the measurement of the depth of the erosion-pavement layer. What is wanted is the depth represented by these coarse particles when they were embedded in the soil layer now gone, that is, the average depth if they were melted down into a solid layer. A check on field measurements may be obtained as follows: Depth is equal to volume divided by area. Areas can be obtained when taking the samples. Volume is equal to weight divided by density. The weight of the coarse particles is readily determined, and it can be assumed that the density of solid mineral material is close to 2.65. If the computed value for volume is substituted in the equation for depth, the depth of the erosion pavement can be computed as a check on the measured depth.

For example, after the Tehachapi flood of 1932 a sample of the erosion pavement was taken from an area of 3 by 3 in. (58.1 cm^2) on a gently sloping knoll. On the basis of a density of 2.65, the volume equaled the weight of the coarse fraction, 16.1 g, divided by 2.65 or 6.1 cm^3. Then depth of the pavement was

$$\frac{6.1}{58.1} = 0.104 \text{ cm} = 0.04 \text{ in.}$$

The depth eroded x then can be obtained from the proportion

$$\frac{x + 0.04}{16.1} = \frac{3}{30.9}$$

where 3 and 30.9 are respectively depth and weight of the coarser particles in the 3 in. underlying the pavement. Hence, the depth eroded was 1.52 in.

149. Sediment Content of Streams and Silting of Reservoirs.—The sediment or silt content of a stream has been suggested as the most sensitive index of changes in the erosion conditions in a drainage basin. As a measure of the suspended sediment the water may be sampled at different depths, but such sampling, unless special equipment is used, does not include the bed load, which may be a large proportion of the eroded material. However, the sediment measured in this way may be either more or less then the amounts indicated by collections from erosion plots on the slopes. Thus in Tarkio Creek in Iowa, 17.3 tons per acre in the stream exceeded the 3.6 tons per acre indicated by five experimental plots, but in Coon Creek, Wisconsin, 3.5 tons per acre in the stream was much less than the 60.9 tons per acre average on plots (57).

The filling of reservoirs by the deposition of sediment constitutes another measure of the erosion taking place in the tributary drainage basins. It is related to the sediment carried by streams, except of course that at any given time a portion of the eroded material is in transit along the stream channels, and only a part has reached the reservoir. Thus the erosion, as measured by the silting of reservoirs, is likely to be less than that obtained by measurements on the land above or in the tributary streams.

Surveys of the silting of reservoirs have made available considerable information as to the amounts and rates at which this process is going on in different parts of the country. A survey of 66 reservoirs by the Soil Conservation Service (57) indicated annual silting at rates ranging from 0.2 to 20 cu yd per acre with a median value of 1.6 cu yd. The annual depletion of the capacity of reservoirs varied from 0.1 to 7.3 per cent of the total capacity, with a median value of 0.9. Thirteen reservoirs in the Piedmont region, with dams averaging 30 ft high, had filled with sediment in 29 years to levels such that their values for storage were completely destroyed.

The two different measures of sedimentation that have been used, namely, the volume rate of filling, expressed in cubic yards per acre per annum, and the annual percentage rate of reduction of storage, obviously are not measures of the same phenomena. The volume rate of filling is the better expression for the rate of erosion on the drainage area. The percentage reduction of storage, however, involving as it does the capacity of the reservoir, is a more useful measure for purposes of evaluating the life and usefulness of the storage. In the data so far available the two measures are not closely correlated.

150. Datum Level Stakes.—Metal stakes driven 2 ft or more into the ground and projecting somewhat above the surface have been used in the measurement of erosion. Successive measurements of the distances from the tops of the stakes to the ground surface provide measures of the changes in surface level. The changes are usually quite small and may be obscured by other factors than erosion. For example, the surface level of a soil high in colloids may rise and fall appreciably with changes in moisture content and it is not always easy to make periodic remeasurements when the moisture content is the same. Removal of soil by erosion is not uniform even over small areas. The surface near a stake may be subject to unusual removal or to local deposition. If a straightedge laid on the surface is used as the surface level, its position is determined by the high points including local deposits of sediment rather than by the depressions which are more important in erosion. If a considerable number of stakes are used, significant results may be obtained where erosion is heavy.

151. Conversion of Units.—Erosion is commonly expressed in units of depth, of volume per unit area, or of weight per unit area. In making comparisons between data from different sources, conversions to a common unit must be made. Depth D, usually in inches, is independent of area.

$$V = 134.4D$$

when V is volume in cubic yards per acre.

$$W = \frac{V\rho_a}{1.3}$$

when W is in metric tons per acre of dry soil. If the volume weight is 1.3, then W is numerically equal to V. Depth and volume per acre are not greatly affected by the moisture content of the soil and, because they do not necessitate oven drying, are preferable to weight per acre as units of amount of soil eroded.

152. Laboratory Test of Erodibility.—In many instances it would be useful to be able to determine in the laboratory to what degree soil might be susceptible to erosion before risking its exposure on a large scale. Soils known from their field behavior to be erodible or nonerodible were tested for many different properties by Middleton (253). He found that three of the properties were somewhat correlated with the field behavior. One was the dispersion ratio, which is the ratio of silt and clay in suspension after limited shaking and settling to the total silt and clay by mechanical analysis. This is a measure of the ease with which the soil is brought into suspension. The second was the ratio of colloid content to moisture equivalent, which is a measure of the percolation, quantity, and water-holding capacity of the colloid. The third was the erosion ratio, which is the quotient of the other two, namely, the dispersion ratio divided by the ratio of colloid to moisture equivalent. In general it appeared that nonerodible soils have a low dispersion ratio of less than 15. They have a high ratio of colloid to moisture equivalent, greater than 1.5. They have a low erosion ratio, less than 10. Although these laboratory tests gave some indication, they have not yet been developed to a point where the degree of erodibility in a specific case can be predicted with assurance.

153. Effect of Forest or Denudation on Erosion.—Previous evidence that destruction of vegetation increased surface runoff would in itself clearly lead to the expectation that erosion would likewise be increased. Moreover, because the cutting and carrying power of water increase with the velocity by powers greater than 1, the erosion would be expected to increase more strikingly than the runoff. Obviously the forest floor and litter forming a mulch or protective covering on the surface is the principal medium through which the forest exercises its influence on erosion.

Fire is a frequent cause of destruction of the litter and forest floor, and the effect of burning in increasing erosion has been demonstrated in many instances. Experiments in tanks by Lowdermilk (225) showed that burning the litter increased the discharge of eroded material from 50 to 3,000 times. Erosion in the undisturbed brush type of ceanothus, manzanita, and buckeye at North Fork, California, did not exceed 0.01 cu yd per acre, whereas after the plots were burned over four times in different years the erosion increased to between 2 and 4 cu yd per acre (226). On one-sixtieth-acre erosion plots in the chaparral of the San Dimas Experimental Forest, erosion was increased

1,000 times by the burning of the vegetation. In Los Angeles County, Eaton (115) concluded that the discharge of debris was increased from ten to thirty times by fire. In the first winter after the San Gabriel fire of 1919, erosion was at the rate of 10 to 120 cu yd per acre, according to Munns (264). After the fire of November, 1933, in the Montrose area, the erosion from unburned drainages was less than 0.1 cu yd per acre. From the burned areas it varied from 48 to 137 cu yd per acre (210). An attempt at generalization by the Flood Control Survey for the Los Angeles River basin is contained in Table 61. For a rain

TABLE 61.—VARIATION IN EROSION WITH AGE OF CHAPARRAL AND MAXIMUM 24-HOUR PRECIPITATION IN LOS ANGELES RIVER DRAINAGE BASIN*

Precipitation, in. 24 hr	Years since fire			
	1	4	17	50+
	Erosion, cu yd/acre			
2	5	1	0	0
5	20	12	1	0
11	180	140	28	3

* Data from an unpublished appendix of the Flood Control Survey Report on the Los Angeles River through the courtesy of the California Forest and Range Experiment Station, U.S. Forest Service.

of 2 in. in 24 hours, erosion is likely to start in burns less than 4 years old. Five inches precipitation in 24 hours will start erosion in chaparral as old as 17 years. The unusual storms of as much as 11 in. of precipitation in 24 hours will cause erosion even in the oldest and most stable areas of chaparral. The increased effectiveness of the vegetation as it recovers, even in the first few years after a fire, is strikingly shown.

Similar results of fire have been recorded elsewhere in the country. In plots of post oak in Oklahoma (46), burning increased the erosion from 0.01 to 0.15 tons per acre. In northern Mississippi two annual burnings in mature oak increased the erosion from 0.05 to 0.83 tons per acre and the same treatment in broomsedge caused an increase from 0.18 to 0.79 tons per acre (245).

Other forms of destruction of the vegetation, when they also involve the disturbance of the upper layer of the mineral soil, may be even more serious than burning. Cultivation leaving the exposed fallow is one such form of disturbance. In northern Mississippi, where the erosion under oak, broomsedge, or Bermuda grass was less than 0.7

ton per acre, a cotton field that had been plowed on the contour eroded 69 tons per acre, a barren, abandoned field eroded 160 tons, and a cotton field plowed up and down the hill eroded 195 tons per acre over a 2-year period (245). In Wisconsin, erosion of cultivated lands removed as much as 1 to 3 in. of topsoil in a single storm (39).

Destruction of the forest by smelter fumes in the area near Kennett, California, resulted in erosion of 427 cu yd per acre, whereas from areas 50 per cent denuded it was 121 cu yd, and from areas only slightly damaged, 49 cu yd per acre (264).

The road banks immediately after construction constitute another form of complete denudation and maximum disturbance of the soil. On plots of over 60 per cent on a mountain road in the San Dimas Experimental Forest the erosion rate reached a maximum of 500 cu yd per acre in the year 1937–1938, when there was 40 in. of precipitation with a maximum intensity of 1.25 in. per hour. It is interesting to compare this figure with that of the previous year, when the precipitation was over 37 in., but the maximum intensity was only 0.4 in. per hour. The erosion amounted to only 48 cu yd per acre.

Marked differences in the rate of silting of reservoirs are associated with differences in the vegetation on the drainage areas. For example, Bridgewater Reservoir on the upper Catawba River, draining a section of the Blue Ridge Mountains in North Carolina, where the land is 75 per cent forested and 10 per cent in pasture, is filling at a rate of less than 0.04 per cent annually. On the other hand, Mountain Island Reservoir, also on the Catawba River, but in the Central Piedmont section, where most of the land is cultivated, is filling at a rate of over 0.5 per cent annually. The reservoirs in the Northeast, many of which have drainage areas 90 per cent forested, are filling at an almost negligible rate, certainly less than 0.01 per cent annually (47).

However, such comparisons of rate of loss of storage between different reservoirs reflect not only the protective efficiency of the vegetation but also the ratio of the original storage capacity of the reservoir to the inflow from its drainage basin. If data for inflow are lacking, the area of the drainage basin may be used as the denominator of the ratio. Thus in the southeastern states, as the original storage capacity in acre-feet per square mile of drainage area increased from 12 to over 200, the annual loss of storage decreased at a decelerating rate from 2.8 to 0.3 per cent (57). In evaluating the influence of vegetation, allowance should be made for differences in this ratio.

The reduction of silting following the establishment of vegetation is exemplified by the records from Lake McMillan on the Pecos River in

New Mexico (57). From 1894 to 1915, when only a sparse growth of saltgrass covered the valley above the reservoir, the annual sedimentation was at the rate of 0.22 cu yd per acre of drainage area. From 1915 to 1932, the introduced French tamarisk spread naturally and formed thickets around the upper end of the reservoir and along the floor plain for 200 miles upstream. As a result, in this period the silting in the reservoir was reduced to 0.034 cu yd per acre per annum or about one-seventh of the earlier rate, and deposition at the upper end and upstream was 5 ft deep.

Conversely, the increased silting with progressive denudation by fire is illustrated by the following figures for the Gibraltar Reservoir near Santa Barbara, California, built in 1920 with a storage capacity of 14,500 acre-ft and a drainage area of 215 square miles.

Period	Annual accumulation, cu yd/acre	Average annual loss of capacity, %	Cumulative area burned, sq miles
1920–1925	2.0	1.2	72.8
1925–1931	6.3	3.7	88.3
1931–1934	7.6	4.8	165.5

Both the rate of sedimentation and the loss of storage capacity increased more than threefold as the area burned was more than doubled.

In sampling the sediment in the flood water from the denuded basin of Squaw Creek in northern California, Munns (264) found from 2.2 to 5.6 per cent of silt in comparison with the flow from Buckeye Creek, 100 per cent forested, which showed less than 0.2 per cent of silt.

After the flood of 1938 the debris deposited in reservoirs by seven streams issuing from the chaparral-covered south-facing slope of the San Gabriel Mountains in southern California constituted from 4 to 13 per cent of the flood discharges. Much larger proportions of solid material have also been reported in that region. After a fire in Rogers Creek, Hoyt and Troxell (177) found from one-half to two-thirds by volume of sand and ash carried by the water. Streams carrying two or three times their volume of solid matter were mentioned by Kraebel (210). Three years after the fire in Pickens Canyon, near Montrose, a rain of 0.8 in. with an intensity of 1.37 in. per hour for 35 minutes, yielded 25,011 cu yd of flow impounded in the debris basin at the mouth of the canyon (194). After the water drained off, the debris deposit had a volume of 22,062 cu yd. Thus, of the total flow, wet solids comprised 88 per cent and water 12 per cent or a ratio of 7.5 to 1.

Obviously the wet solids had pore space which may have amounted to 50 per cent and which was largely filled with water, so that the comparison between dry matter and water would be quite different, but still impressive as to the amount of solid material that flood flows may carry.

In the flood of January, 1934, from the same area, a boulder estimated to weigh 58 tons was left poised on the Foothill Boulevard 8 ft above the level of the channel bottom down which the flood flow moved, a striking example of the power and buoyancy of these muddy flows in transporting large rocks.

Examples of erosion are also available from range lands that have been subject to varying degrees of utilization and abuse. The most notable is that of the two 10-acre pastures at the Great Basin Experiment Station near Manti, Utah. The experiment was started in 1915 and, during the first 5 years, when watershed *A* had 16 per cent cover and *B*, 40 per cent, the ratio of *A* to *B* in silt accumulation was 5.4. During the period 1929–1934 both plots had 40 per cent cover, and the ratio was reduced to 2.9. In 1936, by heavy grazing, plot *B* vegetation had been reduced to 25 per cent and the ratio of *A* to *B* had become 1.5. In all cases the actual amounts of eroded material were less than 6 cu yd per acre annually (125, 348).

The 7-in. rain of late September, 1932, in the Tehachapi Mountains of California almost destroyed the surface of the heavily grazed grassland by the formation of countless gullies nearly to the tops of the ridges. Within the grassland an occasional oak provided a thin forest floor beneath its crown. The gullies came down from above, they started again below, and they were continuous on both sides, but within the projection of the crown, gullies did not form. Similarly, in the adjacent woodland of singleleaf piñon, Digger pine, and blue oak, gullies dissected the surface of the openings between the trees but did not cut the forest floor beneath the crowns. Frequently, where the forest floor covered the slope below well-developed gullies the muddy flows had passed over the surface of the litter without cutting channels and, as the water infiltrated into the soil, left deposits of mineral soil on top of the litter, marking the courses of the miniature mudflows.

Alternate freezing and thawing of the surface layer of the soil is an important source of erosion on steep banks and slopes in Wisconsin, North Carolina, northern Mississippi and doubtless elsewhere. Measurements in Mississippi in two instances showed that the melting of the ice needles on the surface of the unfrozen soil increased erosion in one year from four to eight times and in another year from nine to forty-five times. The influence of vegetation in increasing nightly minimum temperatures and reducing the daily maxima tends to decrease the frequency and intensity of freezing and thawing, and hence also the resulting erosion.

When natural vegetation takes possession of an eroded area there is likely to be a difference in the degree of effectiveness during the early stages of the succession. This process was recorded by Kraebel[1] near San Bernardino in a steep canyon severely burned in 1925. In the first year after the fire the erosion was enormous. In the succeeding 4 years it became progressively smaller. In the fifth and sixth years the erosion again increased to an extent which could not be attributed wholly to the precipitation. The explanation was found in the vegetative succession. In the first year the herbaceous plants occupied the spaces between the sprouting shrubs of the chaparral and protected the surface by contributing both living cover and dead barriers. These plants increased in numbers and vigor in the second, third, and fourth years after the fire, but then, as the sprouting shrubs increased in size, declined to become a negligible influence in the fifth and sixth years. In those years, the shrubs were not yet sufficiently large and dense to provide a closed cover. Hence the secondary acceleration of erosion was attributed to the disappearance of the herbaceous cover before the chaparral had become fully effective.

154. Erosion Surveys.—A procedure for surveys of erosion has been worked out in detail by the Soil Conservation Service (277). Basically these surveys classify and combine four different factors, namely, soil type, land use, slope, and erosion. The soil types are usually those recognized in the soil surveys. The land use distinguishes cultivation, pasture and range lands, idle lands formerly cultivated, and forest lands defined as those with 40 per cent coverage by trees more than 1 year old.

The groupings according to slope on nonagricultural lands are by 5 or 10 per cent divisions. On agricultural lands four groups are recognized on the basis of slope. The minimum erosion includes lands with slopes less than 2 or 5 per cent. The active erosion group where effective control is possible may range from 3 to 7 or from 5 to 25 per cent, depending upon the soil and its erodibility. The third group is too steep for control with clean tillage but may be used for pasture. This

[1] From the Annual Investigative Program of the California Forest Experiment Station, U.S. Forest Service, 1932, through the courtesy of C. J. Kraebel.

may range from 7 to 12 or from 20 to 30 per cent, depending on the soils. The fourth group includes slopes that are too steep for the control of erosion if they are cultivated.

The erosion itself is classified first as to character, whether sheet, gully, wind, geological, or stabilized. These groups in turn are subdivided. Sheet erosion includes: Class 1, where no erosion is apparent; Class 2, where the erosion is slight and less than 25 per cent of the *A* horizon has been removed; Class 3, moderate erosion, where 25 to 75 per cent of the *A* horizon is gone; Class 4, severe erosion, where over 75 per cent or all of the *A* horizon has been removed; Class 5, very severe erosion, extending into the *C* horizon. Class 6 includes special cases such as land slips.

The group of gully erosion comprises Class 7, occasional gullies, three or less per acre; Class 8, frequent gullies, more than three per acre; Class 9, destroyed by gullies. The gullies are also characterized as to their depth and whether tillable or not tillable.

Wind erosion has 10 classes, based on the character of accumulation, whether level or in hummocks or dunes, and on the depth of removal of the upper horizons of the soil. These are defined in terms of depths of accumulation, percentage of area covered, or percentage of the profile removed.

The significance of the sheet-erosion classes in terms of site quality has been indicated for loblolly pine on several soil types in South Carolina, by the following figures (97).

Soil type	Sheet-erosion class			
	2	3a	3b	4
	Avg. site index at 50 years			
Lloyd sandy loam	....	103	....	82
Cecil sandy loam	....	....	88	78*
Helena sandy loam	....	92	82	....
Durham sandy loam	88	83	....	....

*Cecil clay loam, the *A* horizon of sandy loam being eroded away.

The removal of 25 per cent of the *A* horizon results in a reduction of site index by 10 ft in three soil types and 5 ft in the fourth. A difference of 10 ft in site index corresponds to a difference of more than 7,000 fbm per acre in yield of well-stocked stands 50 years old, a considerable deterioration.

155. Erodibility from Erosion-survey Data.—Erodibility has been defined quantitatively (123) as the percentage erosion by area per unit of slope over a specified range of slope. The percentage erosion is derived from erosion-survey data from the area having a certain degree or class of erosion. For each range of slope, areas of each erosion class are weighted by the mid-points of the slope classes to give the mean erosion percentage. When the mean erosion percentages are plotted over the slope percentages, expressed as the mid-points of the classes, the resulting trends are ascending curves of varying slopes.

The area under a curve of erosion over slope, obtained by planimeter, is termed the "erodibility integral." Arbitrary units of this integral may be used, but it has been found convenient to consider that the product of 10 per cent slope by 10 per cent erosion is equal to 1 unit. When the slope is held constant, the erodibility is equal to the erodibility integral times a constant. Thus the erodibility integral can be used as an index as well as erodibility as originally defined.

Using the erodibility integrals with data from the survey of the upper Gila River watershed, the desert soils as a group had integrals two to four times as large as the brown-soil group, with the exception of soils derived from granite, which had the same integral of 9.9. Comparisons between soils within the brown group for different parent materials showed an ascending sequence from limestone through basalt, quartzite, and rhyolite to granite. Similarly, comparisons of textural classes gave erodibility integrals in descending order from sandy loam through loam to clay loam.

In the laboratory the ratio of colloid to moisture equivalent was the only physical criterion that gave a highly significant correlation with erodibility. In all soils tested a high content of calcium or sodium or a low content of iron or potassium was associated with high erodibility. The erodibility was found to increase as the ratio of iron to calcium decreased.

The use of the erodibility integrals makes possible quantitative comparisons of the erosion in the land-use or other classes mapped in an erosion survey. The method should be equally applicable to areas of forest differing in type, age, density, or site quality.

156. Regional Distribution of Erosion.—The survey of erosion in the United States recently made by the Soil Conservation Service provides specific information as to the regions of the country in which erosion is serious both in degree and in area affected. The regions that may be distinguished in this way are shown in Fig. 24 (47). The nine of serious erosion are listed in Table 62 together with the principal causes of the

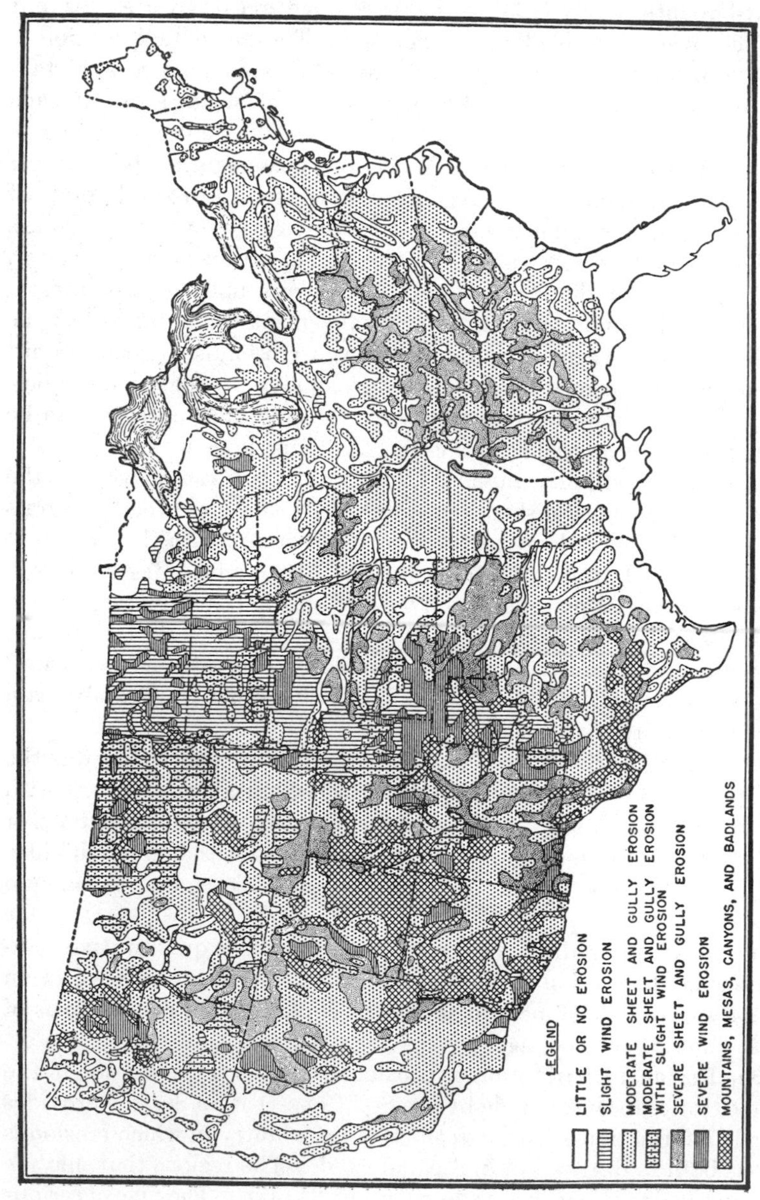

FIG. 24.—Distribution of erosion in the United States. (*Courtesy of H. H. Bennett.*)

erosion. The estimated relative importance of different causes, where more than one has been operative, is indicated by the numerals 1, 2, or 3, 1 indicating the most important and 3 the least important cause. The evidence is quite conclusive that unwise cultivation, often followed by abandonment of the land, and excessive grazing have been two principal causes of serious erosion. The former has been more important in the Piedmont-Appalachian highland and the adjacent

TABLE 62.—PRINCIPAL CAUSES OF ACCELERATED EROSION AND THEIR RELATIVE IMPORTANCE IN DIFFERENT REGIONS

Region	Causes of erosion			
	Over-grazing	Unwise cultivation or abandonment	Repeated fires	Logging
Oklahoma-Texas plains	1*	2*	...	...
Intermountain and southwestern valleys	1	...	...	...
Inner coast ranges of California	1	...	2	...
Piedmont plateau and inner Atlantic plain	3*	1	2	...
Central Missouri-Mississippi Valley	2	1	...	...
Ohio River Valley	2	1	...	...
Appalachian highlands	...	1	2	3
Eastern Oregon-adjacent Idaho	2	1	...	...
Southern California chaparral	...	2	1	...

*1, most important; 2, intermediate; 3, less important.

portions of the Atlantic coastal plain, in Indiana, Ohio, and Kentucky, and also in eastern Washington. The latter has been the principal cause in the Intermountain region, in the Southwest including the plains of Texas and Oklahoma, in the central Mississippi and Missouri Valley, in the territory in and near the Badlands of Montana, and in the foothills of the coast ranges and Sierra Nevada in California. Only in the chaparral region of southern California can repeated fires be considered the principal cause of erosion, and in no region can logging be considered a major cause. Unless the regulation of grazing may be considered a problem in forestry, it is evident that forest practices or failures have not been responsible in any considerable degree for the present conditions of erosion in the United States. On the other hand, forestry has taken an important place in the program to rehabilitate the eroded areas.

157. Erosion by Wind.—Appreciable erosion by air movement is limited to locations where the bare surface of the soil is exposed. Thus wind erosion as a serious problem is found along the sandy beaches of both coasts and the Great Lakes, in parts of the deserts where the sparse vegetation fails to afford protection, and in areas, notably the "Dust Bowl" of Oklahoma and adjacent states and more locally elsewhere, where the vegetative cover has been destroyed or greatly reduced by cultivation or other means.

The intensity of wind erosion becomes more severe with increases in wind velocity, in dryness of the soil, and in fineness of the particles or aggregates at the surface of the soil, at least down to a certain point. Dry beach sand of the textural grade of fine sand on the coast of Oregon began to move when the wind velocity exceeded 9 mph. In the range from 15 to 26 mph the weight W of sand could be expressed in relation to the wind velocity v by the equation

$$W = 0.114v^3$$

where W is pounds of sand per day per foot of width perpendicular to the wind. Velocity was measured at an elevation of 5 ft above the beach (278). Moist sand did not begin to move until the wind velocities exceeded 15 mph.

In the Michigan sand dunes the threshold velocity was 7.8 mph at 5 cm above the surface, and the roughness coefficient was 0.13 cm (211). In the funnel of a blowout, however, the velocity was highest close to the ground and almost uniform from 0.5 to 10 m above the surface.

In wind erosion, particles smaller than 0.1 mm, which is the limit between fine and very fine sand and includes 97 per cent of the loess, are lifted and carried by the wind. Fine and medium sands from 0.12 to 0.5 mm are subject to drifting. Aggregates act as larger particles and do not blow until they are broken by cultivation or trampling. The larger particles of sand and gravel are left on the surface and where they are numerous may form a kind of pavement, which in desert regions has been termed desert pavement.

The recent laboratory and field studies of the movement of desert sands by Bagnold (24) contribute much toward an understanding of the process. Almost all the sands subject to blowing are in the class of fine sand, slightly extended to diameters of 0.3 mm at the upper and 0.08 mm at the lower limit. The movement is a combination of two processes, of which surface creep accounts for about one-fourth and saltation for three-fourths. The latter consists of a series of bounces and the resulting impacts of the sand grains. Sand is not carried in suspension by the wind as is dust, although the turbulence characteristic of winds with velocities of more than 2.2 mph results in upward currents in the wind movement.

In Chap. VIII, Wind, it was shown that as a result of the friction with a rough surface the wind velocity decreased as the logarithm of the height above the surface decreased. The trends for different velocities over any given surface converge at a point K where $v = 0$. Height K is equal to about one-thirtieth the diameter of the grains constituting the surface roughness.

If the drag or shearing force of the wind or any fluid per unit area over a rough surface is represented by τ, then $\tau = \rho v_d^2$ and $v_d = \sqrt{\tau/\rho}$, where ρ is the density of the fluid (air) and v_d is the drag velocity. Because this velocity is proportional to the logarithm of the height above the surface it is also the velocity gradient, so that v_d is proportional to the angle β that $v/\log h$ makes with the height axis. From Prantl's theory of the mixing length, the proportionality factor has been found to be 5.75. Therefore

$$v_d = \frac{\tan \beta}{5.75}$$

$$\tan \beta = \frac{v_1 - v_2}{\log h_2 - \log h_1}$$

subscript 2 representing a higher and 1 a lower level. If the height h_2 is taken as 10 times h_1, so that $\log h_2 - \log h_1 = 1$, then

$$v_d = \frac{v_1 - v_2}{5.75}$$

When the roughness constant K is known, then v at any height may be found from the relation

$$v = 5.75 v_d \log \left(\frac{h}{K}\right)$$

To move particles on the bed the threshold rate of movement must be attained, and it depends upon the Reynolds number, which is equal to $v_d d/\nu$ where d is the mean size of the surface roughness and ν is kinematic viscosity, which for air is 0.14 and for water, 0.01 in cgs units. If $v_d d/\nu$ is greater than 3.5 the surface is "rough," and the threshold velocity v_t is given by the equation

$$v_t = 5.75A \sqrt{\frac{\sigma - \rho}{\rho} gd} \log \left(\frac{h}{K}\right) \qquad (1)$$

where σ is the density of the particles, 2.65 for sands; and A for air is

0.1 and for water, 0.2 mm. This relation holds in air for particles greater than 0.2 mm and in water for those greater than 0.6 mm. It does not apply when $v_d d/\nu$ is less than 3.5.

As the roughness of a surface changes across a boundary by becoming greater, K is increased and v is reduced by the same amount at all heights, whereas the velocity gradient or slope of log h over v remains the same, thus moving the trend line to the left, parallel to its former position. This relation to changing roughness may be applicable to the upper surfaces of forest canopies, provided the trees are evenly and not too widely spaced.

During sand movement the constant K is replaced by K', and the velocity at any height v_h is given by the equation

$$v_h = 5.75 v_d \log \left(\frac{h}{K'}\right) + v_t$$

where v_t is the impact threshold velocity for the sand in question at a height K' above the surface, and v_t is given by equation (1), when $A = 0.08$ and $h = K'$. For air and sand

$$v_t = 680 \sqrt{d} \log \left(\frac{30}{d}\right) \text{ in cgs units}$$

When the threshold velocity v_t is measured at a height K', which is about 1 cm for average dune sand, v_t is 4 m per sec, or 8.9 mph.

Finally, the weight of sand q moved over unit width in unit time is given by the equation

$$q = aC \sqrt{\frac{d}{D}} \frac{\rho}{g(v - v_t)^3}$$

where C is an empirical constant, having a value of 1.8 for natural dune sand and 2.8 for sands of a wide range of texture; d/D is 1 when the diameter d of the natural sand is equal to that of the experimental 0.25-mm sand; ρ is density of air, which is 1.22×10^{-3} g per cm^3; g is acceleration of gravity, which is 981 cm per sec^2; ρ/g is 1.25×10^{-6}; v is wind velocity at any height; and

$$a \text{ is } \left[\frac{0.174}{\log\ (h/K')}\right]^3$$

where 0.174 is the reciprocal of the 5.75 mentioned above as derived from the mixing length. If the height h of measurement is taken as 100 cm and K' as 1 cm, log $(h/K') = 2$ and $a = 6.58 \times 10^{-4}$.

If in the original equation 1.8 is substituted for C

$$q = 5.2 \times 10^{-4}(v - v_t)^3$$

in metric tons per meter width per hour. This relation makes possible the prediction of the amount of dry sand moved by winds of whatever velocity.

Bagnold's results for dune sand have been confirmed and extended by Chepil (83) to other soils and to relations to surface, exposed area, and structure of different soils. The threshold velocity at 6 in. above the surface for particles 0.05 to 0.15 mm in diameter, part of the class of fine sand, was between 8 and 9 mph. For particles of larger diameter it increased with diameter, and for those of smaller diameter it increased as the diameter decreased. The threshold velocity for particles above 0.1 mm varied with the square root of diameter and inversely with the volume weight. Particles below 0.05 mm were highly resistant to erosion by wind. The rate of soil movement by wind varied inversely with the surface roughness K'. However, the value of K' was found to vary not only with the height of the surface projections but also with their character and spacing. For a fine clay dune, initially smooth, K' was 0.9 mm and the threshold velocity was 4 miles per hour (85).

The sizes of particles or aggregates determined by sieving the soil when dry were found to give a satisfactory basis for evaluating erosiveness, even when crusts had formed after rain and exposure. The most erosive soils contained many water-stable aggregates between 0.02 and 0.5 mm in diameter (84). In soils containing mixtures of sizes of particles, the amount of soil eroded by winds of specified velocities varied logarithmically with the ratios of erodible to nonerodible fractions present. A multiple regression equation with four of these ratios as independent variables was derived to express the relation (82).

Aggregates or particles of a size just large enough to withstand movement by wind afford the greatest protection of the soil against erosion. Wind erosion of soils with artificial ridges becomes negligible in from 5 to 15 minutes of blowing. Over bare cultivated fields, the erosion increased with distance from the windward edge up to 1,350 ft as a result of cumulative deposition of the drifting soil. To leeward of stubble strips, reductions in velocity and erosion extended only from 50 to 250 ft (85).

These specific findings confirm two general conclusions. First, any barrier of vegetation or other material reduces the velocity of the wind to leeward as shown in Chap. VIII, and thus also decreases the frequency and amount of erosion by wind. Second, when soil is being blown by the wind, a barrier will also cause deposition of some of the material to leeward.

158. Erosion: Summary.—1. Erosion tends to increase with increase in degree of slope, in velocity and volume of flow, in intensity of rainfall, in size and rate of fall of raindrops, in degree of saturation of soil, in impermeability of soil within about 18 in. of the surface, and in degree of exposure of soil surface as determined by the character and density of the vegetation.

2. As the slope varies by a multiple x the velocity of flow varies as $x^{1/2}$, the cutting power as x, the carrying power as $x^{2.5}$, and the size (volume) of particles carried as x^3. These relations are subject to variation in specific instances.

3. In a given region, climate, natural vegetation, topography, and geology interact to produce a nearly stable soil mantle subject to normal erosion. This normal erosion varies between regions and tends to be less slow in semiarid regions of sparse vegetation. Disturbance of the vegetation and forest floor accelerates erosion.

4. The forms of erosion may be classified as sheet, gully, stream-bank or caving, and landslip or landslide. Incipient gully erosion is often called "rill" or "shoestring" erosion. Headward cutting may also be distinguished as a form of gully erosion.

5. The formation of erosion pavement as a residual layer of coarser particles on the surface after the removal of the finer particles tends to retard erosion and particularly sheet erosion.

6. The chief causes of abnormal or accelerated erosion are disturbance or destruction of the cover by fire, overgrazing, land clearing, improper cultivation, abandonment, construction work, logging, hydraulic mining, smelter fumes, and changing stream channels.

7. Fires or other agencies that destroy most of the vegetation and the forest floor increase the discharge of eroded material by 10 to 6,000 times.

8. Wind erosion is associated only with exposed surfaces of soil. It tends to increase with increase in wind velocity, dryness of the exposed soil, and fineness of subdivision of the surface particles or aggregates, at least down to 0.05 mm diameter.

9. Wind erosion increases with the cube of the velocity in excess of a threshold velocity of about 9 mph for dry fine sand.

10. Particles or aggregates smaller than 0.1 mm in diameter may be lifted and carried by winds of 9 to 25 mph. Grains 0.12 to 0.5 mm are subject to drifting. Larger particles are left and form desert pavement as the finer particles are blown away, thus retarding further removal.

11. Vegetation or any barrier that reduces wind velocity reduces wind erosion and causes deposition of soil to leeward.

CHAPTER XXIII

CONTROL OF EROSION

159. Removal of Causes.—In general, the remedies for accelerated erosion involve the prevention, removal, or regulation of the causes. Because the exposure of bare soil, where wind or water in the form of rain or runoff has an opportunity to move the surface particles, is the immediate cause of most erosion, its control requires first the prevention of exposure of soil. For this purpose, some of the means are well known. Preventing fires and minimizing areas burned is one. Regulating or, in critical areas, excluding grazing is another. Discouraging cultivation of slopes that will erode rapidly is a third. Logging, construction, and mining works could often be located and planned so that a minimum of erosion would be caused. For example, on the Navajo Indian Reservation in New Mexico, logs were skidded along the contours so that the skidding trails formed contour ditches, thus preventing erosion and incidentally conserving water.

If the soil has already been exposed, then the bare areas should be reclothed with vegetation or covered with a mulch of litter, straw, or other similar material. Before outlining these methods of control by vegetation, some of the simpler artificial methods will be considered briefly without attempting to duplicate details of construction, which have been described by Ayres and others (23, 189, 350).

The means of control of erosion by water involve the reduction or diversion of surface runoff or a combination of the two in one or more of three principal ways: First the water may be retained or diverted into channels of low gradient where the velocity and cutting and carrying power will be reduced below the critical point. At the same time the infiltration of water into the soil will be increased by clarifying the water, increasing the head, and enlarging the wetted perimeter. This group of methods of control includes a number of devices that have been extensively and effectively used. Some of them, like pits and terraces, are more applicable to cultivated than to forested lands.

160. Ditches and Trenches.—Contour trenches have been employed with notable success on the range and brush lands of the Wasatch Mountains in Utah (26). In this system the trenches were laid out without gradient, with barriers at intervals along the trench to prevent lateral flow of accumulated water, and at intervals up and down the slope such that the water from any one storm would not be sufficient to overtop the bank on the lower side. In the same class are the interception ditches that are often constructed around the heads of existing gullies to prevent water from flowing down over the exposed bank at the head of the gully. These ditches may be either on the contour or with a very slight slope away from the gully head to prevent the accumulation of water and overflowing of the ditch where the water might again fall into the gully.

Similar systems of ditches or trenches, called "diversion ditches," have been used successfully to carry water from channels or gullies where it was causing serious erosion, laterally along a very low gradient to some point where it could be released on a gentle, well-protected slope, or to ditches or basins prepared to serve as spreading grounds and settling basins. Such a plan on the debris cones of mountain streams in Utah has been described (374) as "the barrier system for control of floods." In this way the water is not only removed as a source of erosion but is also allowed to infiltrate into the soil where it may contribute to the better growth of vegetation or to the recharge of the ground-water table. Where the volumes of water diverted are not large, contour furrows sufficiently long and closely spaced may provide for all flood flows without the nesessity of constructing spreading works.

Soil-collecting trenches across gully channels in badly eroded areas of Mississippi helped to control erosion, although they were devised by Meginnis (247) primarily to provide favorable accumulations of soil for the planting of trees. The trenches were rectangular excavations, 18 in. wide and deep, extending about 3 ft across the drainage channel. The excavated soil was used for a bank on the down slope side of the trench. Although justifiable as a preliminary in the reforestation of difficult sites, they would be less effective than continuous ditches or furrows for erosion control.

The brush wattles placed end to end in trenches along the contours on road fills in California (209) also may be mentioned in this connection, although they are ordinarily used as a first step in the establishment of vegetative control.

161. Check Dams.—Another group of devices to hold water includes the many forms of dams. The larger dams, whether of earth or concrete, require appropriate engineering technique for their design and construction. However, various smaller and simpler structures, usually called "check dams," have been extensively used for erosion control. They have been constructed of many kinds of material, including rock, masonry, concrete, logs or poles, wire, and burlap or brush. Their size is governed in part by their permanency. Thus concrete or masonry check dams well located and constructed to avoid failures may be made 10 or more feet in height. On the other hand the more temporary structures of wood, wire, or brush should be low, usually not over 2 ft, so that in case of failure, serious cutting will not result. All such dams should be provided with a water cushion or some form of apron on the downstream side to prevent cutting by the water as it falls over the spillway of the structure. The spacing of check dams is determined by their height h, the gradient of the channel s_b, and the slope to be allowed for the fill between dams s_f. The distance d between dams can then be computed for slopes expressed in feet per foot as

$$d = \frac{h}{s_b - s_f}$$

Ordinarily, the spacing is such that the top of the spillway of the downstream dam is at or near the level of the base of the dam next above, or higher, if a water cushion instead of an apron is desired. In order to prevent failure from cutting around the end of the dams or against the banks just upstream, it is essential that spillways of ample capacity be provided, sometimes supplemented by wing walls.

Based on some 20 years of experience in the San Gabriel Mountains of California and a century or more in the Alps, Baumann (41) suggested that the following rules might be helpful in the design of systems of check dams for steep mountain canyons.

1. Check dams should be designed for silt and water in the stream bed and for a surcharge and impact due to mud flow, acting simultaneously.

2. Dams should be built of material having high shearing strength and resistance to the weather. If dry masonry is used, the smallest stones outside of fillers and chinks should be at least 1 cu yd in volume or 2 tons in weight. The parts near the crest and the shoulders should always be laid in mortar regardless of the size of the stones, unless other provisions are made to secure adequate bond.

3. Dams should be arched upstream whenever possible and the arch thrust spread by means of flaring abutments or wing walls, or both, unless the arch is anchored into rock or other material of adequate bearing qualities.

4. The overflow section should be wide and level and should be flanked by sloping shoulders of adequate height to prevent scour around the ends.

5. If side slopes are formed by material other than rock or ground of high density and bearing qualities, wing walls should extend far enough above and

below the dam to support adequately the side slopes, to protect them from scour and to prevent piping around the ends.

6. The foundation of one check dam should be protected by the check dam or cross check below, so as to form a cushion the depth of which should not be less than one-third of the fall.

7. If heavy debris flows are expected, no appreciable downstream slope should be provided and there should be no artificial aprons below the spillways. Cushions are much more effective as a protection against scour.

8. Dams and side walls should be properly drained.

9. The side hills should be properly drained where natural seepage flow is insufficient.

10. Check dams must be structures in the true sense of the word and must form an interrelated system. They should be of such height as to effect dissipation of energy during major floods and to stabilize the stream bed and the side slopes so as to permit natural or artificial reforestation of the watershed.

The same principles apply to the smaller check dams of poles, brush, and wire commonly used for gully control in rolling topography, but neglect of or deviations from some of them in the interests of economy entail less serious consequences. Many brush, stake, wire, pole, and loose-rock permeable check dams less than 1½ ft high in small gullies have been effective for several years in preventing cutting and storing sediment. In many cases where they were supplemented by the growth of vegetation and the deposition of litter, cutting has not been renewed even when the dams rotted away. However, the results on many similar areas without check dams indicate that the erosion would have been controlled by the vegetation if the check dams had not been installed. The need for check dams is chiefly in the larger gullies where vegetation is less likely to be effective and where structures of suitable engineering design are required. The high costs of such structures demand consideration of values to be protected and if possible alternative measures such as kudzu in the Gulf States.

Check dams have only limited storage capacity, but they effectively reduce the gradients of channels and consequently the velocity to a point where aggradation replaces erosion. The storage above the dams is quickly filled with sediment and debris. Thereafter, continued deposition gradually reestablishes approximately the original gradient across the spillway levels of the dams. Thus the bed and lower side slopes have been stabilized to the height of the dams. Stabilization of the side slopes to a higher level then requires a new series of check dams. By repetitions of the process, gullies could be completely filled. Often, however, stabilization of the lower slopes may be sufficient to ensure the success of reforestation and thus erosion control for the whole basin.

The Los Angeles Flood Control District over a period of years installed a system of check dams in the canyons along the south face of the San Gabriel Mountains. Most of them were of the wire-rock mattress type. In the flood of Jan. 1, 1934, following a fire in the previous November, most of these check dams were carried away and added appreciably to the damage that was done at the mouths of the canyons. Although a few of the dams stood, notably a series of dry rock checks constructed by experienced men, the general failure of most of the check dams tended to destroy confidence in their effectiveness and resulted in a change to the debris basin as a means of control.

The debris basin is simply an excavation in the alluvial gravel at the mouth of a canyon of sufficient capacity to hold all the solid material brought down in any one storm, and provided with an ample spillway so that the water, after sediment has settled out, can pass on downstream, if necessary through a paved channel. These debris basins are quite large so that their construction is a matter for engineering competence. In locating them it is essential that there be ample space to dispose of the material that is excavated originally and subsequently after each storm or series of storms that may fill them to capacity.

162. Beaver Dams.—In the same group of means of erosion control the possibility of introducing beaver has great promise as an effective and inexpensive method. They build and maintain check dams without cost. Already such introductions have proved valuable in the Rocky Mountains as high as 11,000 ft and near Cashmere, Washington, where Scheffer (309) reported excellent results. In the reach of 2,040 ft in Mission Creek with a total drop of 50 ft, the beavers built 22 dams with an average length of 43 ft and a maximum length of 120 ft. The average height was 2.8 ft, with a maximum of 5 ft and a minimum of 1 ft, and the average spacing between dams was 93 ft, with a maximum of 283 ft and a minimum of 15 ft. These dams stored a total of over 5,800 cu yd of silt and a little less than 1½ acre-ft of water. In Ahtanum Creek, with a flow of 65 sec-ft the beavers built a dam 100 ft long and 10 ft high, with a storage capacity of 2.1 acre-ft. It was estimated that this dam would have cost $2,000 if constructed by usual engineering methods.

Later in Mission Creek all the cottonwood had been cut, and the beavers had to be moved elsewhere to save a flooded road. In any given reach of a stream, their work can be considered permanent only if there is a sufficient supply of trees for food and building material.

Formerly, beaver were found almost throughout the United States, and they can undoubtedly be introduced successfully wherever they can obtain a supply of aspen, cottonwood, or willow. Probably their favorite species is the quaking aspen, which is widely distributed across the whole country in the north and in the higher mountains of the southwest. Along rivers and larger streams, instead of building dams they burrow in the banks below high water levels and thus may undermine vegetative control measures or otherwise increase bank erosion. Where the land slopes and stream gradients are low, as in parts of New England, New York, and the Lake States, the flooding resulting from the beaver dams may become a nuisance on cultivated lands and over roads and trails where they approach the natural drainage ways. However, with this caution in mind, there are still many places where their benefits would exceed the inconveniences, and large additional areas in steep topography where the disadvantages would be nil and where the benefits would be great at an extremely low cost.

163. Other Devices.—Another group of methods of erosion control include those which involve conveying water in artificial channels or protecting eroding surfaces. These include a wide variety of artificial devices, many of which require engineering technique. Gullies or ditches or channels may be lined or paved with masonry or concrete, or culverts or semicylindrical galvanized-iron conduits may be installed to carry the water. Drop-inlet culverts may be included here, although they comprise not only the culvert to convey the water to a lower level but also a small dam downstream from the inlet. The culvert serves in place of a spillway, and it is the dam that functions to control the erosion.

At the heads of gullies where headward cutting and undercutting have left steep drops, various forms of gully head plugs and flumes have been devised to carry water down over the drop and prevent further cutting by preventing the water from having direct contact with the earth banks.

Along the flood channels of larger streams, particularly along the outer sides of bends, to prevent the undercutting and slipping of the banks, various forms of training walls, revetments, ripraps, retards, and jetties have been used. All these come in the class of engineering works and usually involve high costs, which must be balanced against the benefits to be expected.

All the foregoing methods are artificial and more or less temporary, so that they require a certain amount of maintenance in contrast with the vegetative controls, which maintain themselves. Brush and mulch constitute an intermediate group of measures for the control of erosion. They are derived from vegetation and are effective apart from the vegetation but require renewal unless they are promptly reinforced by vegetation.

164. Brush, Litter, and Mulch for the Control of Erosion.—Natural or artificially applied coverings of brush or litter have been shown to be effective in maintaining infiltration at a maximum and in minimizing surface runoff and erosion. Incidentally, mulches also reduce the losses by evaporation and indirectly may improve the survival and growth of trees planted for erosion control. However, this effect had disappeared in the oak–hickory region before the trees were 7 or 8 years old (136). They are effective whether they occur naturally from existing forest cover or when they are applied artificially to bare soils, including road banks. In addition to the beneficial effects in conserving water and preventing erosion they may also be a means of establishing a new cover of vegetation, either from seed included incidentally in the mulch or by providing protection for seedlings that may have been sowed or planted. The extension of the use of brush and litter as a means of erosion control has been a logical outcome of the observations of its beneficial effects.

Although the availability of the material may often be the controlling consideration in selecting what is to be used, there are obvious differences in the kinds, sizes, and amounts which may be applied in any given locality. Where the use of mulch was primarily to aid in the establishment of vegetation for erosion control in the Piedmont region, green pine branches were found to be superior to straw, pine litter, or lespedeza hay. In the same region for checking erosion in gullies the pine litter and lespedeza hay were superior to wheat straw or pine boughs. In an experimental trial in that region, hardwood litter applied at the rate of 23 tons to the acre each fall proved to be slightly more effective than pine needles in reducing both runoff and erosion. A mulch of native weeds and grasses often serves the purpose and will always contain abundant seed. In Arizona, after native grass straw was spread over compacted soil it was disked in and produced a variety of annuals and some perennials. When branches are used the foliage will remain on them longer if they are cut and applied green. A heavy layer of pine litter in one instance resulted in a stand of Virginia and shortleaf pine about 50 per cent stocked. In general, species that naturally produce a heavy forest floor, like Monterey or loblolly pine, are likely to be preferable to those like eucalyptus, which produce little floor.

A variety of sizes, from branches up to 3 or 4 in. in diameter at the large end down to litter, has been used. For erosion control the litter, which lies close to the ground and forms a good contact, is preferable where it will not be washed away. But where pegging or staking are necessary to hold the material in place, the branches form a stabilizing skeleton. For the establishment of reproduction it is better that the mulch be thin and broken and that the foliage be attached to branches which are in part above the ground.

The amount of mulch to be used will also vary with the purpose. For protection of the soil surface exclusively a heavy layer of litter covering 100 per cent of the area is doubtless most effective. On the road cuts in the Appalachian Mountains a mulch of native grass and weeds 4 in. thick has proved satisfactory both as protection and as a source of seed. For the same purpose in Arizona, mats of native chaparral brush are made 6 in. thick, fastened together with wire, and staked in place. Pine and fir brush packed so closely that seedlings were smothered. Experimental work with straw showed that from 4 to 8 tons to the acre, corresponding to depths of 1½ to 3 in. were effective in reducing surface runoff and evaporation (304). For the establishment of seedlings it has been recommended that the mulch should not be too thick and should cover only about 50 per cent of the area.

Various methods of using brush and mulch have been tried, and in general, as experience has been gained, the trend has been from more complete and more expensive to less complete and inexpensive methods. Some years ago, on areas of sheet and rill erosion, broad strips of brush were pegged and wired down along the contour. More recently the strips have been replaced by narrow rows in contour furrows 6 to 12 ft apart, the brush wired down with a single row of pegs. In placing branches, the needles are put next to the soil and the butts on top. On the blow sand near Lake Michigan, oak brush up to 1 in. in diameter, scattered sparsely, was effective in stabilizing the sand and allowing a pine plantation to grow. Oak brush in the southeast up to 2 or 3 in. in diameter was tried, and, although it did not lie close to the ground, it also proved reasonably effective. Brush wattles in contour trenches will be described as a part of one method of control of erosion on road banks.

For the control of erosion in gullies and channel outlets, brush has been used in several different ways. In some instances it has been used to form mats either continuously or at intervals of 5 to 10 or 12 ft along the gully bottom. For these mats, pine brush with butts up to 1 in. in diameter were laid with tops uphill and pegged and wired close to the channel bottom. More commonly the brush has been used to form small check dams usually not over 1 or at the most 2 ft high in the gully bottoms. For this purpose, oak, juniper, eucalyptus, or other brush up to 2 in. in diameter has been staked in the gully bottoms with the tops upstream, usually covered with soil and tramped down firmly. In northern Mississippi, Meginnis (246) developed the pole-frame brush dam and these brush check dams left permeable have been as effective as and have not failed more often than tight checks. The use of wire in their construction is not necessary. In other instances steep gullies have been stabilized by the use of pine limbs up to 3 in. in diameter placed crosswise in the gully at 5 to 10 ft intervals and fastened down with stakes at intervals of 2 to 3 ft. Soil is piled against the brush on the upstream side and tramped down. Brush dams with one or two rows of stakes or posts supplemented with mulch and wire have been widely and successfully used. In these the brush with butts upstream forms a sloping downstream face. Specifications are available in several sources (23, 350). As in all overfall dams the spillways should be sufficiently deep and wide to provide ample capacity for flood flows. Tables give discharge capacities for rectangular spillways of different depths and widths (23, 196).

Low elongated diversion barriers of brush and wire along the contour are used on the gently sloping alluvial fans in the Southwest as a means of spreading flood waters to increase infiltration and the establishment of vegetation.

The effects of mulch on the establishment and growth of plantations vary both with the species and the soils. Obviously a mulch is essential as a preliminary to planting on blow sands. Similarly on steep banks subject to immediate washing it provides protection for the period before vegetation can become established sufficiently to exercise control.

In making recommendations for tree planting on old fields and gullies in the Central states, Ligon (221) included suggestions for the use of mulch on the different soil series. He found mulch unnecessary for the favorable soils derived from limestone and shale but recommended its use on some of the soils from cherty limestone and on the deeply leached soils of Illinoian till and of the inner coastal plain, which are only thinly covered by loess. The latter group includes the sites on which pines are recommended for planting, and in general he concludes that mulches are more likely to benefit pine than locust plantations.

In the uses of brush and litter, part of the effectiveness of growing vegetation for erosion control has been anticipated because litter is always associated with the vegetation.

165. Erosion Control: Summary.—1. The remedies for accelerated erosion should be sought first in the prevention, regulation, or removal of the causes.

2. Erosion may be controlled by decreasing or diverting surface runoff or both in three principal ways:

a. Hold or divert water into basins or low-gradient channels where erosion is negligible.

b. Convey water in artificial channels or protect eroding banks.

c. Cover eroding areas with brush, litter, or vegetation thereby increasing infiltration and retarding surface runoff.

3. The first group includes contour trenches or furrows, diversion and interception ditches, dams, check dams, beaver dams, and debris basins.

4. The second group includes culverts, paved channels, flumes, retaining walls, ripraps, jetties, revetments, retards, tree tops, and tetrahedrons.

5. Brush and litter may effectively reduce sheet, rill, small gully, bank, and wind erosion temporarily but require renewal or reinforcement by vegetation.

CHAPTER XXIV

VEGETATION FOR EROSION CONTROL

The use of vegetation for the control of erosion may be either as a supplement to construction methods or as a self-sufficient procedure in suitable situations. It has distinct advantages. The first cost is low. Little or no maintenance is required. There is the possibility of realizing valuable by-products such as fuel, posts, timber, or landscape beauty. Most important, it is a constructive method, because the vegetation, once established, becomes more and more effective each year as it grows, as a means of controlling the erosion. This statement is subject to qualification if precautions are not taken to prevent renewed cutting and a new cycle of erosion. "When trunk streams are lowered, all tributaries tend to adjust themselves to that level" is a generally accepted principle. Thus gullies in which vegetation is established and erosion stopped may be subject to headward cutting from the outlet and to undermining of the vegetation if a sudden lowering of the level and accelerated erosion in the main stream is not prevented. Vegetation by itself is suitable for all types of sheet erosion for the smaller gullies, and either by itself or as a supplement to mechanical methods in larger gullies, on stream banks, on road slopes, and as a means of reducing the frequency of occurrence of slips and slides, if not of wholly preventing them. The subject of erosion control by vegetation thus involves a wide variety of situations and methods and of species to be used.

Classification of the vegetation has been suggested by Kraebel (209) to include, first, temporary; second, semipermanent; and third, permanent vegetation. The first group would include those plants which would give prompt and effective control of exposed soil by reason of their quick growth, fibrous roots, ability to endure unfavorable site conditions, and abundant and cheap seed. The cereal grains, grasses, mustards, sweet clover, annual lespedezas, and vetch are examples in this group.

The semipermanent vegetation includes the species that are perennial but will be replaced naturally or artificially by other larger and more permanent species. This group includes the willows, the willow-like species of *Baccharis* (*B. viminea* in southern California and *B. glutinosa* in Arizona and New Mexico), blueberry elder, cottonwood, various perennial grasses, and, in the Southeast, *Lespedeza sericea,* Japanese honeysuckle, and kudzu.

The permanent vegetation includes the long-lived species, usually the native woody species, which will occupy the site indefinitely or at least until they are replaced gradually in the process of natural succession.

166. Time to Become Effective.—In the selection and use of species to remedy erosion, the time that will be required for vegetation to reestablish control and minimize the erosion on areas where it has been serious is an important consideration on which some evidence is available. For prompt cover the annual plants mentioned above as temporary vegetation provide the quickest protection. The grasses that spread vegetatively like Bermuda grass become effective sooner than those that reproduce only by seed. Sodding with such species further accelerates control of erosion. Kudzu and sericea give considerable protection in the first year but take 2 or 3 years for maximum usefulness. Longer periods, usually 5 to 10 years, are required for the necessary development of crowns, floor, and roots of the woody species. In northern Mississippi over a period of 2 years under a plantation of black locust and Osage orange 23 years old, the soil eroded at the rate of 1 ton per acre (245). Barren, abandoned fields, presumably in the condition of the area on which the planting was done 23 years previously, yielded 160 tons per acre. Unburned mature oak forest gave only 0.05 ton. Thus the plantation in 23 years had greatly reduced the erosion but not yet to the degree maintained by the natural forest.

In the same locality, after 1 year of measurements on small plots on an abandoned field of 10 per cent slope where the soil was exposed, compacted, and eroding, the oak litter from an adjacent forested plot of the same size was added to the bare soil. The soil losses from the first year's precipitation were reduced to 1/88 of the former amount. The artificial application each autumn of 2-in. layers of forest litter to freshly cultivated sandy clay loam of the Piedmont in North Carolina kept the erosion to less than 0.3 ton per acre, when an unprotected cotton field gave erosion of 15 tons the first year and over 40 tons per acre the second and third years (Table 63). The control by the litter improved in the 2 later years. Hardwood litter was slightly more effective than pine. The annual applications of 23 tons to the acre were much heavier than the usual total forest floor in that region and probably also heavier than would be needed for the purpose.

The forest floor under most forest types becomes effective in controlling erosion some time before the age of crown closure. Obviously a plantation of trees in the first year produces little litter. The number of years for the development of the forest floor as for crown closure varies with the species, density, and site quality. "Southern pines often furnish considerable site protection by the second or third year" according to Meginnis (245). On favorable sites, 6 by 6 ft spacing of black locust resulted in crown closure in 5 or 6 years. The crowns of loblolly and slash pines planted 6 by 8 ft closed in 7 or 8 years, and those of shortleaf pine in 10 years (360). At 8 by 8 ft spacing, loblolly pine requires 10 years for crown closure. In New England the crowns

TABLE 63.—EFFECT OF UNDECOMPOSED FOREST LITTER IN 2-IN. LAYERS, 23 TONS TO THE ACRE APPLIED EACH FALL, ON EROSION OF NEWLY CLEARED CECIL SANDY CLAY LOAM, ON 10 PER CENT SLOPE, AT STATESVILLE, NORTH CAROLINA (47)

Cover	1934	1935	1936
	49.34 in. rainfall	42.80 in. rainfall	60.00 in. rainfall
	Erosion, tons/acre		
Pine needles	0.27	0.18	0.11
Hardwood litter	0.25	0.12	0.09
Cotton plants only	15.15	40.72	40.92

of white pine, jack pine and Norway spruce plantations close in 14 or 15 years on the average, although the time may vary from 8 to over 25 years. Red pine crowns close at 11 years with a range of from 7 to 18 years with 6 by 6 ft spacing and 4 or 5 years later with 8 by 8 ft spacing (145). These two groups of species approach the extremes in rate of juvenile development, the southern pines growing rapidly and the northern species slowly. It is reasonable to expect, therefore, that most species used for erosion control in these or other parts of the United States will lie within these limits of 7 to 25 years in the ages of crown closure. The time for the formation of effective forest floor will be somewhat less than that of crown closure, because at the same age the coverage of floor is greater than that of the crowns. This was shown in the chaparral of southern California where floor coverages of 0.8 and 0.9 were associated with crown densities of 0.7 and 0.8 respectively (200).

Erosion control by means of vegetation involves four sets of interrelated and complex variables. These four are the kinds of erosion and the resulting sites where it must be controlled, the region and its climatic characteristics, the methods to be used, and the species, including their habits and adaptability to the sites. The most useful approach is probably to make the primary classification on the basis of the sites and kinds of erosion and for each of them to take up the methods and species suited to each region. Following the classification previously outlined, the different forms of erosion will be treated separately, namely, sheet and rill erosion, gullies, stream banks and channels, slips and slides, sand dunes and blowouts, and snowslides.

167. Vegetative Control of Sheet and Rill Erosion.—The areas of sheet and rill erosion include fresh burns, road banks, smelter-denuded areas, mine dumps, overgrazed range or pasture lands, and abandoned cultivated areas. In the case of fresh burns, in most parts of the country where growing conditions are reasonably favorable, natural vegetation reclothes the exposed soil within a year or two after the fire and effectively limits further erosion. In the less favorable growing conditions of the foothill and chaparral areas of California where a concentrated rainy season follows immediately after a prolonged fire season, it is essential to provide vegetative protection for the exposed soil as promptly as possible after the first autumn rains. Black mustard and mixed grains (oats, barley, wheat, and rye) have been used for this purpose, because they will germinate and grow during the winter period and thus provide protection during the first season when the erosion is most serious.

The range lands in the west and the pastures in the east, where excessive grazing has reduced the vegetation to a point where erosion has been accelerated, constitute areas of great importance both in the large area involved and in the seriousness of the need for control. The subject, however, involves almost the whole field of range reseeding and pasture improvement and is treated elsewhere in considerable detail (307, 334). A variety of native grasses adapted to the different regions and some introduced species are being used with notable success. These include crested and western wheatgrass, blue grama, side-oats grama, big bluegrass, buffalo grass, bluestem, and switchgrass. All these species in climato-geographic regions to which they are adapted have proved effective in controlling erosion on badly deteriorated areas. Perhaps the chief requisite for their success is that grazing be excluded during the period when they are becoming established.

168. Mustard.—The extensive use of black mustard for the control of erosion in the first season following fires in the chaparral of southern California justifies some details. Seed has been obtainable in large quantities at a cost of 5 to 7 cents per lb. It is sowed at the rate of 5 lb per acre, either by a ground crew, or in inaccessible areas, by airplane at a somewhat higher cost. Such sowing has been found to give from 5 to 6 plants per square foot or 200,000 per acre over 90 per cent of the area. Not over 10 per cent of the seed is carried away in heavy rains, and even then it tends to concentrate in the small gullies where protection is most needed. It germinates and grows promptly after the first fall rains. The taproots extend to a depth of 3 ft, and in addition there is a good development of laterals. The litter, after the first year, may amount to as much as 2¼ tons per acre, and when dry it is less inflammable than that of most other species. The stand density increases as much as 100-fold in the second year, and it reseeds naturally for 3 or 4 years after sowing. Over 50,000 acres have been sown on burns in the national forests of southern California and have provided more satisfactory control than could be obtained by the use of clover or grain (138).

169. Road Banks.—Erosion on road banks may involve both mechanical and vegetative means of control. Vegetation can be used effectively if the slopes are graded down to the angle of repose of the material. Cuts are more difficult sites than fills because of the compact infertile subsoil, as compared with the loose and thoroughly mixed material of the fills. Three different plans adapted to different regions have been developed for the stabilization of road banks. Under the favorable growing conditions in the Gulf States, natural revegetation is so rapid on the fills that special control measures are not needed. On the cuts the following method has been successful: First, the surface is thoroughly roughed. Then about 3 in. of topsoil is spread over the surface. Sods of Bermuda grass 3 to 6 in. in diameter are placed at 1 ft intervals up and down the hill, staggered so that channels will not form between rows of sods. In that region, the method has proved thoroughly effective and more satisfactory than the use of brush or mulch. It has the disadvantage that the grass becomes dry and inflammable during the winter and in some cases may become a nuisance by spreading to adjacent cultivated lands. However, these have not been considered serious objections to the method. Bermuda grass is hardy as far north as Maryland and Kansas and into the southwestern valleys.

A different method has been developed in the Appalachian region of North Carolina and adjacent states, where Bermuda grass is not hardy. The road cuts are first graded to a slope of 1:1 or less. Stakes 18 to 24 in. long and 2 in. in diameter are spaced 12 in. apart over most of the surface. About 6 in. of each stake is left above the surface.

Brush of less than 1 in. in diameter is then hung from all stakes. A mulch of weeds, brush, and litter 4 in. thick is spread over the exposed soil. A second layer of brush is placed over the litter. Finally, poles from the cut brush are laid horizontally on the bank and if necessary cross-staked in order to bring the whole layer close to the bank. Various native plants and grasses tend to come in naturally after this treatment, and the conditions are favorable for the planting of honeysuckle or other species if more permanent control is needed (180).

In California, where winter rains cause the washing, a method has been developed for the control of road fills that involves a combination of mechanical and vegetative control (209). The slope is first graded down to the angle of repose, usually about 34 deg or 67 per cent. Any large boulders that might later get loose are rolled down or carried off. The toe of the fill is stabilized by a permanent barrier to prevent undercutting by concentrations of water from above. In all work of this sort it is essential that the drainage water from the road be carried in culverts or conduits away from the fill itself and, on steep slopes, even down to the natural channel to avoid serious cutting. Beginning at the bottom, contour trenches about 6 in. deep are dug at intervals of 3 to 5 ft slope distance. In these trenches, wattles consisting of bundles of any handy brush 4 to 6 ft long and 4 to 6 in. in diameter are placed end to end and covered with earth. The crew stands on the lower bundles, thus tramping them in at the same time that the next higher row is being placed. The wattles are staked at intervals of 1½ to 3 ft with a combination of live and dead stakes. Usually at least 25 per cent of live stakes are used. Native willows or, in southern California, mulefat have proved successful. The live stakes are ordinarily 3 ft long and not more than 2 in. in diameter and are placed so that about one-third of the length including at least one good bud projects above the ground. In hard ground a crowbar should be used to make the holes, so that the top of the stake will not be split by driving. Scrap lumber or any convenient material can be used for the dead stakes. The live stakes take root and grow and provide a more permanent control as the wattles and dead stakes rot away. The wattles and stakes may be supplemented by sowing mixed grain—wheat, rye, oats, and barley—in contour rows, one just above and one about halfway between the rows of wattles. The grains should be sowed about 1 in. deep and covered. Six to eight ounces of mixed grain of 80 per cent germination per 100 linear feet of trench give 1 seedling per square inch in a strip 4 to 6 in. wide. Some of the grain reseeds itself in the second year, but little appears thereafter. Its chief value is in the first rainy season, when it provides anchorage and cover before the willows or other species have started to grow. The mechanical support provided by the wattles and stakes even before the vegetation has had time to start has proved surprisingly effective in stabilizing large road fills. For example, a fill that, before treatment, eroded seriously with ¼ in. of rain, 2 weeks after treatment showed no erosion at the toe in a rain of 1 in. in 6 hours.

Ten years after the treatment, this fill had 600 living clumps of willows (*Salix laevigata*) per acre, representing about 20 per cent survival, with 2 or 3 stems per clump, averaging 13 ft high. In addition, there were 60 Monterey pines per acre, averaging 16 ft high, partly volunteers and partly planted, and 200 plants per acre of the native *Baccharis pilularis* about 6 ft high. The forest floor on the average was 1 in. deep and covered 0.6 of the surface.

A mulch cover of hay, straw, leaf litter, or the leafy parts of brush or treetops forms an effective protection on road banks for a temporary period until vegetation becomes established. Such a mulch initially should be from 2 to 6 in. deep or from 1 to 10 tons to the acre.

A summary of the species used for the stabilization of road banks in different regions is contained in Table 64. In this and in similar subsequent tables for other sites, the regions are those of natural vegetation (313) for which the following abbreviations are used.

LLP	Longleaf–loblolly pine
OP	Oak–pine
OC	Oak–chestnut–yellow poplar
OH	Oak–hickory
BM	Beech–birch–maple
JP	Jack, red, and white pine
TG	Tall grass
SG	Short grass
SB	Sagebrush
CB	Creosote bush
DG	Desert grass
BG	Bunch grass
C	Chaparral
J	Piñon–juniper
P	Ponderosa pine
R	Redwood
DF	Douglas fir

The species are grouped under three headings, namely, trees, shrubs and vines, and grasses and herbs. Within each group an arrangement has been attempted under which the order is from the more to the less important or widely used. The characteristics of the different species that contribute to their usefulness are not yet sufficiently known to justify tabulation. Consequently, those that are known are given in the text following the first mention of the species on a site to which it is well adapted.

TABLE 64.—SPECIES FOR STABILIZATION OF ROAD BANKS

Species	Region					
	LLP	OP	OC	OH	BG	C
Trees:						
Willows	...	...	...	...	...	x
Black locust	...	x	x	x	...	...
Shrubs and vines:						
Mulefat	...	...	...	...	...	x
Coralberry	...	x	x	x	...	...
Bush honeysuckle	...	x	x	x	...	...
Kudzu	x	...	...	...	...	...
Japanese honeysuckle	x	x	x	x	...	...
Virginia creeper	...	x	x	x	...	...
Nootka rose	...	...	...	...	x	...
Grasses, etc.:						
Bermuda grass	x	...	...	...	...	...
Kikuyu grass	...	...	...	...	...	x
Canada bluegrass	...	x	x	x	...	...
Orchard grass	...	x	x	x	...	...
Mixed grain	...	...	...	...	...	x
White sweet clover	...	x	x	x	...	...
Redtop	...	x	x	x	...	...
Italian rye grass	...	x	x	x	...	...

170. Mining Wastes.—There are two outstanding examples of areas completely denuded by smelter fumes where preliminary trials have been made by way of erosion control and rehabilitation. Both areas represent the most severe gully as well as sheet erosion. In the Copper Basin, near Ducktown, Tennessee, several species have been planted on small areas mostly since 1936 but long enough to provide a tentative indication of their value. Pitch and Virginia pines have shown the highest survival on the most severe sites. Loblolly pine has less high survival but makes the most rapid growth on the less unfavorable sites. A plantation 15 years old and 25 ft high had a forest floor ½ to 1 in. thick which had stabilized the smaller gully channels. Another loblolly pine plantation 10 years old and 20 ft high had a closed canopy, 1 in. of forest floor, and a good stand of Bermuda grass in the openings where trees had failed. Shortleaf pine was the poorest of the four pines in both survival and growth. Black locust did quite well for the first 10 years but thereafter failed suddenly before the trees reached fence-post size. In the deposits above check dams young trees looked more promising. However, here, as in many other localities under the locust, a good stand of broomsedge had become established while none appeared in the adjacent area without locust. Kudzu was tried and after 3 years had almost disappeared. Weeping lovegrass has been seeded on experimental basins by the Tennessee Valley Authority and has formed a dense cover which in combination with stake-wire-brush check dams has controlled the erosion.

In the Kennett smelter area in Shasta County, California, the poison oak and a native manzanita (*Arctostaphylos viscida*) are coming in naturally in numbers. A considerable variety of grasses, herbs, shrubs, and trees have been seeded or planted, but most of them have been failures. After 5 years, cuttings of mulefat showed 20 per cent survival, a few willows survived in the gully bottoms, and root cuttings of the tree of heaven showed 82 per cent survival. Douglas fir and incense cedar were moderately successful. A few volunteer knobcone pines have made the best growth. The two species that on the whole are most promising are ponderosa and Jeffrey pine. Planted nursery stock of these two species survived to the extent of 80 to 95 per cent after 5 or 6 years, and the trees had attained heights of 3 to 5½ ft.[1] These plantings, however, have not been on the most severely eroded parts of the area, and the gullying has not yet been arrested in the plantations. The question remains unanswered whether the trees will gain control or whether they will be undermined in the continued cutting of the gullies.

Somewhat similar conditions of raw subsoil exposed in steep slopes are represented by the spoil banks from the stripping of the surface layers in the coal-mining sections of Pennsylvania, Ohio, and Illinois. Plantings for the stabilization of these dumps have been tried on an experimental scale for a number of years. In Ohio the character of the spoil-bank materials has been suggested by Chapman (81) as a guide to the time the spoils should be allowed to settle before planting. Where the sandstones and shales are mixed with overlying glacial till, the planting is likely to succeed the first year after the mining. On the spoils of sandstone and acid shale it is safer to wait 2 or 3 years. The deep overburdens of sandstone, marly shale, and limestone should be allowed to settle 8 or 10 years before planting. In Pennsylvania

[1] Unpublished data furnished through the courtesy of C. J. Kraebel of the California Forest and Range Experiment Station, U.S. Forest Service.

the results indicated that some of the pines including Scotch, pitch, jack, Virginia, shortleaf, and white, are the most promising species for this purpose. In this locality black locust proved a near failure (232). On the other hand, in parts of Ohio and Illinois, black locust has been the outstandingly successful species. It has grown in height 4 or 5 ft annually after the first year, attained diameters of 2 or 3 in.

TABLE 65.—SPECIES FOR STABILIZATION OF STRIPPED-COAL AND OTHER MINING WASTES

Species	Region		
	OC	OH	P
Trees:			
Pitch pine	x	...	...
Jack pine	x	...	...
Virginia pine	x	...	...
Loblolly pine	x	...	...
Shortleaf pine	x	x	...
Black locust	x	x	...
Eastern redcedar	x	...	...
Scotch pine	x	x	...
Red pine	x	...	...
White pine	x	x	...
Ponderosa pine	...	...	x
Shrubs:			
Coralberry	...	x	...
Grasses:			
Weeping lovegrass	x	...	...

in 3 or 4 years, was not seriously damaged by the borer, and promised to yield a good crop of posts or mine props. In a 5-year-old plantation in Ohio a 2-in. layer of forest floor had formed. The localities where the locust has been successful were underlain with either sandstone and acid shale or limestone. The locusts usually have grown well on soils high in lime content, but where the lime is combined with clay that bakes hard in dry hot weather the locust may fail. In Ohio, white, red, jack, pitch, and shortleaf pines and eastern redcedar have done well on the spoil banks. White and green ash, northern red oak, and yellow poplar did best under a nurse crop of black locust. In southern Illinois, shortleaf pine has been almost as successful as the locust.

Coralberry is invading the stripped coal lands in places in Illinois and may be a useful supplement to tree planting on the most unfavorable sites. It spreads by numerous runners to form thickets. Species used by regions are summarized in Table 65.

171. Old Fields.—Old fields, formerly cultivated and later abandoned usually as a result of progressive erosion, constitute another serious problem in many places in the eastern half of the country. The problem is more acute in those areas where the cultivation has been practiced on sloping land, reaching a climax on some of the steep slopes of the Appalachian Mountains where abandonment follows clearing and cultivation after only 2 or 3 years.

On old fields and on gullied lands also, as a result of early successful trials in certain localities, black locust has been considered the outstanding tree species for erosion control. Until recently it has undoubtedly been the foremost species in the number of localities and the number of trees that have been planted for this purpose. As time has allowed at least a preliminary evaluation of the results, it has become evident that there are many localities where black locust is distinctly undesirable for planting because of its poor growth and susceptibility to the locust borer. These areas where locust has proved a near failure include much of the Piedmont and inner coastal plain. However, within this large area there are notable variations in the behavior of the species. A recent study, the results of which are summarized in Table 66, indicates some of the causative variables. In general it is clear that locust does not do well on the soils with a high water table or where a compact and impervious subsoil is close to the surface. The same conclusion has been reached by Auten and Ligon (22, 221) in the Central states. It responds to fertilization. Its growth becomes much worse as the degree of erosion increases. These results have been confirmed in other studies. In western Kentucky and Tennessee the growth is poor where less than 8 in. of loess overlies gravel and sand, whereas it is good when the layer of loess is more than 12 in. deep. The development of black locust in the Central States on sites covering a range of from 37 to 133 in site index is found by Auten (22) to be related to the plasticity, compactness, and structure of the subsoil. The plantations did poorly on the poorly drained or excessively drained soils and well on the large group of intermediate soil types. No relation was found to the acidity, nitrogen, phosphorus, or organic-matter content of the soil or to the properties of the *A* horizon. Trees in a plantation on soil derived from limestone in Indiana attained heights of 75 ft and diameters up to 15 in. at 31 years (14). The best plantations in New York State are reported by Cope (99) to be growing on limestone formations.

A survey of numerous plantations on most of the eroding soil types in Ohio, Indiana, Kentucky, and Tennessee led Ligon (221) to conclude that "locust grows best on calcareous soils, soils of good drainage, aeration and moisture holding capacity, soils free from compact impervious layers and from a high content of iron or manganese concretions." The favorable soils thus included those derived from the noncherty limestones and some shales, and from loess, and the unfavorable soils included those from sandstones, cherty limestones, and from the deeply leached soils of the Illinoian till and of the coastal plain. Subsequently, Gibbs and Ligon (136) have summarized by the statement, "while a calcareous influence in the soil is not always necessary for good locust growth, practically no failures occur when it is present; other failures are too numerous."

TABLE 66.—AVERAGE HEIGHT INDEX OF PLANTED BLACK LOCUST AT 4 YEARS ON COASTAL PLAIN AND LOESSIAL SOILS FROM NORTH CAROLINA TO MISSISSIPPI, IN RELATION TO DEGREE OF EROSION AND SOIL-PROFILE CHARACTERISTICS (140)

Profile characteristics	Height on noneroded soil, ft	Degree of erosion	Height on eroded soil, ft
More than 6 in. topsoil over plastic subsoil	21	Trace	17
Well-drained friable, colluvial	18	1	11
More than 6 in. topsoil over subsoil with moderate to no compaction	12	2	10
Wet subsoil	7	3	5
Six inches or less of topsoil over plastic subsoil	6	4	1
Mottled subsoil	5		
Subsoil at surface	1		
Piedmont soils	....	Old-field gall spots	4
		Gully sides and ridges	9
		Gully bottoms	19
Effect of fertilizers			
Median for 18 unfertilized (above)	10.5		
Median for 6 fertilized with lime and manure	15		
Median for 11 fertilized with N-P-K.	21		

Experiments with fertilizers on the cherty limestone soils in Indiana showed that locust responded to applications of 2–12–6[1] or 4–24–12 in which the phosphorus was the chief cause of the increased growth. On worn-out fields of Cecil soil near Spartanburg, South Carolina, locust growth improved with application of 5–12–4 fertilizer and still more when this was supplemented with agricultural lime, but the improvement lasted less than eight years.

In northern Mississippi black locust does well on the brown loam and loess soils but poorly on wet, poorly drained, or very sandy soils. In east central Mississippi the average heights of black locust in plantations 4 years old increased at a decreasing rate with increasing depth of the surface soil above a compact clayey horizon from 2 ft for ½ in. depth to 12 ft where the surface soil was 37 in. deep (295).

Cultivation before and after planting and mulching have been tried as means of improving the survival and growth of the locust. In southern Illinois and Ohio benefits were observed on some critical areas. In northern Mississippi a single cultivation resulted after 55 days in a fourfold increase in leaf area and a corresponding increase in height growth (248).

The planting of black locust in mixture with other species has been suggested as a means of reducing the damage by the borer and improving the value and utilization of the products from the plantations. However, in the Central States, Kellogg (193) cautions against the simultaneous planting of conifers with black locust. With few exceptions such mixtures, including those with white, jack, red, ponderosa, Scotch, shortleaf, and Austrian pine and with Norway spruce and Japanese and European larch, were failures. The locust, by reason of its early rapid growth, overtopped and suppressed the conifers, and their growth was greatly retarded if they did not actually die. On the other hand, examples were found where sugar maple, which came in under the locust voluntarily, was doing well in mixture. Abundant reproduction of other hardwoods also may become established under black locust in plantations 6 to 10 years old (80). This suggests that the tolerant hardwoods may be used effectively in mixture with locust.

The reasons for the extensive use of black locust for erosion control are several. First, it was used with notable success for gully reclamation in Tennessee under the guidance of State Forester Maddox in one of the earliest examples of that kind of work in the United States. Second, it is a member of the legume family, and, like many other members of

[1] Two per cent total N, 12 per cent P_2O_5, 6 per cent K_2O.

that family, has the faculty of producing root nodules that fix atmospheric nitrogen and thereby improve the nitrogen content and fertility of depleted soils. The degree of this effect has been indicated in Sec. 100.

Third, black locust has a large and widely spreading root system that is effective in holding the soil. In a 25-year-old plantation, suckers from the roots were found as much as 83 ft beyond the border of the plantation. The species spreads largely by suckers, and Larsen has shown that the rate of spread in this manner is from 3 to 10 ft each year (212). The roots spread widely up or down the slopes, under gullies, through loose shale or hard subsoils. They do not invade the native woods. In general the species does not produce taproots, although, in shelterbelts along the 100th meridian, depths of penetration of 5 to 10 ft and even down to 25 ft have been recorded.

Fourth, black locust produces a litter exceptionally high in nitrogen, and, although as a result the decomposition is rapid, the effect in improving the fertility of impoverished soils is an important one. It may be for this reason that the grasses develop early and densely under stands of locust. In any case the grass sod is often more effective than any direct influence of the locust trees in controlling erosion.

Other advantages of black locust include abundant seed production and rapid seedling growth so that 1–0 stock can be used in planting. Natural reproduction by suckers from the roots is abundant although this is not always an advantage. The species is quite resistant to drought and to frost damage except in the northern plains. If the site is favorable the growth rate and crown development are rapid so that in plantations spaced 6 by 6 ft the crowns close within 5 to 6 years.

Its chief disadvantage is its poor growth and susceptibility to the locust borer on unfavorable sites. This becomes less serious or negligible in stands that grow vigorously or in which locust is mixed with other species.

In the oak–pine region where black locust has often been planted on eroding abandoned fields, the badly forked and borer-infested trees are not much improved by pruning or by cutting them back to renew the growth by sprouts. The trees respond to applications of fertilizer at the time of planting but only for a limited number of years thereafter. It has been suggested that by using fertilizer and thus obtaining a larger development of the trees, a wider spacing of the planting would become feasible, and the saving in the cost of stock and planting resulting from the wider spacing would more than compensate for the cost of the fertilizer. Questions remain, however, as to the duration of the benefit from the fertilizer and how many repetitions of the applications would be needed to assure development of the trees to a size to produce posts.

On the same abandoned fields, which cover so much area in the Piedmont and inner coastal plain on the slopes where erosion has exposed the red *B* horizon in spots locally known as "galls," interspersed with broomsedge and where the black locust is proving to be almost a failure, loblolly, slash, and shortleaf pines have been planted with notable success. All three have done well without fertilization or cultivation on the tight upland soils. Loblolly and slash grow more rapidly than the shortleaf, and the slash pine has grown almost as well as the loblolly as much as 200 miles north of its natural range. However, the survival of the slash pine in the first few years is less high than that of the other pines. Formerly the slash pine was planted on the moister slopes, the loblolly on the medium slopes, and the shortleaf pine on the dry ridges. More recently experience has shown that loblolly pine will grow anywhere that shortleaf will and more rapidly. In central Alabama and northern Louisiana where the compact, red, cemented subsoil is exposed in the galls, slash and loblolly pines have been used on almost all sites. They grow well both in the broomsedge and on the galls. Cultivation after planting improves growth but is not essential. Slash pine is preferable on the moister sites. In northern Mississippi and northward in the corresponding silt loam uplands in western Tennessee, loblolly and shortleaf pines have given good results without special preparation. Near Lindale, Texas, loblolly pine has made good growth, both on an old field of Norfolk sand and on heavier soils.

In general, on the abandoned fields of the southeast where the soils have been strongly leached and the *A* horizon more or less removed by continued erosion, the three pines are proving to be the outstanding tree species for erosion control. The more rapidly growing loblolly and slash pines are being planted and so far have succeeded many miles north and west of their natural ranges, although in this extension there is the danger that serious damage will result in years of climatic extremes. These two species grow rapidly even on poor sites. In 4 to 6 years, with spacing of 6 by 6 ft, the crowns have closed, and ½ in. or more of litter covers the ground completely. Loblolly, with its wider crowns and more branches, lays down more litter than slash pine. Taproots of shortleaf may go down 14 ft in 8 years, although most of the roots are concentrated in the *A* horizon or upper layer where they may spread to 1½ times the height of the tree. Loblolly roots rarely penetrate more than 4 or 5 ft and the laterals extend 25 or 30 ft radially. Shortleaf pine is subject to damage by the tipmoth and by the littleleaf disease. Loblolly and slash pines are damaged by ice storms, and slash frequently has cankers on the stems. Seed and planting stock are readily available.

In the northern portion of the Piedmont and adjacent coastal plain and on the Allegheny Plateau, pitch pine and Virginia pine have also been used effectively for the reforestation of abandoned fields. These two species are naturally pioneers in succession on the most unpromising sites and do correspondingly well on old fields. Within the range of shortleaf and loblolly pine they are probably not more effective for erosion control and are certainly less desirable in respect to growth and utilization. Near Norris, Tennessee, white pine and loblolly pine have made the best showing of any of the pines in plantations 8 years old. Yellow poplar has been the best of the hardwoods.

Still farther north in New England, New York, and the Lake states, reforestation of abandoned fields has been in progress for many years, although not primarily as a measure of erosion control. The red and white pine, which are most commonly planted, are effective in control on those lands where erosion is noticeable. White pine is apt to make a less uniform showing than red. On a good site in Connecticut, the white pine forest floor was 1 to 2 in. thick and had a dry weight of 3.2 to 6.8 metric tons per acre in 25 years (145). Information in regard to planting these species is available in numerous publications. Red pine on a soil with calcareous till near the surface in Indiana was a failure. In the Ohio Valley on eroded soils, ash, oaks, yellow poplar, and black walnut have been unsatisfactory both in survival and growth.

The other region where erosion of abandoned cultivated lands is a serious problem is in the foothills of the coast ranges of central and southern California, where orchards and beans have been grown on steep slopes until the erosion has rendered the continued attempt to raise crops unprofitable. The problem is complicated by the fact that these lands for the most part have never been forested, so that native species of trees are not available. A variety of exotics has been tried, most of them with only moderate success.

The species used most extensively and with considerable success is the blue gum. It grows rapidly on a wide variety of soils and is reasonably effective in the control of erosion. The species has an exceptionally large root system, dense, and spreading widely as far as 80 ft from the trees when unobstructed. The roots provide barriers as the soil is washed away, and also anchorage for the slips and slides that are a frequent form of erosion in parts of the coast ranges. Because of its rapid growth and large root system, it is a serious competitor, and other species in mixed plantings are likely to suffer. In its native habitat in Australia, and the same observation applies wherever it is planted, the species does not build up a forest floor, and there is a question whether it even maintains the productivity of the site. It is easily damaged by fire or freezing weather but after such damage the numerous sprouts make phenomenal growth. Manna gum is preferable to blue gum where freezing temperatures are prevalent. In the interest of improving the soil, it would be highly desirable if some species of legume could be substituted for the eucalyptus.

Among a considerable number, two species that have given promise in experimentation in the last decade are Arizona cypress and velvet ash. They grow naturally on very unfavorable sites in Arizona and New Mexico and have shown high survival and reasonably good growth for the first few years after planting, even in the prolonged dry hot summer season in San Diego County, California.

Coulter pine is another promising species for the lower elevations in central and southern California. It is one of the few species that has succeeded and grown well over a period of years in the chaparral area of Los Angeles County. In that area, however, it was necessary to clear the chaparral more than once during the early development of the plantations. It maintains a much better form than the knobcone or Digger pines, which occupy similar sites, and it lays down a good forest floor.

Monterey pine is another local species near the California coast that has done extremely well in plantations within 50 miles of the coast all along the central section of California. Extensive plantations now 30 years old in the foothills back of Oakland are reaching merchantable size, and groups of trees 60 or 70 years old average as much as 2 ft in diameter. The trees 30 years old are reproducing naturally in abundance. The species is outstanding both in natural and planted stands for the development of a thick layer of forest floor, often 3 to 6 in. in depth. Its root system is not particularly deep or wide, but it is a distinctly promising species for the fog belt of central California.

Abandoned fields constitute an important part of the sites for reforestation in the eastern half of the United States. Those on which the planting might be justified for control of erosion would usually be the poorer sites. The species for planting, therefore, are the same as those used for reforestation on the poorer sites. Many of them are included in the list for gully banks and heads, and Table 67 will serve for suggestions.

TABLE 67.—SPECIES FOR EROSION CONTROL OF GULLY BANKS AND HEADS

Species	Region													
	LLP	OP	OC	OH	TG, N	TG, S	SG, S	SG, N	SB	BG	C	J.	P	DF
Shrubs and vines:														
Lespedeza sericea	x	x	...	...	...	...	...	...	...	...	...	...	...	...
Chokecherry	...	...	...	x	x	x	x	x	...	x	...	...	x	...
Kudzu	x	...	...	...	...	...	...	...	...	...	...	...	...	...
Japanese honeysuckle	x	x	x	x	...	...	...	...	...	...	...	...	...	...
Russian olive	...	...	...	...	...	...	...	...	x	x	...	x	...	...
American plum	...	...	x	x	...	...	...	...	...	...	...	...	x	...
Siberian pea tree	...	...	...	...	x	x	x	x	...	x	...	...	...	...
Skunkbush	...	...	...	...	...	...	...	...	x	...	...	x	...	...
Grasses, etc.:														
Redtop	...	...	x	x	...	...	...	...	...	...	...	...	...	...
Bermuda grass	x	...	...	...	...	...	...	...	...	...	...	...	...	...
Kikuyu grass	...	...	...	...	...	...	...	...	...	...	...	...	...	...
Western wheatgrass	...	...	...	...	...	...	...	...	...	...	x	...	...	...
Crested wheatgrass	...	...	...	...	...	...	...	x	x	x	...	x	x	...
Big bluegrass	...	...	...	...	...	...	...	x	x	x	...	x	...	...
Kentucky and Canada bluegrass	...	...	x	x	...	...	...	...	...	x	...	...	x	...
Little bluestem	...	...	...	...	x	x	...	...	...	...	...	...	...	...
Buffalo grass	...	...	...	...	...	...	x	x	...	...	...	...	...	...
Sudan grass	...	...	...	...	x	x	...	...	...	...	...	...	...	...
Blue grama	...	...	...	...	...	...	x	x	...	...	...	...	...	...
Burnet	...	...	...	...	...	...	...	...	...	...	...	...	...	x

TABLE 67.—SPECIES FOR EROSION CONTROL OF GULLY BANKS AND HEADS—(*Continued*)

Species	Region													
	LLP	OP	OC	OH	TG, N	TG, S	SG, S	SG, N	SB	BG	C	J	P	DF
Trees:														
Black locust	...	...	x	x	...	x	x	...	...	x	x	...	x	...
Shortleaf pine	x	x	x	x	...	...	...	...	...	...	...	...	...	...
Loblolly pine	x	x	...	...	...	...	...	...	...	...	...	...	...	...
Virginia pine	x	x	x	x	...	...	...	...	...	...	...	...	...	...
Eastern redcedar	...	...	x	x	x	x	x	x	...	...	...	...	...	...
Honey locust	...	...	...	x	x	x	x	x	...	...	...	...	...	...
Ponderosa pine	...	...	...	...	...	...	...	...	...	x	...	...	x	...
Douglas fir	...	...	...	...	...	...	...	...	...	x	...	...	x	x
Rocky Mountain juniper	...	...	...	...	...	...	x	x	x	...	...	x	x	...
Siberian elm	...	...	...	...	x	x	x	x	x	x	...	...	x	...
Velvet ash	...	...	...	...	...	...	...	...	x	...	...	x	...	...
Green ash	...	...	...	...	x	x	x	x	...	...	...	...	...	...
Cottonwood	...	...	...	...	x	x	...	...	...	...	...	...	...	...
Hackberry	...	...	...	...	x	x	x	x	...	...	...	...	...	...
Osage orange	...	...	...	...	...	x	x	...	...	...	...	...	...	...
Eucalyptus	...	...	...	...	...	...	...	...	...	...	x	...	...	...
Arizona cypress	...	...	...	...	...	...	...	...	...	...	x	...	...	...
Coulter pine	...	...	...	...	...	...	...	...	...	...	x	...	...	...
Pitch pine	...	...	x	x	...	...	...	...	...	...	...	...	...	...
Red pine	...	...	x	x	...	...	...	...	...	...	...	...	...	...
Desertwillow	...	...	...	...	...	...	x	...	x	...	...	...	...	...

172. Species for Erosion Control of Gully Banks and Heads.—The upper slopes of gully banks at best represent unfavorable sites for vegetation. The slopes are steep and often unstable. They are subject to excessive exposure and drying out, and often they are composed of raw subsoil material. Around the head and edge of a gully just above the bank, conditions may be somewhat more favorable, and often approximate those of the abandoned fields. The lower slopes of gullies, where moisture is often available by capillary rise from the gully bottom or from subsurface drainage, where some of the top soil has accumulated from above, and where seepage water has added both moisture and nutrient constituents, are distinctly favorable. Consequently vegetation that may do well on the lower slopes may fail on the upper slopes. For the same reasons less exacting species must be used on the upper slopes and around the heads of gullies. The species that have been most successfully used on gully banks and heads in the different regions are listed in Table 67.

In the Southeast and elsewhere, black locust has been extensively used but has not been satisfactory on the upper slopes and around the heads of gullies. On the other hand, it has succeeded and grown well on the lower slopes where soil and moisture conditions were more favorable.

Loblolly pine is better adapted to the lower slopes of the gullies although it also grows on the more severe exposures. For the latter, shortleaf or Virginia pine is preferable. The important characteristics of shortleaf pine in this connection are its ability to grow on poor, dry soils and its characteristically deep taproot system. Virginia pine is a pioneer on poor sites throughout its range. It is highly resistant to drought, grows rapidly during the early years, and reproduces abundantly. It produces a good protective covering of forest floor. On the other hand, it is distinctly less desirable than shortleaf or loblolly pine from the standpoint of utilization.

Kudzu, a perennial legume, is being used extensively and effectively for erosion control in gullies of all sizes throughout the southeast (248). It grows well on poor soils, provided that they are not poorly drained or too shallow or too high in lime content. It requires a reasonable amount of moisture and consequently should usually be planted on the lower slopes or around the tops of the gullies whence it spreads rapidly and covers the upper slopes and surrounding areas. The vines may spread as much as 50 ft in a year by the runners, which take root where they touch the ground. On steep banks, if slipping occurs they settle down after the movement and again take control. Its roots penetrate to at least 3 ft. It has considerable incidental value as forage for domestic animals (27). Prolonged heavy grazing kills the plants. The crowns are planted 4 to 8 ft apart in rows 12 to 18 ft apart to give about 500 plants to the acre. They produce a cover in 1 year which increases in density in subsequent years. The thick dense cover of stems and litter withstands erosion by large concentrations of flow and stabilizes the channels of gullies as well as the slopes. It is the opinion of many that kudzu will control erosion without other measures even in the most spectacular gullies. Others, less optimistic, point to the vertical or overhanging banks at the heads of gullies on which nothing can take root. However, if the channel is stabilized, then as the overfall at the gully head slips down, the slope from head to channel must decrease, thus giving the kudzu a chance to cover the slope. A continuation of this process should control the accelerated erosion, provided the concentration of water entering the gully is not sufficient to tear out the kudzu in the channels. In this case, it would be necessary to divert the water by an interception ditch above the gully head. Kudzu is hardy as far north as Virginia, Tennessee, and Oklahoma, but has not succeeded in the southwestern states. It will climb over and damage trees and should not be planted too near them.

Lespèdeza sericea, commonly called "sericea," is an introduced perennial species that is being used extensively and successfully for erosion control in the Southeast as far north as the Ohio River and central Kansas and Oklahoma. It is a deep-rooted legume that forms dense stands of woody stems up to 3 ft in height at maturity. After the first year a stand will have from 20 to 30 plants per square foot. The weight of the roots may be as much as 2 tons of dry matter to the acre. It will grow on the poor exposed soils of upper slopes of gully banks and provides not only a dense cover of foliage but also a good layer of litter. It improves the fertility of the soil and incidentally in the early part of the season provides nutritious grazing or may be cut for hay. It also provides excellent cover and much food for quail (287). If left undisturbed for more than 5 years the sericea gradually dies out and is replaced by broomsedge.

Among the grasses, Bermuda grass has been the outstanding species for erosion control in the Southeast. Its underground rootstalks and creeping stems that take root at the nodes make it highly effective in stabilizing gully banks. It is readily established by seed, stolons, rootstalks, or sods. Once established it will withstand considerable concentrations of flowing water. It is useful as far north as North

Carolina, Tennessee, and Oklahoma and in the warmer parts of New Mexico, Arizona, and California.

Centipede grass is another species that has proved effective in the control of gully banks in the deep south. Like Bermuda grass it spreads by runners that take root at the nodes, but it does not have the underground stolons. It will grow and serve its purpose through severe droughts and on the driest soils, where Bermuda grass and other species are more likely to fail. It is probably not hardy north of the Gulf States.

In the Central States and northward a variety of species has been used effectively for erosion control of gullies. Black locust and shortleaf and Virginia pine are useful, at least to the northern limits of their natural ranges and probably somewhat farther as in the case of black locust, which has been planted as far north as central Michigan and central New York and New England. Redcedar and honey locust, like black locust, are more likely to be satisfactory on soil high in lime content or otherwise of good fertility. On the poorer, drier soils, the pines, to the northward pitch pine and red pine, are the most promising species. Both species are deep rooted but grow rather slowly. Pitch pine sends down vertical roots to depths of 8 or 9 ft and the laterals extend 1 to 1½ times the height in trees up to 25 ft high. Roots are less deep in clayey than in sandy soils. The red pine is distinctly superior from the timber-utilization viewpoint.

Of the shrubs and vines, chokecherry, American plum, coralberry, and Japanese honeysuckle have been used most widely and successfully. The plum is better adapted to soils with high content of lime. Like the chokecherry it forms thickets by sending up suckers from the roots and thus develops a good protective cover in a short time. Coralberry is useful on the poorest sites, where its runners enable it to produce a dense cover. Single runners may extend over 7 ft in a year. Most of the roots are within 1 to 3 ft of the surface. Japanese honeysuckle has been used successfully in the Central States and in Pennsylvania and provides a dense mat of vines, which climb over banks and even over trees and other vegetation. For the latter reason it should probably not be planted in close association with trees or too near to natural woodland. Except for its rapid growth and effective coverage there is little to recommend it.

Among the grasses, Kentucky and Canada bluegrass and redtop have been most promising. The Kentucky bluegrass requires a reasonably fertile soil and in this respect is more limited in its use than the other two species, which grow farther north and on poorer soils. Both the bluegrasses have dense, fibrous root systems, and the Kentucky bluegrass particularly forms a tough sod that withstands considerable flows of water. Seed of these species is readily available, and they have high value for grazing, in addition to that for erosion control.

On the banks and at the heads of gullies in the sagebrush and piñon–juniper of the Great Basin and Southwest, the conditions for establishment of vegetation are extremely unfavorable. With two exceptions the species suggested in Table 67 are native to the region or at least to some part of it. Ordinarily they do not naturally form closed stands, and it is probably essential to their survival that a very large root system has access to a correspondingly large volume of soil in order to support a small top. The indicated need of wide spacing in planting obviously conflicts with the desirability of close spacing as a means of early control of eroding banks. The seriousness of the erosion over much of the region similarly involves a conflicting problem as to the justification for difficult and expensive erosion-control measures where the results at best are dubious and where the justification in the values at stake is questionable.

Among tree species the native junipers, of which Utah juniper is most widely distributed, offer one possibility. The trees grow naturally on the driest, gravelly, rocky, steep slopes with south and west exposures, where no other tree species survive. Their establishment by planting, however, is difficult on such sites because apparently the seedlings, as in the case of natural reproduction, require some shade during the first years. The difficulty is greater on sites where they have not grown naturally. The species grow very slowly and cannot be expected to form a closed canopy. Their value for erosion control through the large root systems is high, and they have additional value because the wood may be utilized for posts.

Another possibility is the velvet ash, which grows naturally in parts of the region. Under the specific conditions of gully banks in the southwestern semidesert country, almost no experience is available as to the success in establishing the species by planting. It may be that these two species could be used in mixtures and experimental trials would be valuable as guides.

Among shrubby species the skunkbush grows naturally in the region. It tends to form thickets of bending stems from suckers. However many plantings have failed. It can be propagated by stem or root cuttings, or from seed. The Russian olive has been widely and successfully planted under the most severe conditions and would presumably succeed. It has a very large root system, and once established is effective in control of erosion.

The grasses, in a region where grazing is the most extensive land use, are undoubtedly the most desirable forms for erosion-control planting. In the southern part of the region and perhaps north of its natural range, curly mesquite is a promising perennial that grows naturally in bunches and reproduces both from seed and from stolons. It is highly resistant to drought and also to grazing. The black grama and hairy grama are two other native perennials that are being used effectively in revegetation. They reproduce by stolons and, at least in places, tend to form sod. Bermuda grass is also naturalized in the southern part of the region and its characteristics previously outlined for the southern pine region make it equally effective in the Southwest in situations where it will survive. It has become naturalized chiefly in the valleys and may not succeed in large gullies where moisture is farther from the surface. It will not withstand freezing temperatures. Lehmann's lovegrass has done well on eroding sites experimentally and may have a much wider use.

In the sagebrush and piñon–juniper regions, western and crested wheatgrasses are most likely to be satisfactory. The former grows rapidly from creeping rootstalks and produces a heavy sod, highly effective in binding the soil. The crested wheatgrass is resistant to drought, reseeds readily, and has a deep and wide root system.

In the ponderosa pine–Douglas fir region of the central and southern Rocky Mountains at middle elevations the two native timber species are also the conifers most likely to be useful for erosion control. Ponderosa pine has a deep taproot system, will grow on severe sites such as gully banks, and produces a reasonably good cover of litter. Douglas fir has a less deep root system and is less likely to survive on the most exposed, driest sites, but once established it forms a denser stand and builds up a heavier forest floor.

The Rocky Mountain juniper may be more satisfactory than the ponderosa pine on the steepest, driest gully banks, and once established its large root system is effective in the stabilization of gully banks. Unfortunately all three of these species grow slowly, particularly in early life, and cannot be expected to form a closed canopy, even when spaced as close as 4 by 4 ft, before they are 20 years old.

Black locust and honey locust have been successfully introduced in various parts of the region for erosion-control purposes. In growth of both tops and roots, and in soil-improving qualities, they are superior to the conifers. It is probably desirable for gully stabilization as in other planting that mixtures by groups of the introduced locusts and native conifers be used with the locusts on the lower slopes.

Among shrubby species the western chokecherry and American plum, native at least in parts of the region, are useful erosion-control species. Both will grow on all but the most unfavorable sites, and they tend to form thickets by the development of suckers from the roots. They will provide cover much more quickly than the tree species, with the possible exception of black locust.

The western wheatgrass, which is native in the northern part of the region, and the big bluegrass, most abundant in the Northwest, serve excellently for erosion control in gullies in the mountain valleys as well as on the plains and plateaus to the north and west. Big bluegrass is less likely to form dense stands, but it has a large root system and will grow on adverse sites. Both species, incidentally, are excellent forage plants. The grasses tend to provide a protective cover much more promptly than the woody species. Where grazing is an important use they should undoubtedly be given preference. Mixtures of grasses, shrubs, and trees would be desirable but are not advisable on gully banks because the competition of the grasses is likely to result in heavy mortality of the trees or shrubs that are planted with them.

In the short-grass plains most of the same species that have been more extensively used for the establishment of shelterbelts are useful for the control of gully banks and heads. Black and honey locust are extensively used and have the advantages elsewhere enumerated. Black locust is likely to be killed back in the winter in the northern part of the plains. Redcedar, probably either the eastern or the Rocky Mountain species, succeeds well but grows slowly.

The other woody species widely used in the shelterbelts and suitable also for gully banks include green ash, cottonwood, hackberry, Osage orange, chokecherry, and the two introduced shrubs—Russian olive and Siberian pea tree. Cottonwood grows rapidly, but requires more moisture than most of the others. It also dies at an early age and has to be replaced. Osage orange is a southern species and should not be planted north of Kansas. The Siberian pea tree although hardly more than a shrub has shown a wide adaptability under the most severe conditions of drought and temperature. As a legume it is definitely beneficial in building up the fertility of the soil, and its root system is both wide and deep. However, it is not adapted to heavy, poorly drained, or alkali soils, and it is subject to severe damage by grasshoppers.

In a grazing region the grasses are the most desirable species for erosion control, and there are several that have proved valuable. In the northern plains, western wheatgrass, previously mentioned, and the crested wheatgrass, a species introduced from Russia, are highly resistant to drought and have deep, wide root systems. The latter, although an introduced perennial, reseeds readily. Buffalo grass is native over much of the region and forms the native short-grass sod, which is so effective in stabilization of the soil. It is drought-resistant and spreads naturally by means of stolons, and the sod is so tough that it can be used as lining for terrace outlets. The roots penetrate to depths of 4 or 5 ft. One experimental plant produced 650 ft of stolons in 84 days. Little bluestem, another native perennial with a wide range, is similarly useful throughout the region. It has a large root system but tends to form only an open stand on poor sites. Blue grama is an associate of buffalo grass over much of its range and, with it, helps to form the erosion-resistant sod. The species has a dense fibrous root system extending to depths of 6 ft as well as high palatability. Finally, Sudan grass has been introduced from north Africa and has been much planted in the semiarid part of the plains region. It is an annual but reseeds readily.

In the bunch grass and lower part of the ponderosa pine–Douglas fir region of eastern Washington and Oregon and adjacent states, erosion control of gullies is locally an important matter, as for example in the Palouse wheat country. The climatic conditions are sufficiently similar that many of the species of the short-grass plains are used also in this region. They include black locust, ponderosa pine, Douglas fir, Russian olive, Siberian pea tree, crested and western wheatgrass, and big bluegrass. The ponderosa pine and Douglas fir in the bunch-grass zone are difficult to establish and have not proved entirely satisfactory. However, so far no other evergreen conifer has been more successful.

The Siberian and Chinese elms (*Ulmus pumila* and *U. parvifolia*) introduced from northern Asia, and often not distinguished from each other, do better than the conifers in the first 5 years under the driest conditions. The Siberian elm has been planted much more commonly than the Chinese. The trees have very large root systems and grow rapidly so that they make a good showing in 5 years. However, doubt has been expressed whether they will live more than 25 years. Their later value for wood products other than firewood is questionable. The severe freeze of November, 1940, killed or damaged many of the trees as far south as northern Oklahoma. The damage was greater on moist sites, on fine-textured soils, and in the exposed north and west rows. The severe competition of the Siberian elm may retard or kill other species growing near it.

West of the Cascade Mountains the burnet has been introduced and found quite effective as an erosion-control plant. It survives on sites where most other species fail, and its deep root system is effective in holding the soil.

In the chaparral region of California the same species are promising for gully control that were suggested for use on abandoned agricultural lands, namely, black locust, eucalyptus, Arizona cypress, Coulter pine, and Kikuyu grass. The grass is unquestionably the most effective species for prompt control. Black locust, as in so many places, does well only where conditions of moisture and fertility are not too unfavorable. An exceptional plantation on deep, alluvial, sandy soil of a large gully in Santa Cruz County attained a height of 20 ft and formed a closed canopy and good layer of litter in 5 years.

In the parts of California that are not subject to freezing, Kikuyu grass is proving to be an unusually effective species in gullies and on road slopes in the most unfavorable situations. It has the faculty of spreading both by runners and by stolons and produces a deep as well as a fibrous root system, which enables it to stay green throughout the long dry summer and fall season. It gradually forms a thick mat of decaying vegetation over the surface through which the green plants continue to grow. Examples on a small scale are available to demonstrate its effectiveness not only in holding steep gully or road slopes but also in forming a mat that resists cutting in channels and gully bottoms. Like other aggressive introduced plants it has the possibility of becoming a nuisance in cultivated lands, and its use for this reason is prohibited in some counties.

173. Species for Erosion Control of Stream Banks and Channels, Gully Bottoms, and Waterways.—Erosion control in channels involves several considerations. Usually the aim is to prevent or minimize scouring of the bottoms and undermining and sloughing of the banks. Above reservoirs, however, the objective may be to cause deposition of sediment and, by means of settling basins or vegetative screens, prevent silting of the reservoirs. Finally, in minimizing the cutting or the silting the channels must not be restricted to an extent that increases unduly the heights and hazards of flood flows.

Species that are useful in situations where they must withstand intermittent flooding and the force of flowing water must have somewhat specialized characteristics. They should develop a dense growth and a wide and fibrous root system in channels that carry considerable flood flows. The effect of aggressive woody vegetation encroaching upon the channel itself and thereby restricting its capacity to carry flood flows must also be considered. Most of the species suitable for

TABLE 68.—SPECIES FOR EROSION CONTROL OF GULLY BOTTOMS, WATERWAYS, AND STREAM CHANNELS

Species	Region									
	LLP	OP	OC	SG	SB	CB	DG	BG	C	J
Trees:										
Sandbar willows	x	x	...	x	x	x	x	...	x	x
Cottonwood	x	x	...	x	x	x	x	...	...	x
Black locust	...	...	x	x	x	...	...	x	x	x
Athel tamarisk	...	...	...	...	...	x	x	...	...	...
Desertwillow	...	...	...	...	...	x	x	...	...	...
Shrubs and vines:										
Mulefat	...	...	...	...	...	...	...	...	x	...
Russian olive	...	...	...	...	...	x	x	...	...	...
Seepwillow	...	...	...	...	...	x	x	...	...	...
Kudzu	x	...	...	...	...	...	...	...	...	...
French tamarisk	...	...	...	...	...	x	x	...	...	...
Lespedeza sericea	x	x	x	...	...	...	...	...	...	...
Paloblanco	...	...	...	...	...	x	x	...	...	...
Grasses:										
Bermuda grass	x	...	...	...	...	x	x	...	x	...
Kentucky bluegrass	...	...	x	...	...	...	...	...	...	...
Canada bluegrass	...	...	x	...	...	...	...	...	...	...
Kikuyu grass	...	...	...	...	...	...	...	...	x	...
Western wheatgrass	...	...	...	x	x	...	...	x	...	x
Crested wheatgrass	...	...	...	...	x	...	...	...	...	x
Buffalo grass	...	...	...	x	...	...	...	...	...	...
Vine mesquite	...	...	...	...	...	x	x	...	...	...
Black grama	...	...	...	...	...	x	x	...	...	...
Giant wildrye	...	...	...	...	...	...	...	...	x	...
Galleta grass	...	...	...	...	...	x	x	...	...	...

this group of sites have already been mentioned as desirable in other situations. The species used in different forest regions are summarized in Table 68.

Willows and cottonwoods of several species grow naturally along streams throughout the United States. Cottonwood is less aggressive and generally less satisfactory than willow. The willows, and particularly the sandbar willows, have been most widely used to protect banks or to form barriers. The introduced golden and crack willows have also been used. Similar methods have been developed and used in Washington, Wisconsin, Ohio, Arizona, and Vermont and essentially include the following steps (5, 260):

First, the bank is graded to a slope of 34 to 45 deg. Second, willow poles are laid up and down the slope 2 ft apart with the butt ends forced into the mud below the level of low water. The poles must be long enough to span the slope and are trimmed of branches. The work should be done when the willows are dormant, and if possible the poles should be placed the day they are cut, to prevent drying. Third, the poles are covered with brush with the tops alternately at the top and bottom of the bank. Any species of brush can be used that will make a mat, which should be about 1 ft thick at the water line and ½ ft at the top. Fourth, the brush is held in place by strips of used woven wire, laced together in horizontal bands. Fifth, the whole mat is held close to the bank by No. 9 wire, fastened at the top to posts 8 ft apart and holding at the low-water mark large rocks or other weights. Willow stakes ½ to 1 in. in diameter are also used for this purpose. Sixth, at each end the mats are fastened to small logs, held in place on the bank by posts.

In large streams it is sometimes necessary to protect the upstream end of the mat by a wire-rock sausage jetty. Livestock must be kept away from the willows by fencing. The poles, stakes, and often the brush itself, if it is also of willows, take root wherever they come in contact with the soil and grow 2 to 5 ft in the first season. On the outsides of bends, where cutting is most active, it has been found necessary to supplement the willow mat by a riprap of stones along the toe of the bank to a height of 5 ft more or less above low-water level. In places it has also been found advantageous to spread loose earth over the mats, to encourage sprouting. Any locally growing species of willow will serve the purpose, although tree willows such as the white willow are necessary for the poles. The sandbar willow or some similar brushy species serves well for the mats and for living cover which will bend over before the force of flood waters or ice.

This method of protection of stream banks has not always been successful, and the failures have resulted from one or more of the following causes: The rock or riprap below the willow mats has been omitted at critical locations. The river has changed its channel thus forming outsides of bends at unforeseen places. Heavy rocks, logs, or other debris thrown against the mats in flood flows have made holes that enlarged and undermined the part remaining. Failure to maintain fences has allowed livestock to browse and damage the willows. Beavers have cut the willows and burrowed in the banks.

The banks of larger rivers may require a structural protection of which three types have been tried along streams in the Southwest. Along the less critical reaches a tree and cable revetment has been used successfully. Whole trees, usually cottonwoods, are placed 6 to 8 ft apart, with the butt ends on the bank and making a 30-deg angle with the bank downstream. They are held in place by two lines of 1-in. cable anchored at both ends (151).

In the more critical situations rail "tetrahedrons" made from four 50- to 75-lb rails in lengths of 7½ to 10 ft have been used. The tetrahedrons are placed in rows out from the bank and, if on a bend, in a catenary curve. They are anchored with 1-in. steel cables fastened to deadmen on the banks both up- and downstream. Sometimes these are supplemented by tree and cable deflectors, which are placed at the anchors to the line, extending from the tetrahedrons to the bank.

Ordinarily in 1 year enough sediment will be deposited between the tetrahedrons and the bank so that willows can be planted in the deposit. Cuttings are used for the planting, some of them large, from 4 to 16 ft long and 1 to 9 in. in diameter, and some small, from 1 to 3 ft long and ½ to 1½ in. in diameter. Vertical planting of the cuttings is best in the deposits out from the bank, and angular planting as previously described is used along the banks themselves. Along the upper parts of the banks, cuttings and layers may form a stand that will give sufficient protection without other means after 2 years.

Willows have certain characteristics that are particularly advantageous in their use for erosion control. They are easily reproduced by cuttings, which are almost always available locally from natural stands. They form dense, pliable thickets. The growth rate is rapid. The native species in any locality are frost-resistant, and most species will withstand flooding and bending for considerable periods of time. They produce a good layer of litter within 6 years after planting. The root system is dense and widely spreading although not deep. Most willows are considered to be restricted to moist sites, and hence lacking in resistance to drought, but the local willows of the canyon bottoms in California grow well on the excessively drained and exposed highway fills for at least 10 years after planting. This may, of course, be a matter of lack of competition of other species for the moisture that is available in the deep deposits. The willow scab fungus (*Fusicladium saliciperdum*), which has spread rapidly across New England and southwestward and has destroyed much of the native willow, is a serious threat to the continued use of willows. Some of the species of willow introduced from Europe and naturalized seem to be reasonably resistant to the disease, notably *Salix babylonica*, *S. pentandra*, *S. purpurea*, and *S. viminalis*.

Black locust has been tried in most regions and does better in the moist soils of the bottoms and lower banks of channels than in most other situations. Even in the southern pine and oak–pine regions it has often succeeded in these situations. It is not suitable, however, for bottoms where it will be subject to much flooding, although in large gullies it has sometimes been closely spaced in rows across the bottom at intervals to serve as pervious check dams.

In the southern pine region, *Lespedeza sericea*, kudzu, and Bermuda grass are the three species that have been shown to resist erosion effectively under considerable volumes and velocities of flow. The sericea can also be used through the oak–pine region and in the oak–chestnut–yellow poplar region, at least as far north as the Ohio River. In the latter region Kentucky and Canada bluegrass are also valuable species for waterways.

In the short-grass plains, buffalo grass and western wheatgrass in addition to being able to survive on the driest sites are also effective in protecting gully bottoms by reason of their dense fibrous root systems and sod-forming habits.

In the desert grassland and desert shrub region of the southwest, Bermuda grass is again useful. Among the native species, vine mesquite, black grama, and galleta grass may be suggested. Vine mesquite produces rapid-growing creeping runners that root at the nodes and form dense mats. The native seepwillow and the Russian olive are useful shrubby species. The latter is resistant to alkali, forms dense stands, and produces fence posts. The native cottonwood and willows may be started from cuttings, grow rapidly, and withstand floods and silting. The crack willow has done even better than the native species. Athel tamarisk is another promising tree for these sites. It can be started from cuttings and is drought-resistant, although some irrigation is essential for rapid growth. The French tamarisk has spread naturally over the flood plains of some of the larger streams and forms dense thickets that entrap sediment but also restrict seriously the carrying capacities of the channels. Its root system is shallow. The above-ground parts are mostly killed by flooding for more than a week, but vegetative regeneration is rapid after the water recedes. It is resistant to alkali. The heavy forest floor is said to be toxic to other species. For this region, Wilson (372) concludes that the best woody species are cottonwood, willow, Russian olive, and athel tamarisk.

Tree planting has also been tried in broad shallow streams in the Southwest as a means of causing the deposition of sediment and thus reducing the rate of silting of reservoirs below, just as did the natural invasion of French tamarisk above Lake McMillan. The trees, usually willow cuttings, are planted during low water in bands across the stream bottom in places where the wash is broad, straight, and shallow and has a low gradient. Along deeper streams and on bends, double strips of willows spaced 8 ft or more between the strips have been planted on the lowest flood plain at an angle of 30 deg with the direction of the current. The purpose in this plan is twofold, first to cause deposition between and below the strips, and second to reduce erosion of the banks by deflecting the current away from them. Changing channels during floods have sometimes washed away the bands of trees. Where tetrahedrons are used to protect the outsides of bends, willows planted close to them form a network of roots that prevent the tetrahedrons from sinking into the alluvial material during floods.

In the chaparral region the mulefat grows naturally along stream channels and can be used in the same ways that willows are used. Among the grasses, Bermuda grass, giant wildrye, and Kikuyu grass are the most promising. Where the latter has been used it forms a thick dense mat of living and decaying vegetation that is highly effective in protecting the bottoms. In one channel about 3 ft across with a maximum depth of 5 in. and a slope of 18 per cent, the grass successfully prevented cutting where it withstood a discharge of 2.3 sec-ft at a velocity of 3.3 ft per second. Serious cutting occurred both above and below the Kikuyu.

In the bunch-grass region of the northwest, western wheatgrass is probably the most promising of the nonwoody species.

174. Species for Stabilization of Sand Dunes and Blowouts.—The species for control of wind erosion of moving sand are subject to undermining of the roots, to burying, to abrasion of the tops, and often to severe deficiencies of soil moisture. They should, therefore, have the ability to recover after partial burying, both deep and spreading roots, rapid height growth in the seedling stages, and rapid crown development with the formation of a cover of litter. For the grasses, capacity to spread laterally by vegetative reproduction is also important. A number of species (Table 69) have been tried with varying degrees of success in the regions where sand blowing is a problem. A mulch of litter, weeds, straw, or boughs of species like white pine or birch is essential for the successful establishment of some species

TABLE 69.—SPECIES FOR STABILIZATION OF SAND DUNES AND BLOWOUTS

Species	Region							
	LLP	BM	JP	TG	SG	C	R	DF
Trees:								
Jack pine	...	x	x	...	...	...	...	...
Ponderosa pine	...	...	...	x	x	...	...	x
Red pine	...	...	x	...	...	...	...	...
Austrian pine	...	x	x	...	...	...	...	x
Maritime pine	...	...	...	...	...	x	x	x
Scotch pine	...	x	x	x	...	...	...	...
Monterey cypress	...	...	...	...	...	x	x	...
Shore pine	...	...	...	...	...	...	x	x
Pitch pine	...	x	...	...	...	...	...	...
Black locust	...	x	...	...	...	...	...	...
Cottonwood	...	x	...	x	...	...	...	...
Willow	...	x	x	...	...	...	...	...
Victoria tea tree	...	...	...	...	...	x	x	...
Siberian elm	...	x	x	...	...	x	...	...
Shrubs:								
Bayberry	x	x	...	...	...	...	...	...
Scotch broom	x	...	...	...	...	...	...	x
Sand cherry	...	x	...	x	...	...	...	...
Tree lupine	...	...	...	...	...	...	...	x
Grasses and herbs:								
European beachgrass	...	x	x	...	...	x	x	x
American beachgrass	...	x	x	...	...	...	...	x
American dunegrass	...	...	...	...	...	...	...	x
European dune wildrye	...	x	x	...	...	...	...	x
Bermuda grass	x	...	...	...	...	...	...	...
Sand bluestem	...	...	...	x	x	...	...	...
Blowout grass	...	...	...	x	x	...	...	...
Prairie sandreed	...	...	...	x	x	...	...	...
Sand dropseed	...	...	...	x	x	...	...	...
Purple beach pea	...	x	x	...	...	...	...	x

and probably improves the growth of all species planted on moving sands. Large planting stock must often be used.

In the pine and hardwood region along the shores of the Atlantic and the Great Lakes, jack pine, red pine, and pitch pine are the three native species of sandy sites. All three have been used with moderate success. The jack pine has the advantage of growing more rapidly during the first few years. The introduced Scotch pine has also done well. Of the deciduous species, cottonwood and willow can be used where moisture is not too far below the surface. The cottonwood particularly grows rapidly and has the characteristic of sending out new roots into the layers of sand that may bury the stems. Siberian elm has also succeeded in areas of moving sand. Sand cherry is native in such situations and has value for stabilization purposes. A number of grasses are pioneers on sand dunes and in blowouts. They include the European beachgrass, American beachgrass, American dunegrass, and, in the Central states, prairie sandreed. The purple beach pea is a legume, naturally adapted to these sites, which forms a protective cover and incidentally builds up the soil for other species.

In the tall-grass and short-grass plains, cottonwood and ponderosa pine are most commonly used for sand stabilization. The slow early growth of the pine usually necessitates the use of a mulch or some sand grass as temporary protection during the first few years. Among the grasses, sand bluestem, sand dropseed, and prairie sandreed are all useful.

In the north Pacific Coast region, one of the outstanding examples of sand-dune stabilization was carried out by the Soil Conservation Service just south of the mouth of the Columbia River near Warrenton, Oregon. The methods have been described in detail by McLaughlin and Brown (236). The essential steps in the plan of attack included the following. Next to the beach and perpendicular to the onshore winds a foredune was established by the use of two parallel rows of pickets 30 ft apart. The pickets were 4½ ft long and were set 1½ ft deep in the sand. Each one was 3 to 4 in. wide and they were spaced 3 to 4 in. apart. American dunegrass was planted between the rows of pickets, and the European beachgrass was planted in strips 50 ft wide on both sides of the foredune. Legumes, including purple beach pea and seaside lupine, were seeded to supplement the grass and build up the fertility of the soil. Over the area back of the foredune, plantings of European or American beachgrass were spaced 18 by 18 in. After 1 or 2 years Scotch broom was planted in belts, each consisting of 6 rows, 10 ft wide, with a band 40 ft wide to windward. The spacing in the rows was 1½ to 2 ft. Tree lupine was seeded directly to supplement the broom.

One or two years later pines were planted 3 to 4 ft apart within the 50- by 100-ft blocks bounded by the broom. The pines that seemed to be adapted to this locality were shore, maritime, Scotch, Table-mountain, and Austrian, of which only the shore pine is native. The Oregon crab apple and red alder were also used.

Experience in the years since the study by McLaughlin and Brown has indicated important simplifications. The European beachgrass forms a foredune close to the beach without the installation of the barrier of pickets. The Scotch broom improves the growth of the trees near it notably, and that benefit is obtained over the whole of an area by planting the broom 6 by 6 ft apart and then, 1 or 2 years later, planting the trees also 6 by 6 ft between the rows of broom. Where the annual precipitation is 70 in. the competition of the broom for moisture is negligible compared with the benefit of the addition of nitrogen to the sand through the root nodules. The purple beach pea has similar value as a "nurse crop."

The shore pine has exceeded other tree species in survival and especially in growth, and consequently other species have been planted only sparingly in the last few years. Next to shore pine, maritime pine and Austrian pine would rank second and third. The other pines, alder, willow, crab apple, and rose have done too poorly to justify continued use. The crowns of shore pine spaced 6 by 6 ft close in 7 to 10 years. At this time the forest floor is ¼ to ½ in. thick and does not cover the ground completely. The broom forms a forest floor more rapidly than the pine.

In the redwood region the classical example of sand stabilization is in Golden Gate Park in San Francisco, where the results have had more than 50 years of testing. European beachgrass, spaced 3 by 3 ft in plowed furrows, was used originally as a sand binder but has long since disappeared. Brush fences, 4 to 6 ft high, at intervals of 100 yd provided protection for the planted trees. In 1900, McLaren (235) reported the following species had succeeded: *Acacia lophantha*, *A. latifolia*, Monterey pine, Monterey cypress, and French tamarisk. The trees were fertilized when about 10 ft high, because their growth had almost stopped. In 1941 the two species that had maintained themselves along the shore close to the beach were Monterey cypress and Victoria tea tree, with a little acacia. Just inland, where the preceding species provided some protection, maritime pine survived and provided good cover, and, still further from the beach, Monterey pine formed a fine forest.

In the southern pine region along the coast, moving sands are also troublesome in places. Bermuda grass has been suggested as the most useful species for the stabilization of the sands. The native sea oats, a pioneer on the sand dunes, survives and stabilizes the sand even where wind and salt spray are most unfavorable. It produces numerous rhizomes and sprouts that climb the sandy slopes and finally cover the crests. The bayberry, native of sandy areas near the coast, and the Scotch broom can be used to supplement the grass. Once the movement of the sand is stabilized, numerous native species are likely to invade the area naturally, or loblolly, slash, or longleaf pines may be planted.

175. Species for Shelterbelts.—Although the terms "shelterbelt" and "windbreak" are often used interchangeably, a distinction can and should be made. A shelterbelt is a long barrier protecting fields. A windbreak is a protective planting around the house, garden, orchard, or feed lot. Shelterbelts of 10 rows, more or less, are called the "basic" or "multiple-use" belts in contradistinction to the one- to three-row belts, which are referred to as "intermediate." Single rows have been called "buffer strips." A complete pattern of shelterbelts for a farm should give protection to all fields. If the trees grow to 65 ft in height and reduce wind velocities for twenty times their height to leeward, the protective effect of a shelterbelt will extend across a 40-acre square. The pattern would therefore include basic belts on the windward sides of the farm and intermediate belts at least on the windward sides and perhaps completely surrounding each "forty."

In the parts of Wisconsin where the sand blows, shelterbelts of five, seven, or nine rows have been planted with jack or Scotch pine in the two or three outside rows and red pine in the inside rows.

Experience with the establishment of shelterbelts in nonforested areas of the Central and Western states extends over a period of more than 50 years. Renewed interest and study has resulted from the initiation of the Prairie States Shelterbelt Project by the U.S. Forest Service in 1934 in the belt along the 100th meridian, where 30,223 shelterbelts with a total length of 18,600 miles were planted between 1935 and 1942.

The plans developed in connection with the shelterbelt project provided for rows of dense low-growing shrubs, usually Siberian pea tree and Russian olive, and conifers on the windward side, with rapidly growing deciduous species in the center, bordered to leeward by longer lived species, if possible those having value for different forms of utilization. Finally, one or more rows of shrubby species on the leeward side completed the total of 10 rows. The standard plan is indicated in Fig. 25. Belts of nine or seven or even fewer rows were frequently used, and some belts included 20 or more rows. Sometimes the leeward shrub rows were omitted. Belts of less than six rows did not develop desirable forest conditions. Green ash, Siberian elm, cottonwood, and honey locust have been the tree species most frequently planted. Boxelder, honey locust, black locust, catalpa, green ash, eastern redcedar and American plum had the highest percentages of survival, all of them over 70 per cent. The mean annual growth in height of the leading species ranked them as follows: cottonwood, 2.9 ft; black locust, Siberian elm, honey locust, boxelder, and catalpa, 1.6 ft. Height growth increases from north to south. Thus,

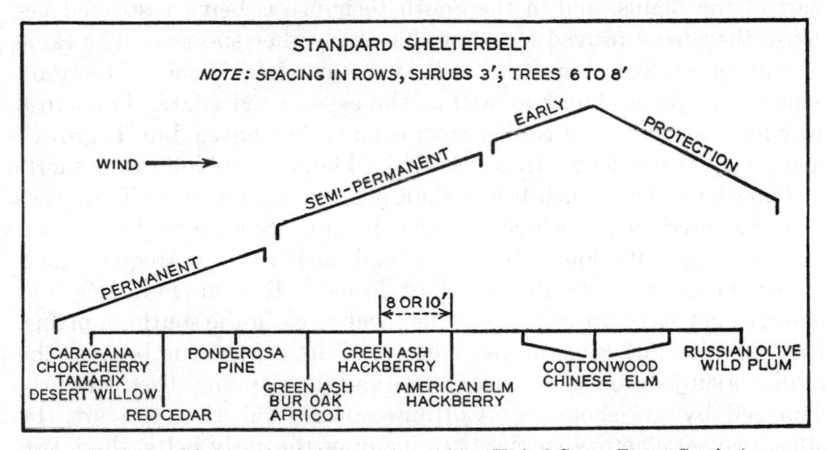

FIG. 25.—The standard shelterbelt. (*After United States Forest Service.*)

the average height of all belts measured at an average age of 6.5 years was 16 ft in North Dakota, 20 ft in Nebraska, and 24 ft in Texas. In mean annual spread of crown, black locust led with 1.5 ft, followed by cottonwood, Siberian elm, catalpa, and boxelder with 1.2 ft. On the whole, for the period from 1935 to 1942, 78 per cent of the prairie shelterbelts were rated as satisfactory by Munns and Stoeckeler (267).

During 1930–1937 in the southern high plains roots of redcedar and Siberian elm penetrated to depths of 25 and 27 ft, respectively. The longest laterals of the elm extended 43 ft, and of the cedar, 20 ft. However, the elm suffered much damage north of Oklahoma in 1940 by cold, and there is a question whether its continued extensive planting is justified. Black locust and Osage orange had the same penetration and lateral extensions of 21 and 14 ft, respectively. The green ash penetrated only 6 ft but had a spread of 43 ft (59). In general the deep-rooted species to depths of 10 to 20 ft include ponderosa pine, hackberry, honey locust, bur oak, Russian mulberry, and Osage orange. Green ash, American elm, black locust, Russian olive, and Siberian pea tree have roots 5 to 10 ft deep. The cottonwood roots ordinarily are within 5 ft of the surface (146). On the basis of the distance to which the root competition of different trees adversely affects the growth of adjacent crops, Osage orange should not be planted in an outside row toward the crop, whereas Russian mulberry is one of the best species for that location.

The hackberry and Osage orange are not hardy in the northern part of the plains, and in the South their use is being restricted because they have proved less desirable than other species. The same is true of apricot, bur oak, and American elm. Rocky Mountain juniper has done almost as well as the eastern redcedar. Ponderosa pine has been the other conifer most commonly planted, but its growth and survival are low. In Texas and Oklahoma, loblolly and shortleaf pine have done much better than ponderosa, and they will deservedly be used more widely in the future. Sycamore has grown rapidly, holds its lower branches, and merits more frequent use. Honey locust is preferable to black locust. Russian mulberry and desertwillow have proved best in the shrub rows in the southern plains. Russian olive and Siberian pea tree are of doubtful value because the former is subject to a gummosis disease and the latter has been seriously damaged by grasshoppers. Cottonwood and Siberian elm are the other two satisfactory species. In many of the early belts, these two fastest growing species were planted in adjacent rows. As a result in most cases the cottonwood has been retarded or eliminated by the competition of the elm. The two species should not be planted in adjacent rows, which usually means not in the same belts.

In general, experience indicates that for successful development of trees the soils should be permeable and at least 6 ft deep. In the bottomlands and elsewhere in Texas and southern Oklahoma, the broad-leaved species are likely to be seriously damaged by the cotton root rot. This and the desire of farmers for narrower shelterbelts has led to plans for all-conifer belts of three or five rows. In such belts, redcedar is planted in the windward row and loblolly, shortleaf, or Austrian pines in the other rows.

Spacing between the rows has been 8 or 10 ft and, between the trees in a row, has varied from 2 to 4 ft for the shrubs and from 4 to 8 ft for the trees. The crowns of the rapidly growing species close in 5 to 10 years. In one-fifth of the belts a forest floor ½ to 1½ in. thick had formed within 10 years.

Further west in the sagebrush region many of the same species are being used that have proved successful in the plains. The trees

TABLE 70.—SPECIES FOR SHELTERBELTS

Species	Region								
	JP	TG,N	TG, S	SG,N	SG,S	SB	CB	BG	C
Trees:									
Green ash	...	x	x	x	x	x	...	...	...
Cottonwood	...	x	x	x	x	x	...	...	...
Blue gum	...	...	...	...	...	...	...	...	x
Siberian elm	...	x	x	x	x	x	...	...	...
Boxelder	...	x	x	...	...	x	...	...	...
Black locust	...	...	x	...	...	x	...	x	...
Honey locust	...	...	x	...	x	x	x	...	...
Catalpa	...	...	x	...	...	...	...	...	...
Osage orange	...	...	x	...	x	...	...	...	...
Athel tamarisk	...	...	...	...	...	...	x	...	x
Loblolly pine	...	...	x	...	...	...	...	...	...
Eastern redcedar	...	x	x	x	x	x	...	...	...
Rocky Mt. juniper	...	x	x	x	x	x	...	...	...
Shortleaf pine	...	...	x	...	...	...	...	...	...
Ponderosa pine	...	x	...	x	...	x	...	x	...
Jack pine	x	x	...	...	...	...	...	...	...
Red pine	x	...	...	...	...	...	...	...	...
American elm	...	x	...	x	...	x	...	...	...
Russian mulberry	...	...	x	...	x	x	x	...	...
Sycamore	...	...	x	...	...	...	...	...	...
Scotch pine	x	...	...	...	...	...	...	...	...
Austrian pine	...	...	x	x	...	...	...	...	...
Arizona cypress	...	...	...	...	...	...	x	...	x
Red gum (eucalyptus)	...	...	...	...	...	...	...	...	x
Desert gum	...	...	...	...	...	...	x	...	x
Shrubs:									
Russian olive	...	x	x	x	x	x	x	x	...
Siberian pea tree	...	x	...	x	...	x	...	...	...
Chokecherry	...	x	...	x	...	x	...	...	...
American plum	...	x	...	x	...	x	...	...	...
Desertwillow	...	...	x	...	x	...	x	...	...
Golden willow	...	...	...	...	...	x	...	...	...
French tamarisk	...	...	x	...	x	...	...	...	...

usually receive the same irrigations as the crops, although Siberian elm and Russian olive survive without irrigation. The promising species are indicated in Table 70.

A still more restricted group of the same species is being planted for windbreaks in the bunch-grass region of the Northwest, including black locust, ponderosa pine, Siberian pea tree, and Russian olive. The latter is subject to serious damage by an unidentified disease (254).

In the chaparral and adjacent semidesert regions of California and the Southwest, Arizona cypress is one of the promising species. It is being planted in California in place of the Monterey cypress, which has been so effective but is now being destroyed by a leaf disease. Three kinds of eucalyptus—the blue gum, red gum, and desert gum—are planted commonly and successfully. The latter two are better adapted to the drier sites or to situations where alkali is concentrated. Red and desert gum have been found to withstand concentrations of sodium carbonate as high as 0.08 per cent, and blue gum as high as 0.04 per cent. The root development of the gums is enormous. Records are available of penetration of 16 ft in 1½ years, and total depths of 60 ft in southern California. The spread of the roots may extend 75 to 100 ft laterally. Another introduced species used effectively in the semidesert part of the region is the athel tamarisk.

A summary of species for different regions is given in Table 70.

176. Species for the Stabilization of Landslides and Slips.—Frequently landslides are of such magnitude that they are beyond the possibility of control by vegetation, as shown by the existence of slides in areas of long undisturbed natural vegetation, even of old-growth redwood. At the same time it may be that many more slides and slips would have occurred in areas susceptible to that form of soil movement if they had not been protected by vegetation. By the same reasoning it seems likely that the establishment of deep-rooted vegetation on areas with little or no plant growth may minimize the occurrence of slides and slips. In the coast ranges of California near San Francisco Bay where many slips occur in wet years on the clay soils of the heavily grazed grasslands, it is a notable fact that slips do not occur on these same soils in the extensive plantations of eucalyptus more than 25 years old.

Obviously, for the stabilization of these deep earth movements, species with large deep root systems are essential. The suggestions included in Table 71 are made with that as the primary consideration. The regions for which suggestions are included are those in which slides and slips are prevalent.

In the oak–chestnut region of West Virginia, Ohio, and similar parts of New York, New Jersey, and North Carolina, black locust is a desirable species for those soils sufficiently high in lime content and fertility. Redcedar is another species of limey soils, and in deep soils its root system penetrates deeply. Shortleaf pine is native over much of the region, and its deep taproot should give stability to a considerable depth of soil.

On the tall-grass prairies and short-grass plains from Texas to Montana, and notably in portions of Iowa and Kansas, small slips occur frequently. The grasses tend to be less deeply rooted than some tree species, and therefore the latter are preferable for the stabilization of slips. Honey locust and Siberian elm are two species with large

TABLE 71.—SPECIES FOR STABILIZATION OF SLIPS AND SLIDES

Species	Region					
	OC	TG	SG	BG	C	P
Black locust	x	x	...	x	x	x
Honey locust	...	...	x	...	...	...
Siberian elm	...	...	x	x	x	x
Shortleaf pine	x	...	...	...	...	...
Eastern redcedar	x	x	x	...	...	...
Osage orange	...	x	x	...	...	...
Blue gum	...	...	...	...	x	...
Ponderosa pine	...	...	...	x	...	x

and deep roots. Redcedar roots penetrate to considerable depths in the plains region. Black locust is desirable in the tall-grass region except at the northern end. Osage orange is a deep taprooted species suitable for the southern half of the plains region.

In the foothill grasslands of the California coast ranges, blue gum is undoubtedly the most promising species for this type of stabilization. Black locust can be established successfully and should also be valuable. Siberian elm and velvet ash have been less often planted, but their deep root systems and their adaptability to severe sites suggest their use in some locations.

In the region of the Columbia plateau and in the Wasatch Mountains in Utah, ponderosa pine is the outstanding taprooted native species. Black locust and Siberian elm also have value in this region. Other areas, including the Cascade Mountains of Washington, the San Juan Mountains of Colorado, certain valleys in Wyoming, the New England Coast, the St. Lawrence River Valley, and the mountains of New Hampshire and Vermont, are also subject to frequent slides. In these areas the planting of trees for stabilization has not been attempted, and there is no basis for specific recommendations. The native species with deep or tap roots are the most likely to be effective.

177. Stabilization of Avalanches and Snowslides.—Whether or not erosion properly includes these mass movements of snow, the retarding influence of forests is related to the influence on the mass movements of soil considered in the previous section and may be taken up appropriately at this point.

In mountainous areas of steep slopes where heavy snows accumulate, snowslides and avalanches may be both frequent and destructive. The locality in the United States where they occur often and have forced attention is in the northern Cascade Mountains, where Munger (263) reported the results in the particularly destructive winter of 1910. He distinguished two types of avalanches, one, the canyon slides that originate at the heads of steep, narrow canyons and concentrate in their movement along the canyon bottoms. These slides tend to recur in the same canyon with sufficient frequency that woody vegetation does not have time to establish itself, and hence the canyon slides are ordinarily little related to or influenced by the forest vegetation.

The second type, called slope slides, may originate on any smooth slope where the degree of slope is high and where the exposure to the sun results in surface melting and the consequent lubrication of part or all of the underlying mass of snow. These slope slides are less frequent than the canyon slides and are less likely to recur in the same locations. They occur most often on recent burns where the destruction of the timber leaves the snow exposed to the sun. Eleven avalanches of the slope-slide type in the winter of 1910 in the Wenatchee National Forest covered 19 per cent of the area mapped. Of these, 73 per cent were on deforested slopes and 27 per cent on forested, although the ratio of forested to deforested for the area as a whole was 3:2. Records of 10,000 avalanches in Switzerland subsequent to 1872 showed that only 6 per cent of them originated on areas covered with timber or brush, and 60 per cent on rocky ground.

The influence of the forest seems to consist first in protecting the snow from the sun, and second in providing anchorage for the snow mantle by the stems of the trees. It is at or near the place of origin that forest is most effective. A snowslide once under way is quite capable of destroying almost any forest that may be in its path. Munger concluded that the trees must be more than 15 ft high to resist the progress of snowslides.

In general, there are four ways in which the likelihood and frequency of slope slides may be reduced. The first is the maintenance and protection from fire and other damaging agencies of the existing forest. The second is the exclusion of grazing from recent burns where vegetation might otherwise become established more rapidly. Third, if any cutting of the timber is permissible it should be under a light selection system, particularly on slopes of more than 50 per cent gradient. In Switzerland the selection cuttings are arranged in horizontal strips. Finally, areas denuded by fire or previous slides should be reforested as promptly as possible. The planting of slide areas where the surface soil has been scoured and where the slopes are steep and sometimes unstable involves more difficulty than the planting of similar undisturbed or burned areas. Most of the snowslides occur at high elevations in the fir types, and lacking specific experience the native species growing on exposed areas in the locality should be used.

Experience with the control of snowslides in Switzerland indicates that structures on extensive steep rocky faces are impractical. Elsewhere, the structures should begin at the points where the avalanches start. Walls of stone or rows of short piles have proved effective. The piles are 4 to 5 ft long, 5 to 6 in. in diameter, driven perpendicularly to project not more than 3 ft above ground, and spaced 2 ft apart in the rows. The rows should be along the contours. Distance between the rows varies with the slope. "Breaks" from gentler to steeper slopes are favorable locations. Permeable dry masonry walls, 3 ft high on the upper side are made on a shelf levelled 3 ft wide to a natural foundation. The ends must be well anchored, and the roofs formed by large slabs. Reforestation with seedlings of species growing naturally on the site spaced 3½ by 3½ ft is recommended.

178. Application of Principles of Forest Planting.—Erosion control by planting trees or other vegetation involves not only the choice of species that has been emphasized in this chapter but also most of the subject of seeding and planting with minor modifications in application to eroding sites. Collection of seed, nursery practice, class of stock, seasons and methods of planting or seeding, spacing, and mixtures are governed by the usual principles of artificial reforestation. Mixtures are preferable to pure stands. The closer the spacing and the larger the stock, the sooner will protection become effective. On eroding sites the more time-consuming methods of planting and the selection of the season of most favorable moisture conditions are likely to be essential to success. Failures in the first planting should be replanted promptly. Exclusion of livestock for at least the first few years after planting is necessary.

179. Vegetation for Erosion Control: Summary.—1. Vegetation, once established, provides a self-perpetuating and increasingly effective permanent control. For its establishment, preparatory measures are often necessary such as grading banks to angles of repose, preventing undercutting, conveying concentrations of water to natural channels, covering surfaces with mulch or brush, and excluding livestock.

2. Species for the control of erosion should be pioneers in succession and should have as many as possible of the following characteristics: ecological adaptability to the local sites, rapid juvenile growth, vegetative and aggressive regeneration, ability to grow on adverse sites, large fibrous root systems, and resistance to damage by insects, diseases, and grazing animals.

3. Most of the principles and practices of forest planting and seeding are applicable with minor modifications to control of erosion.

CHAPTER XXV

WATERSHED MANAGEMENT AND THE PROTECTION FOREST

A protection forest has been defined as an area wholly or partly covered with woody growth, managed primarily for its beneficial effects on water or soil movement rather than for wood or forage production. This general concept has been familiar in forestry literature for many years. Some of the noncommercial forest areas in the United States, including much of the subalpine forest and the chaparral areas in southern California, have been classified as protection forests. Actually they have been serving the objectives mentioned in the definition but without benefit of management other than protection from fire. With that protection the natural vegetation has seemed to provide all that was needed by way of watershed protection. The use of the term "watershed protection" has carried that implication.

180. Watershed Management and Its Objectives.—More recently watershed management has been suggested and used as a more suitable term. It has been defined as the administration and regulation of the aggregate resources of a drainage basin for the production of water and the control of erosion, stream flow, and floods (329). For the control of erosion, or the stabilization of moving sands, or the reduction of flood damage, natural vegetation protected from fire usually serves the purpose. When the objective of producing a maximum yield of usable water is introduced, however, it cannot be assumed that the unmanaged natural vegetation will be the most effective in contributing to the greatest amount of water.

Four phases of watershed management may be recognized. One is the recognition phase, which involves a survey to determine extent, location, and severity of deterioration on critical or misused areas. Second is the restoration phase, which includes the correction of the unstable conditions causing erosion and floods by vegetation or engineering methods. The third is the protection phase, which involves not only protection from fires or damaging agencies but the maintenance of existing conditions, provided they are acceptable for the uses to which the area is subject. The fourth is the improvement phase, when practices are initiated to increase the yield of water. This phase may involve various measures of different degrees of effectiveness as far as the yield of water is concerned. A storage reservoir with capacity to impound the maximum runoff may be contrasted with a drainage basin where the soil serves as the reservoir, and maximum infiltration is the objective to assure the most efficient use of the soil reservoir. The recognition and protection phases of watershed management ordinarily involve no conflicts with other uses, but the restoration and improvement phases may often require some modifications of existing forest and range-management practices.

Maximum yield of water has been shown to depend in part upon the character of the vegetation in that the losses by interception, evaporation, and transpiration may be minimum. It is reasonably certain that in most areas these losses will not be minimum when the area is occupied by mature, luxuriant climax forest. In other less extreme instances the losses resulting from the presence of large, dense vegetation are likely to be too great to be harmonized with a maximum yield of water. The replacement of existing vegetation by smaller species and less dense stands that are more economical in their use of water logically has a place in watershed management. Many protection forests are inaccessible or unmerchantable, so that cuttings and thinnings have not been attempted, and the expense might not be justified by any prospective increase in value of water yield. Logging operations in merchantable timber probably increase water yield but they are not planned with that objective. Destruction of existing vegetation by fire, although it may increase the water yield temporarily, at the same time renders that yield unusable and hence does not serve the purpose.

Where maximum water yield is not a consideration, as in much of the humid part of the country, the other objectives of protection forest management including prevention of erosion, control of avalanches and snowslides, fixation of shifting sand, and control of floods are likely to be attained by the maintenance of a dense forest canopy and the underlying well-developed layer of forest floor. The climax forest in any given region will almost certainly fulfill these objectives, although it is also likely that preclimax stages in succession, including well-stocked stands of conifers some years after the age when the crown canopy closes, may be equally effective. Similarly, the dense chaparral in southern California or the buffalo grass in the plains region may be as effective as the larger forms of woody vegetation.

181. Ways and Means.—The actual execution of plans for protection-forest management involves activities following from the effects that have been outlined previously of burning, cutting, logging, and grazing. They include, therefore, first, effective fire protection. In the chaparral region of southern California, where intensive protection in recent years has reduced the percentage of area burned to about 1, it has been suggested that for purposes of flood control the protection should be intensified until an average of less than one-quarter of 1 per cent of the area is burned annually.

Areas that are burned or otherwise denuded should be treated immediately to forestall erosion. Erosion-control measures for the road banks that might result from more intensive fire protection would be essential. If cutting of the timber is permissible at all in an area where forest influences are important, the lighter the cutting and the less ground surface that is exposed in contiguous areas, the better the results are likely to be. In general the selection system of cutting is most likely to meet these requirements.

Logging methods are quite definitely related to the objectives of protection forest management. Among the different methods, steam skidding, and particularly the high-line skidding, causes denudation of the largest proportion of the logged area. Skidding of logs directly down the slope gouges channels in which water concentrates and cuts with maximum effectiveness. The longer the skidding trails, the longer they are used, and the steeper the grade, the greater is the erosion. On the Navajo Indian Reservation contour skidding in ponderosa pine has been used successfully as a means of conserving water and avoiding erosion damage. The skidding trails served as contour ditches. Tractor and truck logging exposes a minimum of bare soil and is more desirable than donkey logging from the viewpoint of protection forests. Trucks are less destructive than tractors, and the big wheels of some years ago, like the arches used with tractors, by lifting the forward ends of the logs off the ground, cause less disturbance of the surface. Chutes are preferable to skidding trails or roads.

The slash resulting from logging may be used to advantage for the protection of exposed or scoured surfaces such as skid trails. When it is lopped and scattered it is beneficial. On the other hand, when it is burned the areas covered by the fires are completely denuded. The aim should be to keep these fire-denuded areas small, and particularly so on slopes. For this purpose, burning in piles is far preferable to burning in windrows or broadcast. If windrow burning is necessary for other reasons, obviously rows along the contours are less likely to cause damage than those up and down the slope.

On range lands the damage to vegetation and the exposure of bare soil at places where livestock concentrate, as around water holes or salt, constitute clear examples of the need for regulation of grazing use to reduce erosion. The remedy lies evidently in management which will minimize the concentration of the animals at certain points. Some concentration is almost inevitable near water and salt, but at least these necessities can often be located on areas that are nearly flat and are therefore little subject to washing. Locations with rock or gravelly material at the surface should be selected in preference to soils subject to cutting.

When provision for maximum yield of water becomes an objective in protection-forest management it becomes necessary to consider ways of reducing the losses of water by interception, transpiration, and evaporation to a minimum. One obvious method used successfully in certain stream channels in southern California is to pipe the water out of the channel so that the riparian vegetation will not be able to draw upon the subsurface supply in the alluvial gravels. Considering the high use of water by this stream-bottom vegetation, this method may save considerable amounts. After several years of operation near San Bernardino it appears that the trees in the canyon bottom are still able to obtain sufficient moisture for survival. Because the sum of interception and transpiration exceeds evaporation in most plant communities, the aim should be to reduce those two losses, which are functions of the crowns and roots of the plants. In forests this may be done by cutting to reduce the density of the stands and to limit the size to which the trees are allowed to grow. In reducing the density by cutting, sufficient crowns should be reserved to protect the ground surface with a layer of forest floor. Similar results might be obtained by converting to species of small size, shallow roots, and light crowns. Grasses or shrubs often intercept and transpire less than tree species. It is known for a few species that there are wide differences in the rate of use of water by transpiration. Obviously, species economical in the use of water are preferable to those that are lavish.

In regions where the snow is an important source of water, the density of the forest should be reduced and the accumulation of snow increased by cutting to provide openings or wells of diameters at least equal to the height of the trees and sufficiently close together that a considerable proportion of the total area is occupied by the openings. Clear cutting in strips as wide as the height of the trees and roughly perpendicular to the prevailing direction of the snow-bearing winds should have most of the same advantages. The limit to the number and spacing of the openings or strips in any given forest type will probably be found in the resistance to windthrow of the trees left between the wells or strips.

182. Progress and Status.—Accomplishments in the establishment and management of protection forests are more substantial than most foresters or others realize. Acquisition of forest lands primarily for watershed management includes many areas beginning with the Angeles and Tonto and other national forests in the Southwest and including most of the national forests in the Appalachian and northeastern mountains. Many communities in the eastern states have acquired lands surrounding their water-supply reservoirs. Up to 1933 there were about 340 such municipal forests with a total area of 290,000 acres in the New England and Middle Atlantic States. Private and semi-public water and power companies have also acquired lands around their reservoirs. In the West, many of the water developments are within the national forests where the tributary lands are managed by the U.S. Forest Service.

The management of these protection forests consists in protection from fire, regulation of use in the interests of sanitation, and frequently in reforestation of eroding areas. For example, the Mahoning Valley Sanitary District in Ohio has planted 4,500,000 trees on its 4,500 acres since 1931. Progress beyond these stages into the measures for maximum yields of water is only in initial stages on experimental areas, which, however, include experimental forests of several square miles.

183. Protection Forest: Summary.—1. The objectives of management of protection forests may include one or more of the following:

a. Prevention of erosion, avalanches, and snow slides.
b. Fixation of shifting sand.
c. Control of floods.
d. Protection of water supply for regularity, purity, and maximum quantity.

2. The first three (*a*, *b*, and *c*) objectives are attained by the establishment and maintenance of a dense, complete canopy and the resulting forest floor. Pioneer species are usually easiest to establish but ultimately a mixture of the climax species of the region will most effectively meet these requirements.

3. The management of such forests should provide for:

a. Effective fire protection.
b. Immediate reestablishment of cover on any areas burned or otherwise denuded.

c. Light cuttings preferably under the selection system.

d. Slash disposal by lopping and scattering or by placing in gullies, skid trails, or over exposed soil.

e. Logging methods like wheels, chutes, or trucks, which will denude a minimum area.

f. Range management for minimum concentration of stock.

4. Protection of water supply and particularly for maximum water yield involves considerations supplementing, and to some extent conflicting with, the foregoing, namely, the minimizing of interception, transpiration, and evaporation from soil or snow. For the attainment of this objective the following means should be considered:

a. Favor species of small size; light, slowly decomposing foliage; and of low water requirement; in stands of only sufficient density to provide just enough forest floor to maintain the soil at high infiltration capacity. Perhaps preclimax conifers with frequent light cuttings on a very short rotation, or brush or grass species may most nearly meet these requirements.

b. Remove water by piping or otherwise from stream bottoms where the roots of the vegetation reach the water table.

APPENDIX

SYMBOLS USED AND THEIR MEANINGS

A area
a cross-sectional area; altitude
B barometric pressure at actual elevation
B_0 barometric pressure at sea level
C constant; runoff coefficient; heat leakage through walls of containers; centigrade
D diameter; depth of soil
D_s depth of snow
d depth of water; water equivalent of snow
E evaporation, depth of water
e vapor pressure
e_a vapor pressure, actual or saturated at t of dew point
e_s saturated vapor pressure
e_w saturated e at t of water surface
F forest; Fahrenheit; total infiltration
f rate of infiltration
G sediment in discharge or eroded material
g acceleration of gravity
H_a absolute humidity
H_d saturation deficit
H_q specific humidity
H_r relative humidity
h height
I interception
I_t index of turbulence
i intensity of rainfall
K or k constant
L latent heat of vaporization of water
l length or distance
l_o length of overland flow
M moisture content
M_e moisture equivalent
M_p % M by weight
M_f field moisture capacity
M_v % M by volume
M_w wilting point
m slope of line
n roughness coefficient
O open
P precipitation
P_e precipitation excess
P_s precipitation per shower

p wetted perimeter; pressure difference
Q flow in volume per unit of time
Q_g ground-water flow or percolation
Q_s surface runoff, total or cumulative
q_s rate of surface runoff
R solar radiation
R_b back radiation
r radius; hydraulic radius; rate of recharge
S storage; heat stored in unit time in a column of unit area and depth
s slope of land or channel
T transpiration depth
t temperature; time
t_a temperature of air
t_w temperature of water
t' wet-bulb temperature
V volume
v velocity
W weight
w width
X unknown deep seepage
x length, horizontal
y specific yield
δ surface detention
ϵ base of natural logarithms
η coefficient of viscosity
ρ density
ρ_a apparent density or volume weight
ρ_s density of snow

COMMON AND SCIENTIFIC NAMES OF SPECIES

Alder—*Alnus incana* (L.) Moench.
Alpine fir—*Abies lasiocarpa* (Hook.) Nutt.
American beachgrass—*Ammophila breviligulata* Fern.
American beech—*Fagus grandifolia* Ehrh.
American dunegrass—*Elymus mollis* Trin.
American plum—*Prunus americana* Marsh.
Apricot—*Prunus armeniaca* L.
Arizona cypress—*Cupressus arizonica* Greene.
Athel tamarisk—*Tamarix aphylla* (L.) Karst.
Austrian pine—*Pinus nigra austriaca* Schneid.
Bald cypress—*Taxodium distichum* (L.) Rich.
Balsam fir—*Abies balsamea* (L.) Mill.
Basswood—*Tilia americana* L.
Bayberry—*Myrica carolinensis* Mill.
Bermuda grass—*Cynodon dactylon* (L.) Pers.
Big bluegrass—*Poa ampla* Merrill.
Big bluestem—*Andropogon furcatus* Muhl.
Bigcone spruce—*Pseudotsuga macrocarpa* (Vasey) Mayr.
Bigtree—*Sequoia gigantea* (Lindl.) Decne.

Black cherry—*Prunus serotina* Ehrh.
Black grama—*Bouteloua eriopoda* (Torr.) Torr.
Black gum—*Nyssa sylvatica* Marsh.
Blackjack oak—*Quercus marilandica* Muenchh.
Black locust—*Robinia pseudoacacia* L.
Black medick—*Medicago lupulina* L.
Black mustard—*Brassica nigra* (L.) Koch.
Black oak—*Quercus velutina* Lam.
Black spruce—*Picea mariana* (Mill.) B.S.P.
Black walnut—*Juglans nigra* L.
Blowout grass—*Redfieldia flexuosa* (Thurb.) Vasey.
Blueberry elder—*Sambucus glauca* Nutt.
Blue grama—*Bouteloua gracilis* (H.B.K.) Lag.
Blue gum—*Eucalyptus globulus* Labill.
Blue oak—*Quercus douglasii* Hook. and Arn.
Blue spruce—*Picea pungens* Engelm.
Bluestem—*Andropogon* spp.
Boxelder—*Acer negundo* L.
Brake fern—*Pteridium aquilinum* (L.) Kuhn. var. *pubescens* Underw.
Broomsedge—*Andropogon* sp.
Buffalo grass—*Buchloë dactyloides* (Nutt.) Engelm.
Bur clover—*Medicago hispida* Gaertn.
Bur oak—*Quercus macrocarpa* Michx.
Burnet—*Sanguisorba occidentale* Nutt.
Bush honeysuckle—*Diervilla lonicera* Mill.
California black oak—*Quercus Kelloggii* Newb.
California buckeye—*Aesculus californica* (Spach) Nutt.
California buckwheat—*Eriogonum fasciculatum* Benth. var. *foliolosum* (Nutt.) Stokes.
California chinquapin—*Castanopsis sempervirens* (Kell.) Dudl.
California scrub oak—*Quercus dumosa* Nutt.
Canada bluegrass—*Poa compressa* L.
Canary pine—*Pinus canariensis* C. Smith.
Canyon live oak—*Quercus chrysolepis* Liebm.
Carpet grass—*Axonopus compressus* (Swartz) Beauv.
Catalpa—*Catalpa* sp.
Centipede grass—*Eremochloa ophiuroides* (Munro) Hack.
Chamise—*Adenostema fasciculatum* H. & A.
Chestnut—*Castanea dentata* (Marsh.) Borkh.
Chestnut oak—*Quercus montana* Willd.
Chinese elm—*Ulmus parvifolia* Jacq.
Chokecherry—*Prunus virginiana* L.
Christmasberry—*Photinia arbutifolia* Lindl.
Coralberry—*Symphoricarpos orbiculatus* Moench.
Coulter pine—*Pinus coulteri* D. Don.
Crack willow—*Salix fragilis* L.
Creosote bush—*Covillea tridentata* (DC.) Vail.
Crested wheatgrass—*Agropyron cristatum* (L.) Beauv.
Cucumbertree—*Magnolia acuminata* L.
Curly mesquite—*Hilaria belangeri* (Steud.) Nash.

Desert gum—*Eucalyptus rudis* Endl.
Desertwillow—*Chilopsis linearis* (Cav.) Sweet.
Digger pine—*Pinus sabiniana* Dougl.
Douglas fir—*Pseudotsuga taxifolia* (Poir.) Britton.
Dyer's greenweed—*Genista tinctoria* L.
Eastern cottonwood—*Populus deltoides* Bartr.
Eastern hemlock—*Tsuga canadensis* (L.) Carr.
Eastern redbud—*Cercis canadensis* L.
Eastern redcedar—*Juniperus virginiana* L.
Eastern white pine—*Pinus strobus* L.
Elm—*Ulmus americana* L.
Engelmann spruce—*Picea engelmanni* Parry.
European beachgrass—*Ammophila arenaria* (L.) Link.
European beech—*Fagus sylvatica* L.
European dune wildrye—*Elymus arenarius* L.
European hornbeam—*Carpinus betulus* L.
European larch—*Larix decidua* Miller.
Fireweed—*Epilobium angustifolium* L.
Flowering dogwood—*Cornus florida* L.
Foxtail dalea—*Dalea alopecurioides* Willd.
French tamarisk—*Tamarix gallica* L.
Galleta grass—*Hilaria jamesii* (Torr.) Benth.
Gambel oak—*Quercus gambelii* Nutt.
Giant wildrye—*Elymus condensatus* Presl.
Golden willow—*Salix alba vitellina* Stokes.
Gray birch—*Betula populifolia* Marsh.
Green ash—*Fraxinus pennsylvanica* var. *lanceolata* (Borkh.) Sarg.
Hackberry—*Celtis occidentalis* L.
Hairy grama—*Bouteloua hirsuta* Lag.
Hairy vetch—*Vicia villosa* Roth.
Hazelnut—*Corylus* spp.
Hoaryleaf Ceanothus—*Ceanothus crassifolius* Torr.
Hog peanut—*Amphicarpa monoica* (L.) Ell.
Honey locust—*Gleditsia triacanthos* L.
Hophornbeam—*Ostrya virginiana* (Mill.) K. Koch.
Incense cedar—*Libocedrus decurrens* Torr.
Interior live oak—*Quercus wizlizeni* A. DC.
Italian rye grass—*Lolium multiflorum* Lam.
Jack pine—*Pinus banksiana* Lamb.
Japanese honeysuckle—*Lonicera japonica* Thunb.
Japanese larch—*Larix leptolepis* Sieb. & Zucc.
Jeffrey pine—*Pinus jeffreyi* Grev. & Balf.
Kentucky bluegrass—*Poa pratensis* L.
Kentucky coffeetree—*Gymnocladus dioicus* L. (Koch.)
Kikuyu grass—*Pennisetum clandestinum* Hochst.
Knobcone pine—*Pinus attenuata* Lemm.
Kudzuvine—*Pueraria thunbergiana* (Lieb. & Zucc.) Benth.
Lead plant—*Amorpha canescens* Pursh.
Lehmann's lovegrass—*Eragrostis Lehmanniana* Nees.

Lilac—*Syringa vulgaris* L.
Limber pine—*Pinus flexilis* James.
Little bluestem—*Andropogon scoparius* Michx.
Loblolly pine—*Pinus taeda* L.
Lodgepole pine—*Pinus contorta* var. *latifolia* Engelm.
Longleaf pine—*Pinus palustris* Mill.
Manzanita—*Arctostaphylos* spp.
Maritime pine—*Pinus pinaster* Ait.
Monterey cypress—*Cupressus macrocarpa* Hartw.
Monterey pine—*Pinus radiata* D. Don.
Mountain hemlock—*Tsuga mertensiana* (Bong.) Carr.
Mountain mahogany—*Cercocarpus betuloides* Nutt.
Mulefat—*Baccharis viminea* DC.
Nootka rose—*Rosa nutkana* Presl.
Northern catalpa—*Catalpa speciosa* Warder.
Northern red oak—*Quercus borealis* Michx. f.
Northern white cedar—*Thuja occidentalis* L.
Norway spruce—*Picea excelsa* Link.
Orchard grass—*Dactylis glomerata* L.
Oregon crab apple—*Malus fusca* (Raf.) Schneid.
Osage orange—*Maclura pomifera* (Raf.) Schneid.
Paloblanco—*Forestiera neo-mexicana* Gray.
Paper birch—*Betula papyrifera* Marsh.
Partridge pea—*Cassia Chamaecrista* L.
Pin oak—*Quercus palustris* Muenchh.
Pitch pine—*Pinus rigida* Mill.
Plains cottonwood—*Populus sargentii* Dode.
Poison oak—*Rhus diversiloba* Torr. & Gray.
Ponderosa pine—*Pinus ponderosa* Laws.
Port Orford cedar—*Chamaecyparis lawsoniana* (A. Murr.) Parl
Post oak—*Quercus stellata* Wangenh.
Prairie sandreed—*Calamovilfa longifolia* (Hook.) Scribn.
Purple beach pea—*Lathyrus japonicus* Willd.
Quaking aspen—*Populus tremuloides* Michx.
Raspberry—*Rubus idaeus* var. *strigosus* Michx.
Red alder—*Alnus rubra* Bong.
Redbud—*Cercis canadensis* L.
Red clover—*Trifolium pratense* L.
Red fir—*Abies magnifica* A. Murr.
Red gum—*Eucalyptua rostrata* Schlecht.
Red maple—*Acer rubrum* L.
Red pine—*Pinus resinosa* Ait.
Red spruce—*Picea rubra* Link.
Redtop—*Agrostis alba* L.
Redwood—*Sequoia sempervirens* (D. Don) Endl.
Rocky Mountain juniper—*Juniperus scopulorum* Sarg.
Russian mulberry—*Morus alba tatarica* Loud.
Russian olive—*Elaeagnus angustifolia* L.
Sagebrush—*Artemisia tridentata* Nutt.

Saltgrass—*Distichlis* sp.
Sand bluestem—*Andropogon hallii* Hack.
Sandbar willow—*Salix exigua* Nutt.
Sandbar willow—*S. fluviatilis* Nutt.
Sandbar willow—*S. hindsiana* Benth.
Sandbar willow—*S. interior* Rowlee.
Sandbar willow—*S. sessilifolia* Nutt.
Sand cherry—*Prunus pumila* L.
Sand dropseed—*Sporobolus cryptandrus* (Torr.) A. Gray.
Sassafras—*Sassafras albidum* (Nutt.) Nees.
Scarlet oak—*Quercus coccinea* Muenchh.
Scotch broom—*Cytisus scoparius* (L.) Lk.
Scotch pine—*Pinus sylvestris* L.
Sea oats—*Uniola paniculata* L.
Seaside lupine—*Lupinus littoralis* Dougl.
Seepwillow—*Baccharis glutinosa* Pers.
Sericea—*Lespedeza sericea* Benth.
Shagbark hickory—*Carya ovata* (Mill) K. Koch
Shore pine—*Pinus contorta* Dougl.
Shortleaf pine—*Pinus echinata* Mill.
Siberian elm—*Ulmus pumila* L. (frequently called "Chinese elm").
Siberian pea tree—*Caragana arborescens* Lam.
Side-oats grama—*Bouteloua curtipendula* (Michx.) Torr.
Silver fir—*Abies pectinata* DC.
Singleleaf piñon—*Pinus monophylla* Torr. and Frém.
Sitka spruce—*Picea sitchensis* (Bong.) Carr.
Skunkbush—*Rhus trilobata* Nutt.
Slash pine—*Pinus caribaea* Morelet.
Sourwood—*Oxydendrum arboreum* (L.) DC.
Southern red oak—*Quercus falcata* Michx.
Southern white cedar—*Chamaecyparis thyoides* (L.) B.S.P.
Spring vetch—*Vicia sativa* L.
Sudan grass—*Sorghum vulgare* var. *sudanense* (Piper) Hitchc.
Sugar maple—*Acer saccharum* Marsh.
Sugar pine—*Pinus lambertiana* Dougl.
Sweet birch—*Betula lenta* L.
Sweetgum—*Liquidambar styraciflua* L.
Switchgrass—*Panicum virgatum* L.
Sycamore—*Platanus occidentalis* L.
Table-mountain pine—*Pinus pungens* Lamb.
Tamarack—*Larix laricina* (Du Roi) K. Koch.
Tatarian honeysuckle—*Lonicera tatarica* L.
Trailing wild bean—*Strophostyles helvola* (L.) Britton.
Tree lupine—*Lupinus arboreus* Sims.
Tree of heaven—*Ailanthus altissima* (Mill.) Swingle.
Utah juniper—*Juniperus utahensis* (Engelm.) Lemm.
Velvet ash—*Fraxinus velutina* Torr.
Victoria or Australian tea tree—*Leptospermum laevigatum* Muell.
Virginia creeper—*Parthenocissus quinquefolia* (L.) Planch.

Virginia pine—*Pinus virginiana* Mill.
Weeping lovegrass—*Eragrostis curvula* Nees.
Western chokecherry—*Prunus demissa* (Nutt.) Walp.
Western hemlock—*Tsuga heterophylla* (Raf.) Sarg.
Western larch—*Larix occidentalis* Nutt.
Western red cedar—*Thuja plicata* Donn.
Western soapberry—*Sapindus drummondii* H. & A.
Western wheatgrass—*Agropyron smithii* Rydb.
Western white pine—*Pinus monticola* Dougl.
Western white spruce—*Picea glauca albertiana* (S. Brown) Sarg.
White ash—*Fraxinus americana* L.
White clover—*Trifolium repens* L.
White fir—*Abies concolor* (Gord. & Glend.) Hoopes.
White oak—*Quercus alba* L.
White spruce—*Picea glauca* (Moench) Voss.
White sweet clover—*Melilotus alba* Desr.
White willow—*Salix alba* L.
Wild indigo—*Baptisia tinctoria* (L.) R. Br.
Wild lupine—*Lupinus perennis* L.
Yellow birch—*Betula lutea* Michx. f.
Yellow poplar—*Liriodendron tulipifera* L.
Yellow sweet clover—*Melilotus officinalis* (L.) Lam.

REFERENCES

1. ADAMS, W. R.: Studies in tolerance of New England forest trees. X. The change in the environmental factors caused by thinnings in pine plantations, *Vt. Agr. Expt. Sta. Bull.* 310 (1930).
2. ———: Studies in tolerance of New England forest trees. VIII. Effect of spacing in a Jack pine plantation, *Vt. Agr. Expt. Sta. Bull.* 282 (1928).
3. ———: Studies in tolerance of New England forest trees. XII. Effect of thinning in plantations on some of the physical factors of the site and on the development of young northern white pine and Scotch pine, *Vt. Agr. Expt. Sta. Bull.* 390 (1935).
4. ALDERFER, R. B., and F. G. MERKLE: Structural stability and permeability of native forest soils compared with cultivated areas of the same soil type, *Proc. Soil Sci. Soc. Am.*, **6**:98–103 (1941).
5. ALTPETER, L. S.: Use of vegetation in control of streambank erosion in northern New England, *J. Forestry*, **42**:99–107 (1944).
6. ALWAY, F. J.: Effect of burning the forest floor upon the productivity of Jack pine land, *Proc. First Intern. Congr. Soil Sci.*, **3**:514–524 (1928).
7. ——— and J. KITTREDGE, JR.: The forest floor under stands of aspen and paper birch, *Soil Sci.*, **35**:307–312 (1933).
8. ———, ———, and W. J. METHLEY: Composition of the forest floor layers under different forest types on the same soil type, *Soil Sci.*, **36**:387–398 (1933).
9. ———, T. E. MAKI, and W. J. METHLEY: Composition of the leaves of some forest trees, *Am. Soil Survey Assoc. Bull.* 15, 81–84 (1934).
10. ——— and C. O. ROST: Effect of forest fires upon the composition and productivity of the soil, *Proc. First Intern. Congr. Soil Sci.*, **3**:546–576 (1928).
11. ——— and R. ZON: Quantity and nutrient contents of pine leaf litter, *J. Forestry*, **28**:715–727 (1930).
12. ANDERSON, H. W.: The effect of freezing on soil moisture and on evaporation from a bare soil, *Trans. Am. Geophys. Union*, **27**:863–870 (1946).
13. ANONYMOUS: Determination of water supply from snow fields, *Eng. News-Record*, **83**:766–767 (1919).
14. ———: Notes on black locust, *J. Forestry*, **30**:499–500 (1932).
15. AREND, J. L.: Infiltration rates of forest soils in the Missouri Ozarks as affected by woods burning and litter removal, *J. Forestry*, **39**:726–728 (1941).
16. ARTHUR, J. M., and W. D. STEWART: Transpiration of tobacco plants in relation to radiant energy in the visible and infra-red, *Contrib. Boyce Thompson Inst.*, **5**:489–501 (1933).
17. ASHE, W. W.: Financial limitation in the employment of forest cover in protecting reservoirs, *U.S. Dept. Agr. Bull.* 1430 (1926).
18. AUGUSTINE, M. T.: Infiltration runs on frozen ground, *Proc. Soil Sci. Soc. Am.*, **6**:435 (1941).
19. AUTEN, J. T.: The effect of forest burning and pasturing in the Ozarks on the water absorption of forest soils, *U.S. Forest Serv., Central States Forest Expt. Sta. Note* 16 (1934). Mimeo.

20. ———: Porosity and water absorption of forest soils, *J. Agr. Research*, **46**:997–1014 (1933).
21. ———: Relative influence of sassafras, black locust, and pines upon old-field soils, *J. Forestry*, **43**:441–446 (1945).
22. ———: Soil profile studies in relation to site requirements of black locust and black walnut, *U.S. Forest Serv., Central States Forest Expt. Sta., Note* 21 (1936).
23. Ayres, Q. C.: "Soil Erosion and Its Control," McGraw-Hill Book Company, Inc., New York (1936). 365 pp.
24. Bagnold, R. A.: "The Physics of Blown Sand and Desert Dunes," Methuen & Co., Ltd., London (1941). 265 pp.
25. Bailey, R. W.: Land-erosion—normal and accelerated—in the semiarid west, *Trans. Am. Geophys. Union*, Part II:240–250 (1941).
26. ——— and A. R. Croft: Contour-trenches control floods and erosion on range-lands, *U.S. Emergency Conserv. Work, Forestry Pub.* 4, 1–22 (1937).
27. Bailey, R. Y.: Kudzu for erosion control in the southeast, *U.S. Dept. Agr. Farmers Bull.* 1840 (1939).
28. Baker, D. M., and H. Conkling: "Water Supply and Utilization," John Wiley & Sons, Inc., New York (1930). 495 pp.
29. Baker, F. S.: Some field experiments upon evaporation from snow surfaces, *Monthly Weather Rev.*, **45**:363–366 (1917).
30. ———: "The Theory and Practice of Silviculture," McGraw-Hill Book Company, Inc., New York (1934). 502 pp.
31. Barnes, B. S.: Fluctuation of ground water flow, *Proc. Cent. Snow Conf.*, **1**:174–175 (1942).
32. Barrett, L. I.: Water in its relation to forestry, pp. 16–18 in "Michigan's Water Problems," *Water Conserv. Conf.*, Lansing, Mich. (1944).
33. Barrows, H. K.: A study of the effect of temperature upon different reactions and processes, *J. Boston Soc. Civil Engrs.*, **24**:57–77 (1937).
34. Bates, C. G.: Forest types in the central Rocky Mountains as affected by climate and soil, *U.S. Dept. Agr. Bull.* 1233 (1924).
35. ———: Physiological requirements of Rocky Mountain trees, *J. Agr. Research*, **24**:97–164 (1923).
36. ———: Windbreaks: their influence and value, *U.S. Forest Serv. Bull.* 86 (1911).
37. ——— and A. J. Henry: Forest and streamflow at Wagon Wheel Gap, Colorado. Final report. *Monthly Weather Rev. Suppl.*, **30**:1–79 (1928).
38. ——— and J. H. Stoeckeler: Snowdrift control on highways by planting of trees and shrubs, *U.S. Forest Serv., Lake States Forest Expt. Sta.*, 1–14 (1941). Mimeo.
39. ——— and O. R. Zeasman: Soil erosion—a local and national problem, *Wisc. Agr. Expt. Sta. Research Bull.* 99 (1930).
40. Bauer, H. L.: Moisture relations in the chaparral of the Santa Monica Mountains, California, *Ecol. Monographs*, **6**:409–454 (1936).
41. Baumann, P.: The function and design of check dams, *Civil Eng.*, **6**:355–358 (1936).
42. Baver, L. D.: "Soil Physics," John Wiley & Sons, Inc., New York (1940).
43. Beall, H. W.: The penetration of rainfall through hardwood and softwood forest canopy, *Ecology*, **15**:412–415 (1934).

44. Belgrand, E.: De l'influence des forêts sur l'écoulement des eaux pluviales, *Soc. Mét. de France, Annuaire,* **1**:176–193 (1853); **2**:81–87 (1854).
45. Belotelkin, K. T.: Soil-freezing and forest-cover, *Trans. Am. Geophys. Union,* Part I:173–175 (1941).
46. Bennett, H. H.: Relation of erosion to vegetative changes, *Sci. Monthly,* **35**: 385–415 (1932).
47. ———: "Soil Conservation," McGraw-Hill Book Company, Inc., New York (1939). 993 pp.
48. Bialoglowski, J.: Effect of humidity on transpiration of rooted lemon cuttings under controlled conditions, *Proc. Am. Soc. Hort. Sci.,* **33**:166–169 (1936).
49. Billings, W. D.: The structure and development of old field shortleaf pine stands and certain associated physical properties of soil, *Ecol. Monographs,* **8**:437–499 (1938).
50. Blaney, H. F., C. A. Taylor, and A. A. Young: Rainfall penetration and consumptive use of water in Santa Ana River Valley and coastal plain, *Calif. Dept. Pub. Works, Div. Water Resources, Bull.* 33, 1–162 (1930).
51. Bodman, G. B.: The forest floor developed under conditions of summer rainfall deficiency in a Californian pine and fir forest, *Am. Soil Survey Assoc. Bull.* 16, 97–101 (1935).
52. Brater, E. F.: The unit-hydrograph principle applied to small watersheds, *Proc. Am. Soc. Civil Engrs.,* **65**:1191–1215 (1939).
53. Broadfoot, W. M., and W. H. Pierre: Forest soil studies: I. Relation of rate of decomposition of tree leaves to their acid-base balance and other chemical properties, *Soil Sci.,* **48**:329–348 (1939).
54. Brooks, C. E. P.: Influence of forests on rainfall, *Quart. J. Roy. Meteorol. Soc., London,* **54**:1–13 (1928).
55. Brooks, C. F.: Need for universal standards for measuring precipitation, snowfall and snow cover, *Int. Comn. Snow Glaciers Trans. Bull.,* 23, 7–58 (1936).
56. ——— and A. J. Connor: "Climatic Maps of North America," Harvard University Press, Cambridge, Mass. (1936). 3 pp., 26 maps.
57. Brown, C. B.: The control of reservoir silting, *U.S. Dept. Agr. Misc. Pub.* 521 (1943).
58. Browning, G. M., and F. M. Milam: Effect of different types of organic materials and lime on soil aggregation, *Soil Sci.,* **57**:91–106 (1944).
59. Bunger, M. T., and H. J. Thomson: Root development as a factor in the success or failure of windbreak trees in the southern high plains, *J. Forestry,* **36**: 790–803 (1938).
60. Burger, H.: Blattmenge und Zuwachs an Fichten im Plenterwald, *Schweiz. Z. Forstw.,* **89**:275–278 (1938).
61. ———: Holz, Blattmenge und Zuwachs. IV. Ein 80-jähriger Buchenbestand, *Mitt. Schweiz. Anstalt forstl. Versuchsw.,* **21**:307–348 (1940).
62. ———: Holz, Blattmenge und Zuwachs. II. Die Douglasie, *Mitt. Schweiz. Anstalt forstl. Versuchsw.,* **19**:21–72 (1935).
63. ———: Holz, Blattmenge und Zuwachs. I. Die Weymouthsföhre, *Mitt. Schweiz. Anstalt forstl. Versuchsw.,* **15**:243–292 (1929).
64. ———: Holzarten auf verschiedenen Bodenarten, chemische Zusammensetzung, Phänologie und Wachstum der darauf erzogenen Holzarten, *Mitt. Schweiz. Anstalt forstl. Versuchsw.,* **15**:50–128 (1930).

65. ———: Influence du pâturage sur l'écoulement des eaux et l'érosion, *J. Forest. Suisse,* **83**:97–102 (1932).
66. ———: Der Kronenaufbau gleichalteriger Nadelholzbestände, *Mitt. Schweiz. Anstalt forstl. Versuchsw.,* **21**:5–57 (1939).
67. ———: Die transpiration unserer Waldbäume, *Z. Forst. u. Jagdw.,* **57**:473–82 (1925).
68. Burke, M. F.: Rainfall on and runoff from San Gabriel Mountains during flood of March, 1938, *Trans. Am. Geophys. Union,* Part I:8–15 (1939).
69. Burns, G. P.: Studies in tolerance of New England forest trees. XIII. The effect of root development on height and diameter growth, *Vt. Agr. Expt. Sta. Bull.* 422 (1937).
70. ——— and E. S. Irwin: Studies in tolerance of New England forest trees. XIV. Effect of spacing on the efficiency of white and red pine needles as measured by the amount of wood production on the main stem, *Vt. Agr. Expt. Sta. Bull.* 499 (1942).
71. Burns, G. R.: A portable instrument for measuring solar radiation in forests, *Vt. Agr. Expt. Sta. Bull.* 261 (1927).
72. Burrage, C. H.: Forestry and land use in a flood-control survey, *J. Forestry,* **41**:873–876 (1943).
73. Busse, W.: Baumkrone und Schaftzuwachs, *Forstwiss. Centr.,* **52**:310–18 (1930).
74. California Department of Public Works, Division of Engineering and Irrigation: Flow in California streams, *Calif. Dept. Pub. Works, Div. Water Resources, Bull.* 5 (1923).
75. Campbell, F. B., and H. A. Bauder: A rating-curve method for determining silt-discharge of streams, *Trans. Am. Geophys. Union,* Part II:603–7 (1940).
76. Chandler, R. F., Jr.: The amount and mineral nutrient content of freshly fallen leaf litter in the hardwood forests of central New York, *J. Am. Soc. Agron.,* **33**:859–871 (1941).
77. ———: Amount and mineral nutrient content of freshly fallen needle litter of some Northeastern conifers, *Proc. Soil Sci. Soc. Am.,* **8**:409–411 (1943).
78. ———: A study of certain calcium relationships and base exchange properties of forest soils, *J. Forestry,* **35**:27–32 (1937).
79. Chandler, W. H.: The dry-matter residue of trees and their products in proportion to leaf area, *Proc. Am. Soc. Hort. Sci.,* **31**:39–56 (1934).
80. Chapman, A. G.: The effects of black locust on associated species with special reference to forest trees, *Ecol. Monographs,* **5**:37–60 (1935).
81. ———: Forest planting on strip-mined coal lands with special reference to Ohio, *U.S. Forest Serv., Central States Forest Expt. Sta. Tech. Paper* 104, 1–25 (1944).
82. Chepil, W. S.: Measurement of wind erosiveness of soils by dry sieving procedure, *Sci. Agr.,* **23**:154–160 (1942).
83. ———: Relation of wind erosion to the dry aggregate structure of a soil, *Sci. Agr.,* **21**:488–507 (1941).
84. ———: Relation of wind erosion to the water-stable and dry clod structure of soil, *Soil Sci.,* **55**:275–287 (1943).
85. ——— and R. A. Milne: Wind erosion of soil in relation to roughness of surface, *Soil Sci.,* **52**:417–432 (1941).

86. ——— and ———: Wind erosion of soils in relation to size and nature of the exposed area, *Sci. Agr.*, **21**:479–487 (1941).
87. CHURCH, J. E.: The conservation of snow: its dependence on forests and mountains, *Sci. American Sup.*, **74**:145, 152–155 (1912).
88. ———: Evaporation at high altitudes and latitudes, *Trans. Am. Geophys. Union*, Part II:326–343 (1934).
89. CLARK, O. R.: Interception of rainfall by prairie grasses, weeds, and certain crop plants, *Ecol. Monographs*, **10**:243–277 (1940).
90. CLARKE-HAFSTED, K.: The spacing of rain gages and the measurement of flood-producing rains, *Trans. Am. Geophys. Union*, Part II:557–558 (1942).
91. CLYDE, G. D.: Snow-melting characteristics, *Utah Agr. Expt. Sta. Tech. Bull.* 231 (1931).
92. COILE, T. S.: Composition of the leaf litter of forest trees, *Soil Sci.*, **43**:349–355 (1937).
93. ———: Distribution of forest tree roots in North Carolina Piedmont soils. *J. Forestry*, **35**:247–257 (1937).
94. ———: Soil changes associated with loblolly pine succession on abandoned agricultural land of the Piedmont plateau, *Duke Univ. School of Forestry Bull.* 5 (1940).
95. CONNAUGHTON, C. A.: The accumulation and rate of melting of snow as influenced by vegetation, *J. Forestry*, **33**:564–569 (1935).
96. COOK, H. L., and F. B. CAMPBELL: Characteristics of some meadow strip vegetations, *Agr. Eng.*, **20**:345–348 (1939).
97. COOPER, W. E.: Forest site determination by soil and erosion classification, *J. Forestry*, **40**:709–712 (1942).
98. COOPER, W. S.: The broad sclerophyll vegetation of California, *Carnegie Inst. Wash. Pub.*, **319**:1–124 (1922).
99. COPE, J. A.: Concerning black locust in New York, *J. Forestry*, **27**:825–831 (1929).
100. ———: Loblolly pine in Maryland, *Md. Dept. Forestry*, 1–96 (1923).
101. COPELAND, E. B.: Transpiration by chaparral and its effect upon the temperature of leaves, *Univ. Calif. Pub. Botany*, **17**:1–21 (1932).
102. COSTER, C. H.: The transpiration of various types of vegetation on Java (Dutch with English summary), *Tectona*, **30**:1–124 (1937).
103. COVILLE, F. V.: The formation of leafmold, *J. Wash. Acad. Sci.*, **3**:77–89 (1913).
104. CRADDOCK, G. W., and C. K. PEARSE: Surface run-off and erosion on granitic mountain soils of Idaho as influenced by range cover, soil disturbance, slope, and precipitation intensity, *U.S. Dept. Agr. Circ.* 482 (1938).
105. CROFT, A. R.: Evaporation from snow, *Am. Meteorol. Soc. Bull.* **25**, 334–337 (1944).
106. CUMMINGS, N. W.: Evaporation from water surfaces; status of present knowledge and need for further investigations, *Trans. Am. Geophys. Union*, Part II: 507–509 (1936).
107. ———: The reliability and usefulness of the energy equations for evaporation, *Trans. Am. Geophys. Union*, **27**:81–94 (1946).
108. DAY, G. M.: Topsoil changes in coniferous plantations, *J. Forestry*, **38**:646–649 (1940).
109. DE FOREST, H.: Rainfall interception by plants: an experimental note, *Ecology*, **4**:417–419 (1923).

110. DENGLER, A.: Kronengrösse, Nadelmenge und Zuwachsleistung von Altkiefern, *Z. Forst. u. Jagdw.*, **69**:321–336 (1937).
111. DEN UYL, D.: The zone of effective windbreak influence, *J. Forestry*, **34**:689–695 (1936).
112. DREIBELBIS, R. F., and F. A. POST: An inventory of soil-water relationships on woodland, pasture, and cultivated soils, *Proc. Soil Sci. Soc. Am.*, **6**:462–473 (1941).
113. DUNCAN, W. H.: A study of root development in three soil types in the Duke Forest, *Ecol. Monographs*, **11**:141–164 (1941).
114. DUNFORD, E. G., and C. H. NIEDERHOF: Influence of aspen, young lodgepole pine, and open grassland types upon factors affecting water yield, *J. Forestry*, **42**:673–677 (1944).
115. EATON, E. C.: Flood and erosion control problems and their solution, *Trans. Am. Soc. Civil Engrs.*, **101**:1302–1330 (1936).
116. EBERMAYER, E.: "Die gesammte Lehre der Waldstreu, mit Rucksicht auf die chemische Statik des Waldbaues, unter Zugrundlegung der in den königlichen Staatsforsten Bayerns angestellten Untersuchungen," Verlag Julius Springer, Berlin (1876). 416 pp.
117. ELLISON, W. D.: Some effects of raindrops and surface-flow on soil erosion and infiltration, *Trans. Am. Geophys. Union*, **26**:415–431 (1945).
118. ENGLER, A.: Researches on the influence of the forest upon the disposition of terrestrial and atmospheric waters, *U.S. Forest Serv., Div. Silvics., Trans.* 100, mimeo. Original in German (1919).
119. ESER, K.: Untersuchungen über den Einfluss der physikalischen und chemischen Eigenschaften des Bodens auf dessen Verdunstungsvermögen, *Forsch. Geb. Agriculturphysik*, **7**:1–125 (1884).
120. FERGUSON, J. A.: Influence of locust on the growth of catalpa, *J. Forestry*, **20**:318–319 (1922).
121. FERNOW, B. E., and others: Forest influences, *U.S. Forestry Div. Bull.* 7 (1902).
122. FINNEY, E. A.: Snow control on the highways, *Mich. Eng. Expt. Sta. Bull.* 57 (1934).
123. FLETCHER, J. E., and E. L. BEUTNER: Erodibility investigations on some soils of the upper Gila watershed, *U.S. Dept. Agr. Tech. Bull.* 794 (1941).
124. FONS, W. L.: Influence of forest cover on wind velocity, *J. Forestry*, **38**:481–487 (1940).
125. FORSLING, C. L.: A study of the influence of herbaceous plant cover on surface run-off and soil erosion in relation to grazing on the Wasatch plateau, Utah, *U.S. Dept. Agr. Tech. Bull.* 220 (1931).
126. FOSTER, E. E.: The effects of bedrock on runoff of Wisconsin streams, *Proc. Eng. Soc. Wis.*, 162 (1926).
127. FOURCADE, H. G.: A note on multiple rain-gages, *Trans. Am. Geophys. Union*, **26**:267–268 (1945).
128. ———: Some notes on the effects of the incidence of rain on the distribution of rainfall over the surface of unlevel ground, *Trans. Roy. Soc. S. Africa*, **29**:235–254 (1942).
129. FREE, G. R., G. M. BROWNING, and G. W. MUSGRAVE: Relative infiltration and related physical characteristics of certain soils, *U.S. Dept. Agr. Tech. Bull.* 729 (1940).

130. Galloway, H. M.: Origin and properties of alkaline raw humus, *Soil Sci.*, **49**: 295–301 (1940).
131. Garstka, W. U.: The calcium content of Connecticut forest litter, *J. Forestry*, **30**:396–405 (1932).
132. ———: Hydrology of small watersheds under winter conditions of snow-cover and frozen soil, *Trans. Am. Geophys. Union*, 1944, Part VI:838–71 (1945).
133. Gast, P. R.: A thermoelectric radiometer for silvical research, with preliminary results on the relation of insolation tó the growth of white pine, *Harvard Forest Bull.* 14 (1930).
134. Geiger, R.: "The Climate of the Layer of Air Near the Ground," 1927, U.S. Soil Conservation Service (translation from the German by John Leighly, 1942).
135. ———: Untersuchungen über das Bestandesklima, *Forstwiss. Centr.*, **48**:337–349 (1926).
136. Gibbs, J. A., and W. S. Ligon: Correlation of sites and species in tree planting, *Proc. Soil Sci. Soc. Am.*, **6**:413 (1941).
137. Gibbs, W. M., and H. W. Batchelor: Effect of tree products on bacteriological activities in soil: II. Study of forest soils, *Soil Sci.*, **24**:351–364 (1927).
138. Gleason, C. H.: Directions for sowing mustard for erosion control in burned areas of southern California, *U.S. Forest Serv., Calif. Forest and Range Expt. Sta., Forest Research Note* 37 (1944). Mimeo.
139. Grah, R. F., and C. C. Wilson: Some components of rainfall interception, *J. Forestry*, **42**:890–898 (1944).
140. Grant, T. J.: A method of judging quality of planting sites for black locust, *U.S. Soil Conserv. Service*, SCS-TP-34 (1940). Mimeo.
141. Griffen, A. A.: Influence of forests upon the melting of snow in the Cascade range, *Monthly Weather Rev.*, **46**:324–327 (1918).
142. Gustafson, A. F.: Composition of black locust leaf mold and leaves, and some observations on the effects of the black locust, *J. Am. Soc. Agron.*, **27**:237–239 (1935).
143. Hamilton, E. L.: Rainfall-measurement as influenced by storm-characteristics in southern California mountains, *Trans. Am. Geophys. Union*, Part III: 502–518 (1944).
143a. Harris, J. A., J. Kuenzel, and W. S. Cooper: Comparison of the physical factors of habitats, *Ecology*, **10**:47–66 (1929).
144. Harrold, L. L., and F. R. Dreibelbis: An accounting of the daily accretion, depletion, and storage of soil-water as determined by weighing monolith lysimeters, *Trans. Am. Geophys. Union*, **26**:283–292 (1945).
145. Hawley, R. C., and H. J. Lutz: Establishment, development, and management of conifer plantations in the Eli Whitney forest, New Haven, Connecticut, *Yale Univ. School Forestry Bull.* 53 (1943).
146. Hayes, F. A., and J. H. Stoeckeler: Soil and forest relationships of the shelterbelt zone, *Sect. 12 of U.S. Forest Serv. Spec. Pub.*, "Possibilities of Shelterbelt Planting in the Plains Region," 111–153 (1935).
147. Hayes, G. L.: A method of measuring rainfall on windy slopes, *Monthly Weather Rev.*, **72**:111–114 (1944).
148. Hendricks, B. A.: Effect of forest litter on soil temperature, *Chron. Botan.*, **6**: 440–441 (1941).

149. ———: Effect of grass litter on infiltration of rainfall on granitic soils in a semi-desert shrub grass area, *U.S. Forest Serv., Southwestern Forest and Range Expt. Sta. Research Note* 96 (1942).
150. HENDRICKSON, B. H.: The choking of pore-space in the soil and its relation to runoff and erosion, *Trans. Am. Geophys. Union,* Part II:500–505 (1934).
151. HERION, G. A.: River bank protection, *Soil Conserv.,* **5**:6–9 (1939).
152. HEYWARD, F., and R. M. BARNETTE: Effect of frequent fires on chemical composition of forest soils in the longleaf pine region, *Fla. Agr. Expt. Sta. Tech. Bull.* 265 (1934).
153. ——— and ———: Field characteristics and partial chemical analyses of the humus layer of longleaf pine forest soils, *Fla. Agr. Expt. Sta. Bull.* 302 (1936).
154. HIRATA, T.: Contributions to the problem of the relations between the forest and water in Japan, *Imp. Forestry Expt. Sta. Meguro, Tokyo* (1929). 41 pp.
155. HOFMANN, J. V.: The natural regeneration of Douglas fir in the Pacific Northwest, *U.S. Dept. Agr. Bull.* 1200 (1924).
156. HOLCH, A. E.: Development of roots and shoots of certain deciduous tree seedlings in different forest sites, *Ecology,* **12**:259–299 (1931).
157. HOLZMAN, B.: Sources of moisture for precipitation in United States, *U.S. Dept. Agr. Tech. Bull.* 589 (1937).
158. HOOVER, M. D.: Effect of removal of forest vegetation upon water yields, *Trans. Am. Geophys. Union,* Part VI:969–975 (1944).
159. HOPPE, E.: Regenmessung unter Baumkronen, *Mitt. forstl. Versuchsw. Österr.,* **21**:1–75 (1896).
160. HORTON, R. E.: Analysis of runoff plot experiments with varying infiltration-capacity, *Trans. Am. Geophys. Union,* Part IV:693–711 (1939).
161. ———: An approach toward a physical interpretation of infiltration-capacity *Proc. Soil Sci. Soc. Am.,* **5**:399–418 (1940).
162. ———: Determination of infiltration-capacity for large drainage-basins, *Trans. Am. Geophys. Union,* Part II:371–385 (1937).
163. ———: Erosional development of streams and their drainage basins: Hydrophysical approach to quantitative morphology, *Bull. Geol. Soc. Am.,* **56**:275–370 (1945).
164. ———: The interpretation and application of runoff plot experiments with reference to soil erosion problems, *Proc. Soil Sci. Soc. Am.,* **3**:340–349 (1938).
165. ———: Measurement of rainfall and snow, *J. New Eng. Waterworks Assoc.,* **33**:14–71 (1919).
166. ———: The melting of snow, *Monthly Weather Rev.,* **43**:599–605 (1915).
167. ———: Rainfall interception, *Monthly Weather Rev.,* **47**:603–623 (1919).
168. ———: Rainfall interpolation, *Monthly Weather Rev.,* **51**:291–304 (1923).
169. ———: The role of infiltration in the hydrologic cycle, *Am. Geophys. Union Rpt. and Pap.,* 446–460 (1933).
170. ———: Surface run-off control, Chap. 11 of "Headwaters Control and Use," Upstream Eng. Conf., Washington, D.C. (1937).
171. ———: Surface run-off phenomena. Part I—Analysis of the hydrograph, *Horton Hydrological Lab. Pub.* 101 (1935). 73 pp.
172. ———: Transpiration by forest trees, *Monthly Weather Rev.,* **51**:571–581 (1923).
173. ———: Water-losses in high latitudes and at high elevations, *Trans. Am. Geophys. Union,* Part II:351–380 (1934).

174. Hough, F. B.: Connection between forests and climate, in his "Report upon Forestry," 1:221–283 (1877).
175. Houk, I. E.: Rainfall and runoff in the Miami Valley, *Miami (Ohio) Conserv. Dist. Tech. Rpt.*, 8 (1921).
176. Hoyt, W. G., and others: Studies of relations of rainfall and runoff in the United States, *U.S. Geol. Survey, Water Supply Paper* 772 (1936).
177. ——— and H. C. Troxell: Forests and streamflow, *Trans. Am. Soc. Civil Engrs.*, **99**:1–111 (1934).
178. Humphreys, W. J.: "Physics of the Air," 3d ed., McGraw-Hill Book Company, Inc., New York (1940). 654 pp.
179. Hursh, C. R.: Litter keeps forest soil productive, *S. Lumberman*, **133**:219–221 (1928).
180. ———: Roadbank stabilization at low cost, *U.S. Forest Serv. Appal. Forest Expt. Sta. Tech. Note* 38 (1939).
181. ——— and C. A. Connaughton: Effects of forests upon local climate, *J. Forestry*, **36**:864–866 (1938).
182. ——— and P. W. Fletcher: The soil profile as a natural reservoir, *Proc. Soil Sci. Soc. Am.*, **7**:480–486 (1942).
183. Israelsen, O. W.: "Irrigation Principles and Practices," John Wiley & Sons, Inc., New York (1932). 422 pp.
184. Jaenicke, A. J., and M. H. Foerster: The influence of a western yellow pine forest on the accumulation and melting of snow, *Monthly Weather Rev.*, **43**: 115–126 (1915).
185. Jarvis, C. S.: Maximum stream-flow with reference to flood formulas, *Trans. Am. Geophys. Union*, Part II:409–416 (1937).
186. ——— and others: Floods in the United States, *U.S. Geol. Survey, Water Supply Paper* 771 (1936).
187. Jemison, G. M.: The significance of the effect of stand density upon the weather beneath the canopy, *J. Forestry*, **32**:446–451 (1934).
188. Jenny, H.: "Factors of Soil Formation," McGraw-Hill Book Company, Inc., New York (1941). 281 pp.
189. Jepson, H. G.: Prevention and control of gullies, *U.S. Dept. Agr. Farmers Bull.* 1813 (1939).
190. Johnson, W. M.: Infiltration capacity of forest soil as influenced by litter, *J. Forestry*, **38**:520 (1940).
191. ———: The interception of rain and snow by a forest of young ponderosa pine, *Trans. Am. Geophys. Union*, Part II:566–569 (1942).
192. Kadel, B. C.: Instructions for the installation and operation of class "A" evaporation stations, *U.S. Weather Bur. Circ.* L (1919).
193. Kellogg, L. F.: Failure of black locust-coniferous mixtures in the Central States, *U.S. Forest Serv., Central States Forest Expt. Sta. Note* 15 (1934). Mimeo.
194. Kenyon, E. C., Jr.: The functions of debris-dams and the loss of reservoir-capacity through silting, *Trans. Am. Geophys. Union*, Part I: 16–20 (1939).
195. Kienholz, R.: Frost depth in forest and open in Connecticut, *J. Forestry*, **38**: 346–350 (1940).
196. King, H. W.: "Handbook of Hydraulics," 3d ed., McGraw-Hill Book Company, Inc., New York (1939). 617 pp.

197. Kittredge, J.: The annual accumulation and creep of litter and other surface materials in the chaparral of the San Gabriel mountains, Calif., *J. Agr. Research*, **58**:537–541 (1939).
198. ———: A comparison of forest floors from plantations of the same age and environment, *J. Forestry*, **38**:729–732 (1940).
199. ———: Estimation of the amount of foliage of trees and stands, *J. Forestry*, **42**:905–912 (1944).
200. ———: The forest floor of the chaparral in San Gabriel mountains, California, *J. Agr. Research*, **58**:521–535 (1939).
201. ———: Forest and water: aspects which have received little attention, *J. Forestry*, **34**:417–419 (1936).
202. ———: The magnitude and regional distribution of water losses influenced by vegetation, *J. Forestry*, **36**:775–778 (1938).
203. ———: Some quantitative relations of foliage in the chaparral, *Ecology*, **26**:70–73 (1945).
204. ———, H. J. Loughead, and A. Mazurak: Interception and stem-flow in a pine plantation, *J. Forestry*, **39**:505–522 (1941).
205. Kohnke, H.: A method for studying infiltration, *Proc. Soil Sci. Soc. Am.*, **3**:296–303 (1938).
206. ———, F. R. Dreibelbis, and J. M. Davidson: A survey and discussion of lysimeters and a bibliography on their construction and performance, *U.S. Dept. Agr. Misc. Pub.* 372 (1940).
207. Korstian, C. F.: Factors controlling germination and early survival in the oaks, *Yale Univ. School of Forestry Bull.* 19 (1927).
208. Kozlowski, T. T., and F. X. Schumacher: Estimation of stomated foliar surface of pines, *Plant Physiol.*, **18**:122–127 (1943).
209. Kraebel, C. J.: Erosion control on mountain roads, *U.S. Dept. Agr. Circ.* 380 (1936).
210. ———: Study of the New Year's storm by the California Forest Experiment Station, *Proc. Flood Control Conf., Conserv. Assoc. of Los Angeles Co.*, Mar. 23, 1934, 3–11 (1934). Mimeo.
211. Landsberg, H., and N. A. Riley: Wind influences on the transportation of sand over a Michigan sand dune, *Proc. Second Hydraulics Conf. 1942, Iowa Univ. Eng. Bull.* 27, 342–352 (1943).
212. Larsen, J. A.: Natural spreading of planted black locust in Southeastern Ohio, *J. Forestry*, **33**:616–619 (1935).
213. Laws, J. O.: Measurements of the fall-velocities of waterdrops and raindrops, *Trans. Am. Geophys. Union*, Part III:709–721 (1941).
214. ———: Recent studies in raindrops and erosion, *Agr. Eng.*, **21**:431–433 (1940).
215. ——— and D. A. Parson: The relation of raindrop-size to intensity, *Trans. Am. Geophys. Union*, Part II:452–459 (1943).
216. Lee, C. H.: Total evaporation for Sierra Nevada watersheds by the method of precipitation and runoff differences, *Trans. Am. Geophys. Union*, Part I:50–66 (1941).
217. ———: Water supply from rainfall on valley floors, Discussion, *Proc. Am. Soc. Civil Engrs.*, **56**:787–800 (1930).
218. Leighly, J.: A note on evaporation, *Ecology*, **18**:180–199 (1937).
219. Leighton, M. O.: Floods, *Natl. Conserv. Comn. Rpt.*, **2**:95–107 (1909).
220. Li, Tsi-Tung: Soil temperature as influenced by forest cover, *Yale Univ. School of Forestry Bull.* 18 (1926).

221. LIGON, W. S.: Influence of soil type and other site factors on the success of tree plantings for erosion control, *J. Forestry*, **38**:226–227 (1940).

222. LINKE, F.: Niederschlagmessungen unter Bäumen, *Meteorol. Z. Braunschweig*, **33**:141 (1916); **38**:277 (1921).

223. LIVINGSTON, B. E., and F. SHREVE: Distribution of vegetation in the U.S. as related to climatic conditions, *Carnegie Inst. Wash. Pub.* 284 (1921).

224. LOWDERMILK, W. C.: Further studies of factors affecting surficial run-off and erosion, *Proc. Intern. Congr. Forest Expt. Sta., Stockholm*, **1929**:606–28 (1930).

225. ———: Influence of forest litter on run-off, percolation and erosion, *J. Forestry*, **28**:474–491 (1930).

226. ——— and P. B. ROWE: Still further studies on absorption of rainfall in its relation to surficial runoff and erosion, *Trans. Am. Geophys. Union*, Part II:509–515 (1934).

227. LUNT, H. A.: The effect of forest litter removal upon the structure of the mineral soil, *J. Forestry*, **35**:33–36 (1937).

228. ———: Effect of weathering upon dry matter, and composition of hardwood leaves, *J. Forestry*, **33**:607–609 (1935).

229. ———: Profile characteristics of New England forest soils, *Conn. Agr. Expt. Sta. Bull.* 342, 743–836 (1932).

230. LYFORD, W. H., JR.: Mineral composition of freshly fallen white pine and red maple leaves, *N.H. Agr. Expt. Sta. Tech. Bull.* 77 (1941).

231. MCHARGUE, J. S., and W. R. ROY: Mineral and nitrogen content of some forest trees at different times in the growing season, *Botan. Gaz.*, **94**:381–393 (1932).

232. MCINTYRE, A. C.: Trees and erosion control, *Am. Soil Survey Assoc. Bull.* 16, 110–114 (1935).

233. ——— and C. D. JEFFRIES: The effect of black locust on soil nitrogen and growth of catalpa, *J. Forestry*, **30**:22–28 (1932).

234. MACKINNEY, A. L.: Effect of forest litter on soil temperature and soil freezing in autumn and winter, *Ecology*, **10**:312–321 (1929).

235. MCLAREN, J.: The reclamation of drifting sand dunes in Golden Gate Park, in "Forest and Water," by A. Kinney, 240–241 (1900).

236. MCLAUGHLIN, W. T., and R. L. BROWN: Controlling coastal sand dunes in the Pacific Northwest, *U.S. Soil Conserv. Serv. Circ.* 660 (1942).

237. MCVICKER, J. S.: The influence of soil type on the composition and yield of white oak leaves, abstracts of Ph.D. theses, Univ. Illinois (1942).

238. MALLIK, A. K.: The depth of the surface layer of the soil taking part in the diurnal exchange of moisture with the air layers near the ground, *Indian J. Agr. Sci.*, **10**:164–171 (1940).

239. MARSH, G. P.: "The Earth as Modified by Human Action," Charles Scribner's Sons, New York (1907).

240. MARTIN, E. V.: Studies of evaporation and transpiration under controlled conditions, *Carnegie Inst. Wash. Pub.* 550 (1943).

241. ———: Effect of solar radiation on transpiration of *Helianthus annuus*, *Plant Physiol.*, **10**:341–354 (1935).

242. MARVIN, C. F.: Psychrometric tables for obtaining the vapor pressure, relative humidity and temperature of the dewpoint, *U.S. Weather Bur. No.* 235 (1915).

243. MEAD, D. W.: "Elements of Hydrology," McGraw-Hill Book Company, Inc., New York (1919). 647 pp.

244. MEANS, T. H.: Fog precipitated by trees, *J. Forestry*, **25**:1015–1016 (1927).

245. MEGINNIS, H. G.: Effect of cover on surface run-off and erosion on the loessial uplands of Mississippi, *U.S. Dept. Agr. Circ.* 347 (1935).

246. ———: The pole-frame brush dam—a low-cost mechanical aid in reforesting gullied land, *U.S. Forest Serv., South. Forest Expt. Sta. Occas. Paper* 76 (1938).

247. ———: Soil-collecting trenches as substitutes for temporary check dams in reforesting gullies, *J. Forestry*, **37**:764–769 (1939).

248. ———: Using soil-binding plants to reclaim gullies in the South, *U.S. Dept. Agr. Farmers Bull.* 1697 (1933).

249. MEINZER, O. E., and others: "Hydrology," Natl. Res. Council. Physics of the Earth Series, Vol. IX, McGraw-Hill Book Company, Inc., New York (1942).

250. MELIN, E.: Biological decomposition of some types of litter from North American forests, *Ecology*, **11**:72–101 (1930).

251. MEYER, A. F.: "The Elements of Hydrology," John Wiley & Sons, New York, 2d ed., 1928. 522 pp.

252. MEYER, B. S.: The daily periodicity of transpiration in the tulip poplar, *Ohio Jour. Sci.*, **32**:104–114 (1932).

253. MIDDLETON, H. E.: Properties of soils which influence soil erosion, *U.S. Dept. Agr. Tech. Bull.* 178 (1930).

254. MILLER, H. W.: A new disease of Russianolive in the Pacific Northwest, *J. Forestry*, **44**:118–120 (1946).

255. MINCKLER, L. S.: Transpiration of trees and forests, *J. Forestry*, **37**:336–339 (1939).

256. MITCHELL, J. A.: Interception of rainfall by the forest, *J. Forestry*, **28**:101–102 (1930).

257. MITCHELSON, A. T., and D. C. MUCKEL: Spreading water for storage underground, *U.S. Dept. Agr. Tech. Bull.* 578 (1937).

258. MOORE, B., H. M. RICHARDS, H. A. GLEASON, and A. B. STOUT: Hemlock and its environment, *N.Y. Botan. Gard. Bull.* 12, 325–350 (1924).

259. MOORE, W. L.: Report on "The Influence of Forests on Climate and on Floods," U.S. Weather Bureau, Washington, D.C. (1910).

260. MOREHEAD, L. B.: Willow mats halt streambank erosion, *Soil Conserv.*, **5**:127, 131 (1939).

261. MOREY, H. F.: The application of our knowledge of the organic layers of the soil profile to flood control, *Hydrology Conf., State College, Pa., Proc. School Eng. Tech. Bull.* 27, 143–151 (1941).

262. MOROZOV, G. F.: "Die Lehre vom Walde," translated from Russian into German by Ruoff and Buchholz, Neudamm (1928).

263. MUNGER, T. T.: Avalanches and forest cover in the northern Cascades, *U.S. Dept. Agr., Forest Serv. Circ.* 173 (1911).

264. MUNNS, E. N.: Erosion and flood problems in California, *Report to the Legislature on Senate concurrent resolution No. 27 (Legis. of 1921) by the Calif. State Bd. Forestry* (1923).

265. ——— and others: Watershed and other related influences and a watershed protective program, *Sen. Resol. 175, 72d Congr. Sen. Doc.* 12 (1933).

266. ——— and I. H. Sims: Forests in flood control; *supplemental report to the Committee on Flood Control, House of Representatives, 74th Congr., 2d Sess., on H.B. 12517. U.S. Forest Serv.* (1936).

267. ——— and J. H. Stoeckeler: How are the Great Plains shelter-belts? *J. Forestry*, **44**:237–257 (1946).

268. Musgrave, G. W.: Field research offers significant new findings, *Soil Conserv.*, **3**:210–214 (1938).

269. ——— and G. R. Free: Preliminary report on a determination of comparative infiltration-rates on some major soil-types, *Trans. Am. Geophys. Union*, Part II:345–349 (1937).

270. Neal, J. H.: The effect of the degree of slope and rainfall characteristics on runoff and soil erosion, *Mo. Agr. Expt. Sta. Research Bull.* 280 (1938).

271. New Hampshire Agricultural Experiment Station, Formation of forest soils: rate of deposition of litter, *N.H. Agr. Expt. Sta. Bull.* 262 (1932).

272. Nichols, M. L., and H. D. Sexton: A method of studying soil erosion, *Agr. Eng.*, **13**:101–103 (1932).

273. Nicholson, J. W.: The influence of forests on climate and water supply in Kenya, *Kenya Colony, Forest Dept. Pamphlet* 2 (1929).

274. Niederhof, C. H., and E. G. Dunford: The effects of openings in a young lodgepole pine forest on the storage and melting of snow, *J. Forestry*, **40**:802–805 (1942).

275. ——— and R. Stahelin: Climatic conditions within and adjacent to a forest plantation in the Nebraska sandhills, *J. Forestry*, **40**:244–249 (1942).

276. ——— and H. G. Wilm: Effect of cutting mature lodgepole-pine stands on rainfall interception, *J. Forestry*, **41**:57–61 (1943).

277. Norton, E. A.: "Soil Conservation Survey Handbook," *U.S. Dept. Agr. Misc. Pub.* 352 (1939).

278. O'Brien, M. P., and B. D. Rindlaub: The transportation of sand by wind, *Civil Eng.*, **6**:325–327 (1936).

279. Ohmasa, M., and K. Mori: The amount of fall and decomposition of the leaf litter of the forests of Japan, *Imp. Forestry Expt. Sta. Bull.* 3, 39–107 (1937).

280. Paschall, A. H., R. T. A. Burke, and L. E. Baver: Aggregation studies on the Muskingum, Chester and Lansdale silt loams, *Am. Soil Survey Assoc. Bull.* 16 (1935).

281. Pearse, C. K., and S. B. Woolley: The influence of range plant cover on the rate of absorption of surface water by soils, *J. Forestry*, **34**:844–847 (1936).

282. Pearson, G. A.: A meteorological study of parks and timbered areas in the western yellow pine forests of Arizona and New Mexico, *Monthly Weather Rev.*, **41**:1615–1629 (1913).

283. ———: Studies in transpiration of coniferous tree seedlings, *Ecology*, **5**:340–347 (1924).

284. Pers, R.: Relations entre les données pluviométriques et les précipitations totales recueillies par un bassin: Introduction à l'étude théoretique de la stéréopluviométrie & note complémentaire sur la stéréopluviométrie, *La Météorologie*, 75, N.S. 8:101–113 (1932).

285. Phillips, J. F. V.: Rainfall interception by plants, *Nature* (*London*), **116**:837 (1926); **118**:837–838, 915 (1928).

286. Pickels, G. W.: Run-off investigations in central Illinois, *Univ. Illinois Eng. Expt. Sta. Bull.* 232 (1931).

287. PIETERS, A. J.: *Lespedeza sericea* and other perennial lespedezas for forage and soil conservation, *U.S. Dept. Agr. Circ.* 534 (1939).

288. PLICE, M. J.: Acidity, antacid buffering, and nutrient content of forest litter in relation to humus and soil, *N.Y. (Cornell) Agr. Expt. Sta. Mem.* 166 (1934).

289. PLUMMER, F. G.: Chaparral, *U.S. Forest Serv. Bull.* 85 (1911).

290. POWERS, W. L.: Characteristics of forest soils of the northwestern United States, *Soil Sci.*, **34**(1):1–10 (1932).

291. RABER, O.: Water utilization by trees, with special reference to the economic forest species of the North Temperate Zone, *U.S. Dept. Agr. Misc. Pub.* 257 (1937).

292. REE, W. O.: Hydraulic tests of Kudzu as a conservation channel lining, *Agr. Eng.*, **22**:27–29 (1941).

293. RENNER, F. G.: Conditions influencing erosion on the Boise River watershed, *U.S. Dept. Agr. Tech. Bull.* 528 (1936).

294. RIESBOL, H. S.: Report on exploratory study of rain-gage shields and enclosures at Coshocton, Ohio, *Trans. Am. Geophys. Union*, Part II:474–482 (1940).

295. ROBERTS, E. G.: Soil depth and height growth of black locust, *J. Forestry*, **37**: 583–584 (1939).

296. ROESER, J. J.: Transpiration capacity of coniferous seedlings and the problem of heat injury, *J. Forestry*, **30**:381–395 (1932).

297. ROHWER, C.: Evaporation from free water surfaces, *U.S. Dept. Agr. Tech. Bull.* 271 (1931).

298. Rome, State Library: Un codice Veneziano del "1600" per la acque e le foreste. Includes in Italian "A Venetian project for the reorganization of the regimen of the forests and lagoons at the beginning of the seventeenth century" (1934).

299. ROWE, P. B.: The construction, operation, and use of the North Fork infiltrometer, *U.S. Flood Control Coord. Com. Misc. Pub.* 1, *Calif. Forest and Range Expt. Sta. Misc. Pub.* 1 (1940).

300. ———: A method of hydrologic analysis in watershed-management, *Trans. Am. Geophys. Union*, Part II:632–650 (1943).

301. ———: Some factors of the hydrology of the Sierra Nevada foothills, *Trans. Am. Geophys. Union*, Part I:90–100 (1941).

302. ———, D. M. ILCH, and R. BOLLAERT: An infiltration study of a denuded and a forest covered soil, *U.S. Forest Serv., Calif. Forest and Range Expt. Sta., Research Note* 14 (1937).

303. RUBNER, K.: Der Nebelniederschlag in Wald und seine Messung, *Tharandter Forstl. Jahrb.*, **83**:121–149 (1932).

304. RUSSEL, J. C.: The effect of surface cover on soil moisture losses by evaporation, *Proc. Soil Sci. Soc. Am.*, **4**:65–71 (1939).

305. SAMPSON, A. W.: Plant succession on burned chaparral lands in northern California, *Calif. Agr. Expt. Sta. Bull.* 685 (1944).

306. ——— and R. SAMISH: Growth and seasonal changes in composition of oak leaves, *Plant Physiol.*, **10**:739–751 (1935).

307. ——— and L. H. WEYL: Range preservation and its relation to erosion control on western grazing lands, *U.S. Dept. Agr. Bull.* 675 (1918).

308. SCHANTZ-HANSEN, T.: Ecological changes due to thinning jack pine, *Minn. Agr. Expt. Sta. Tech. Bull.* 124 (1937).

309. SCHEFFER, P. N.: The beaver as an upstream engineer, *Soil Conserv.*, **Jan.**:178–181 (1938).

310. SCHOLZ, H. F.: Forest cover keeps frost line at shallow depth, *J. Forestry*, **36**:78–79 (1938).

311. SCHUBERT, A.: Untersuchungen über den Transpirationsstrom der Nadelhölzer und den Wasserbedarf von Fichte und Lärche, *Tharandter Forstl. Jahrb.*, **90**:821–883 (1940).

312. SCHUBERT, J.: Wald und Niederschlag in Westpreussen und Posen und die Beeinflussung der Regen- und Schneemessung durch den Wind, *Z. Forst. u. Jagdw.*, **38**:728–735 (1906).

313. SHANTZ, H. L., and R. ZON: Natural vegetation, in "Atlas of American Agriculture, Part I, Sect. E." U.S. Government Printing Office, Washington, D.C. (1924).

314. SHAPIRO, A. A., and H. DE FOREST: A comparison of transpiration rates in chaparral, *Ecology*, **13**:290–295 (1932).

315. SHARPE, C. F. S.: "Landslides and Related Phenomena," Columbia University Press, New York (1938). 137 pp.

316. SHAW, C. F.: The normal moisture capacity of soils, *Soil Sci.*, **23**:303–317 (1927).

317. SHIRLEY, H. L.: Light intensity in relation to plant growth in a virgin Norway pine forest, *J. Agr. Research*, **44**:227–244 (1932).

318. SHREVE, F.: Soil temperature as influenced by altitude and slope exposure, *Ecology*, **5**:128–136 (1924).

319. ———: Soil temperatures in redwood and hemlock forests, *Bull. Torrey Botan. Club*, **54**:649–656 (1927).

320. SHULITS, S.: Rational equation of river-bed profile, *Trans. Am. Geophys. Union*, Part III:622-630 (1941).

321. ——— and W. E. CORFITZEN: Bed-load transportation and the stable-channel problem, *Trans. Am. Geophys. Union*, Part II:456–467 (1937).

322. SIMS, I. H.: Litter deposition and accumulation in the pine-oak type of the southern Appalachians, *J. Forestry*, **30**:90–91 (1932).

323. SIMSON, A. G.: The interception of summer rains by forest cover, *U.S. Forest Serv., Pacific Northwest Forest Expt. Sta., Forest Research Note* 5 (1931). Mimeo.

324. SINCLAIR, J. D., and A. W. SAMPSON: Establishment and succession of vegetation on different soil horizons, *Hilgardia*, **5**:155–174 (1931).

325. SMITH, H. L., and L. B. LEOPOLD: Infiltration studies in the Pecos River watershed, New Mexico and Texas, *Soil Sci.*, **53**:195–204 (1942).

326. SMITH, L. F.: Factors controlling the early development and survival of Eastern White Pine (*Pinus strobus* L.) in Central New England, *Ecol. Monographs*, **10**:373–420 (1940).

327. SMITH, R. M., and D. R. BROWNING: Persistent water-unsaturation of natural soil in relation to various soil and plant factors, *Proc. Soil Sci. Soc. Am.*, **7**:114–119 (1942).

328. Society of American Foresters: Forest terminology, *J. Forestry*, **15**:68–101 (1917).

329. ———, Committee on forestry terminology: "Forestry Terminology, a Glossary of Technical Terms Used in Forestry," Society of American Foresters, Washington, D.C. (1944).

330. SPAETH, J. N., and C. H. DIEBOLD: Some interrelationships between soil characteristics, water tables, soil temperature, and snow cover in the forest and adjacent open areas in south central New York, N.Y., (*Cornell*) *Agr. Expt. Sta. Mem.* 213 (1938).

331. STEWART, G. R.: A study of soil changes associated with the transition from fertile hardwood forest land to pasture types of decreasing fertility, *Ecol. Monographs*, **3**:107–145 (1933).

332. STICKEL, P. W.: The measurement and interpretation of forest-fire weather in the Western Adirondacks, *N.Y. State Col. Forestry Tech. Pub.* 34 (1931).

333. STOCKER, O.: Klimamessungen auf kleinstem Raum an Wiesen- Wald- und Heidepflanzen, *Deut. Bot. Ges. Ber.*, **41**:145–150 (1923).

334. STODDART, L. A., and A. D. SMITH: "Range Management," McGraw-Hill Book Company, Inc., New York (1943).

335. STOECKELER, J. H., and E. J. DORTIGNAC: Snowdrifts as a factor in growth and longevity of shelterbelts in the Great Plains, *Ecology*, **22**:117–125 (1941).

336. STOREY, H. C., and E. L. HAMILTON: A comparative study of rain-gages, *Trans. Am. Geophys. Union*, Part I:133–140 (1943).

337. ——— and H. G. WILM: A comparison of vertical and tilted rain-gages in estimated precipitation on mountain watersheds, *Trans. Am. Geophys. Union*, Part III:518–524 (1944).

338. SVERDRUP, H. U.: The eddy conductivity of the air over a smooth snow field, *Geofysiske Publikasjoner*, Part XI:1–69 (1936).

339. TAYLOR, C. A.: Transpiration and evaporation losses from areas of native vegetation, *Trans. Am. Geophys. Union*, Part II:554-559 (1934).

340. THOM, C. C., and H. F. HOLTZ: Factors influencing the water requirements of plants, *Wash. Agr. Expt. Sta. Bull.* 146 (1917).

341. THORNTHWAITE, C. W.: Atmospheric turbulence and the measurement of evaporation, *Proc. Second Hydraulics Conf., Iowa Univ. Studies, Eng. Bull.* 27 (1943).

342. ——— and B. HOLZMAN: Measurement of evaporation from land and water surfaces, *U.S. Dept. Agr. Tech. Bull.* 817 (1942).

343. TIREN, L.: Ueber die grösse der Nadelfläche einiger Kiefernbestände, *Statens Skogsförsöksanst (Sweden), Meddel.*, **23**:295–336 (1927).

344. TOLMAN, C. F.: "Ground Water," McGraw-Hill Book Company, Inc., New York (1937). 593 pp.

345. TOUMEY, J. W., and E. J. NEETHLING: Insolation, a factor in the natural regeneration of certain conifers, *Yale Univ. School of Forestry Bull.* 11 (1924).

346. TRANSEAU, E. N.: The relation of plant societies to evaporation, *Botan. Gaz.*, **45**:217–231 (1908).

347. TROXELL, H. C.: The diurnal fluctuation in the ground water and flow of the Santa Ana River and its meaning, *Trans. Am. Geophys. Union*, Part II :496–504 (1936).

348. U.S. Department of Agriculture: Influences of vegetation and watershed treatments on run-off, silting, and stream flow, *U.S. Dept. Agr. Misc. Pub.* 397 (1940).

349. ———: "Soils and Men," *U.S. Dept. Agr. Yearbook* 1938.

350. U.S. Forest Service: "Handbook of Erosion Control Engineering on the National Forests," Government Printing Office, Washington, D.C. (1936).

351. ———: "A National Plan for American Forestry," *73d Congr. 1st Sess., Sen. Doc.* 12, 2 vols. (1933).

352. ——— Lake States Forest Experiment Station: Effect of shade in reducing soil surface temperatures, *Lake States Forest Expt. Sta. Tech. Note* 100 (1935).

353. ——— Rocky Mountain Forest and Range Experiment Station: Forest influences, *Rocky Mt. Forest and Range Expt. Sta. 8th Ann. Rpt.* (1943). Mimeo.

354. U.S. National Research Council: "Low Dams," Government Printing Office, Washington, D.C. (1938). 431 pp.

355. Van Burkalow, A.: Angle of repose and angle of sliding friction, *Bull. Geol. Soc. Am.*, **56**:669–707 (1945).

356. Vanoni, V. A.: Dynamics of entrainment of erosional debris and sedimentation (discussion of paper by C. B. Brown), *Trans. Am. Geophys. Union*, Part II:313–315 (1941).

357. Veihmeyer, F. J.: Some factors affecting the irrigation requirements of deciduous orchards, *Hilgardia*, **2**:125–284 (1927).

358. ——— and A. H. Hendrickson: The relation of soil moisture to cultivation and plant growth, *Proc. First Intern. Congr. Soil Sci.*, 1927, **3**:498–513 (1928).

359. ——— and C. N. Johnston: Soil-moisture records from burned and unburned plots in certain grazing areas of California, *Trans. Am. Geophys. Union*, Part I:72–83 (1944).

359a. Wahlenberg, W. G.: Effect of Ceanothus brush on western yellow pine plantations in the northern Rocky Mountains, *J. Agr. Research*, **41**:601–612 (1930).

360. Wakeley, P. C.: Artificial reforestation in the southern pine region, *U.S. Dept. Agr. Tech. Bull.* 492 (1935).

361. Waksman, S. A.: "Principles of Soil Microbiology," 2d ed., The Williams & Wilkins Company, Baltimore (1932). 894 pp.

362. Weaver, J. E.: Evaporation and plant succession in southeastern Washington and adjacent Idaho, *Plant World*, **17**:273–94 (1914).

363. ——— and G. W. Harmon: Quantity of living plant materials in prairie soils in relation to run-off and soil erosion, *Neb. Univ. Conserv. Bull.* 8 (1935).

364. Wicht, C. L.: An approach to the study of rainfall interception by forest canopies, *J. S. African Forestry Assoc.*, **6**:54–70 (1941).

365. Wilm, H. G.: The application and measurement of artificial rainfall on types FA and F infiltrometers, *Trans. Am. Geophys. Union*, Part II:480–486 (1943).

366. ———: Determining net rainfall under a conifer forest, *J. Agr. Research*, **67**: 501–512 (1943).

367. ———: The effect of timber cutting in a lodgepole pine forest on the storage and melting of snow, *Trans. Am. Geophys. Union*, Part I:153–155 (1944).

368. ———: Methods for the measurement of infiltration, *Trans. Am. Geophys. Union*, Part III: 678–686 (1941).

369. ——— and M. H. Collet: The influence of a lodgepole pine forest on storage and melting of snow, *Trans. Am. Geophys. Union*, Part II:505–508 (1940).

370. ———, A. Z. Nelson, and H. C. Storey: An analysis of precipitation measurements on mountain watersheds, *Monthly Weather Rev.*, **67**:163–172 (1939).

371. ——— and C. H. Niederhof: Interception of rainfall by mature lodgepole-pine, *Trans. Am. Geophys. Union*, Part III:660–666 (1941).

372. Wilson, R. E.: Tree planting and soil erosion control in the Southwest, *J. Forestry*, **42**:668–672 (1944).

373. Wilson, W. T.: An outline of the thermodynamics of snow-melt, *Trans. Am. Geophys. Union*, Part I:182–195 (1941).

374. Winsor, L. M.: The barrier system for control of floods in mountain streams, *U.S. Dept. Agr. Misc. Pub.* 165 (1933).

375. Wittich, W.: Untersuchungen über den Einfluss der Kahlschlages auf den Bodenzustand, *Mitt. aus Forstw. u. Forstwiss.*, **1**:438–506 (1930).

376. Wood, O. M.: The interception of precipitation in an oak-pine forest, *Ecology*, **18**:251–255 (1937).

377. Woodruff, C. M.: The movement and evaporation of soil water in relation to pF, *Proc. Soil Sci. Soc. Am.*, **6**:120–126 (1941).

378. Woodward, L.: Infiltration-capacities of some plant-soil complexes on Utah range watershed-lands, *Trans. Am. Geophys. Union*, Part II:468–474 (1943).

379. Yarnell, D. L.: Rainfall intensity—frequency data, *U.S. Dept. Agr. Misc. Pub.* 204 (1935).

380. Young, L. J.: Growth and cultural experiments on the Saginaw forest, *Mich. Acad. Sci. Arts and Letters Papers*, Part IX:541–594 (1929).

381. Zon, R.: Forests and water in the light of scientific investigation, *U.S. Nat. Waterways Comm. Final Rpt., App.* V:205–302, *62d Congr., 2d Sess., Sen. Doc.* 469 (1927).

INDEX

F

S

T

U

V

W

A CATALOGUE OF SELECTED DOVER BOOKS IN ALL FIELDS OF INTEREST

A CATALOGUE OF SELECTED DOVER BOOKS IN ALL FIELDS OF INTEREST

AMERICA'S OLD MASTERS, James T. Flexner. Four men emerged unexpectedly from provincial 18th century America to leadership in European art: Benjamin West, J. S. Copley, C. R. Peale, Gilbert Stuart. Brilliant coverage of lives and contributions. Revised, 1967 edition. 69 plates. 365pp. of text.
21806-6 Paperbound $3.00

FIRST FLOWERS OF OUR WILDERNESS: AMERICAN PAINTING, THE COLONIAL PERIOD, James T. Flexner. Painters, and regional painting traditions from earliest Colonial times up to the emergence of Copley, West and Peale Sr., Foster, Gustavus Hesselius, Feke, John Smibert and many anonymous painters in the primitive manner. Engaging presentation, with 162 illustrations. xxii + 368pp.
22180-6 Paperbound $3.50

THE LIGHT OF DISTANT SKIES: AMERICAN PAINTING, 1760-1835, James T. Flexner. The great generation of early American painters goes to Europe to learn and to teach: West, Copley, Gilbert Stuart and others. Allston, Trumbull, Morse; also contemporary American painters—primitives, derivatives, academics—who remained in America. 102 illustrations. xiii + 306pp. 22179-2 Paperbound $3.00

A HISTORY OF THE RISE AND PROGRESS OF THE ARTS OF DESIGN IN THE UNITED STATES, William Dunlap. Much the richest mine of information on early American painters, sculptors, architects, engravers, miniaturists, etc. The only source of information for scores of artists, the major primary source for many others. Unabridged reprint of rare original 1834 edition, with new introduction by James T. Flexner, and 394 new illustrations. Edited by Rita Weiss. 6⅝ x 9⅝.
21695-0, 21696-9, 21697-7 Three volumes, Paperbound $13.50

EPOCHS OF CHINESE AND JAPANESE ART, Ernest F. Fenollosa. From primitive Chinese art to the 20th century, thorough history, explanation of every important art period and form, including Japanese woodcuts; main stress on China and Japan, but Tibet, Korea also included. Still unexcelled for its detailed, rich coverage of cultural background, aesthetic elements, diffusion studies, particularly of the historical period. 2nd, 1913 edition. 242 illustrations. lii + 439pp. of text.
20364-6, 20365-4 Two volumes, Paperbound $6.00

THE GENTLE ART OF MAKING ENEMIES, James A. M. Whistler. Greatest wit of his day deflates Oscar Wilde, Ruskin, Swinburne; strikes back at inane critics, exhibitions, art journalism; aesthetics of impressionist revolution in most striking form. Highly readable classic by great painter. Reproduction of edition designed by Whistler. Introduction by Alfred Werner. xxxvi + 334pp.
21875-9 Paperbound $2.50

AMERICAN FOOD AND GAME FISHES, David S. Jordan and Barton W. Evermann. Definitive source of information, detailed and accurate enough to enable the sportsman and nature lover to identify conclusively some 1,000 species and sub-species of North American fish, sought for food or sport. Coverage of range, physiology, habits, life history, food value. Best methods of capture, interest to the angler, advice on bait, fly-fishing, etc. 338 drawings and photographs. l + 574pp. 6⅝ x 9⅜.
22383-1 Paperbound $4.50

THE FROG BOOK, Mary C. Dickerson. Complete with extensive finding keys, over 300 photographs, and an introduction to the general biology of frogs and toads, this is the classic non-technical study of Northeastern and Central species. 58 species; 290 photographs and 16 color plates. xvii + 253pp.
21973-9 Paperbound $4.00

THE MOTH BOOK: A GUIDE TO THE MOTHS OF NORTH AMERICA, William J. Holland. Classical study, eagerly sought after and used for the past 60 years. Clear identification manual to more than 2,000 different moths, largest manual in existence. General information about moths, capturing, mounting, classifying, etc., followed by species by species descriptions. 263 illustrations plus 48 color plates show almost every species, full size. 1968 edition, preface, nomenclature changes by A. E. Brower. xxiv + 479pp. of text. 6½ x 9¼.
21948-8 Paperbound $5.00

THE SEA-BEACH AT EBB-TIDE, Augusta Foote Arnold. Interested amateur can identify hundreds of marine plants and animals on coasts of North America; marine algae; seaweeds; squids; hermit crabs; horse shoe crabs; shrimps; corals; sea anemones; etc. Species descriptions cover: structure; food; reproductive cycle; size; shape; color; habitat; etc. Over 600 drawings. 85 plates. xii + 490pp.
21949-6 Paperbound $3.50

COMMON BIRD SONGS, Donald J. Borror. 33⅓ 12-inch record presents songs of 60 important birds of the eastern United States. A thorough, serious record which provides several examples for each bird, showing different types of song, individual variations, etc. Inestimable identification aid for birdwatcher. 32-page booklet gives text about birds and songs, with illustration for each bird.
21829-5 Record, book, album. Monaural. $2.75

FADS AND FALLACIES IN THE NAME OF SCIENCE, Martin Gardner. Fair, witty appraisal of cranks and quacks of science: Atlantis, Lemuria, hollow earth, flat earth, Velikovsky, orgone energy, Dianetics, flying saucers, Bridey Murphy, food fads, medical fads, perpetual motion, etc. Formerly "In the Name of Science." x + 363pp.
20394-8 Paperbound $2.00

HOAXES, Curtis D. MacDougall. Exhaustive, unbelievably rich account of great hoaxes: Locke's moon hoax, Shakespearean forgeries, sea serpents, Loch Ness monster, Cardiff giant, John Wilkes Booth's mummy, Disumbrationist school of art, dozens more; also journalism, psychology of hoaxing. 54 illustrations. xi + 338pp.
20465-0 Paperbound $2.75

JIM WHITEWOLF: THE LIFE OF A KIOWA APACHE INDIAN, Charles S. Brant, editor. Spans transition between native life and acculturation period, 1880 on. Kiowa culture, personal life pattern, religion and the supernatural, the Ghost Dance, breakdown in the White Man's world, similar material. 1 map. xii + 144pp.
22015-X Paperbound $1.75

THE NATIVE TRIBES OF CENTRAL AUSTRALIA, Baldwin Spencer and F. J. Gillen. Basic book in anthropology, devoted to full coverage of the Arunta and Warramunga tribes; the source for knowledge about kinship systems, material and social culture, religion, etc. Still unsurpassed. 121 photographs, 89 drawings. xviii + 669pp. 21775-2 Paperbound $5.00

MALAY MAGIC, Walter W. Skeat. Classic (1900); still the definitive work on the folklore and popular religion of the Malay peninsula. Describes marriage rites, birth spirits and ceremonies, medicine, dances, games, war and weapons, etc. Extensive quotes from original sources, many magic charms translated into English. 35 illustrations. Preface by Charles Otto Blagden. xxiv + 685pp.
21760-4 Paperbound $4.00

HEAVENS ON EARTH: UTOPIAN COMMUNITIES IN AMERICA, 1680-1880, Mark Holloway. The finest nontechnical account of American utopias, from the early Woman in the Wilderness, Ephrata, Rappites to the enormous mid 19th-century efflorescence; Shakers, New Harmony, Equity Stores, Fourier's Phalanxes, Oneida, Amana, Fruitlands, etc. "Entertaining and very instructive." *Times Literary Supplement.* 15 illustrations. 246pp. 21593-8 Paperbound $2.00

LONDON LABOUR AND THE LONDON POOR, Henry Mayhew. Earliest (c. 1850) sociological study in English, describing myriad subcultures of London poor. Particularly remarkable for the thousands of pages of direct testimony taken from the lips of London prostitutes, thieves, beggars, street sellers, chimney-sweepers, street-musicians, "mudlarks," "pure-finders," rag-gatherers, "running-patterers," dock laborers, cab-men, and hundreds of others, quoted directly in this massive work. An extraordinarily vital picture of London emerges. 110 illustrations. Total of lxxvi + 1951pp. 6⅝ x 10.
21934-8, 21935-6, 21936-4, 21937-2 Four volumes, Paperbound $14.00

HISTORY OF THE LATER ROMAN EMPIRE, J. B. Bury. Eloquent, detailed reconstruction of Western and Byzantine Roman Empire by a major historian, from the death of Theodosius I (395 A.D.) to the death of Justinian (565). Extensive quotations from contemporary sources; full coverage of important Roman and foreign figures of the time. xxxiv + 965pp. 21829-5 Record, book, album. Monaural. $3.50

AN INTELLECTUAL AND CULTURAL HISTORY OF THE WESTERN WORLD, Harry Elmer Barnes. Monumental study, tracing the development of the accomplishments that make up human culture. Every aspect of man's achievement surveyed from its origins in the Paleolithic to the present day (1964); social structures, ideas, economic systems, art, literature, technology, mathematics, the sciences, medicine, religion, jurisprudence, etc. Evaluations of the contributions of scores of great men. 1964 edition, revised and edited by scholars in the many fields represented. Total of xxix + 1381pp. 21275-0, 21276-9, 21277-7 Three volumes, Paperbound $7.75

INCIDENTS OF TRAVEL IN YUCATAN, John L. Stephens. Classic (1843) exploration of jungles of Yucatan, looking for evidences of Maya civilization. Stephens found many ruins; comments on travel adventures, Mexican and Indian culture. 127 striking illustrations by F. Catherwood. Total of 669 pp.
20926-1, 20927-X Two volumes, Paperbound $5.00

INCIDENTS OF TRAVEL IN CENTRAL AMERICA, CHIAPAS, AND YUCATAN, John L. Stephens. An exciting travel journal and an important classic of archeology. Narrative relates his almost single-handed discovery of the Mayan culture, and exploration of the ruined cities of Copan, Palenque, Utatlan and others; the monuments they dug from the earth, the temples buried in the jungle, the customs of poverty-stricken Indians living a stone's throw from the ruined palaces. 115 drawings by F. Catherwood. Portrait of Stephens. xii + 812pp.
22404-X, 22405-8 Two volumes, Paperbound $6.00

A NEW VOYAGE ROUND THE WORLD, William Dampier. Late 17-century naturalist joined the pirates of the Spanish Main to gather information; remarkably vivid account of buccaneers, pirates; detailed, accurate account of botany, zoology, ethnography of lands visited. Probably the most important early English voyage, enormous implications for British exploration, trade, colonial policy. Also most interesting reading. Argonaut edition, introduction by Sir Albert Gray. New introduction by Percy Adams. 6 plates, 7 illustrations. xlvii + 376pp. 6½ x 9¼.
21900-3 Paperbound $3.00

INTERNATIONAL AIRLINE PHRASE BOOK IN SIX LANGUAGES, Joseph W. Bátor. Important phrases and sentences in English paralleled with French, German, Portuguese, Italian, Spanish equivalents, covering all possible airport-travel situations; created for airline personnel as well as tourist by Language Chief, Pan American Airlines. xiv + 204pp. 22017-6 Paperbound $2.00

STAGE COACH AND TAVERN DAYS, Alice Morse Earle. Detailed, lively account of the early days of taverns; their uses and importance in the social, political and military life; furnishings and decorations; locations; food and drink; tavern signs, etc. Second half covers every aspect of early travel; the roads, coaches, drivers, etc. Nostalgic, charming, packed with fascinating material. 157 illustrations, mostly photographs. xiv + 449pp. 22518-6 Paperbound $4.00

NORSE DISCOVERIES AND EXPLORATIONS IN NORTH AMERICA, Hjalmar R. Holand. The perplexing Kensington Stone, found in Minnesota at the end of the 19th century. Is it a record of a Scandinavian expedition to North America in the 14th century? Or is it one of the most successful hoaxes in history. A scientific detective investigation. Formerly *Westward from Vinland.* 31 photographs, 17 figures. x + 354pp. 22014-1 Paperbound $2.75

A BOOK OF OLD MAPS, compiled and edited by Emerson D. Fite and Archibald Freeman. 74 old maps offer an unusual survey of the discovery, settlement and growth of America down to the close of the Revolutionary war: maps showing Norse settlements in Greenland, the explorations of Columbus, Verrazano, Cabot, Champlain, Joliet, Drake, Hudson, etc., campaigns of Revolutionary war battles, and much more. Each map is accompanied by a brief historical essay. xvi + 299pp. 11 x 13¾. 22084-2 Paperbound $6.00